Chapman & Hall/CRC
Studies in Informatics Series

SERIES EDITOR
G. Q. Zhang
Case Western Reserve University
Department of EECS
Cleveland, Ohio, U.S.A.

PUBLISHED TITLES

Stochastic Relations: Foundations for Markov Transition Systems
Ernst-Erich Doberkat

Conceptual Structures in Practice
Pascal Hitzler and Henrik Schärfe

Context-Aware Computing and Self-Managing Systems
Waltenegus Dargie

Introduction to Mathematics of Satisfiability
Victor W. Marek

Chapman & Hall/CRC
Studies in Informatics Series

Introduction to Mathematics of Satisfiability

Victor W. Marek

CRC Press
Taylor & Francis Group
Boca Raton London New York

CRC Press is an imprint of the
Taylor & Francis Group, an **informa** business

A CHAPMAN & HALL BOOK

Chapman & Hall/CRC
Taylor & Francis Group
6000 Broken Sound Parkway NW, Suite 300
Boca Raton, FL 33487-2742

© 2009 by Taylor and Francis Group, LLC
Chapman & Hall/CRC is an imprint of Taylor & Francis Group, an Informa business

No claim to original U.S. Government works

Printed in the United States of America on acid-free paper
10 9 8 7 6 5 4 3 2 1

International Standard Book Number: 978-1-4398-0167-3 (Hardback)

This book contains information obtained from authentic and highly regarded sources. Reasonable efforts have been made to publish reliable data and information, but the author and publisher cannot assume responsibility for the validity of all materials or the consequences of their use. The authors and publishers have attempted to trace the copyright holders of all material reproduced in this publication and apologize to copyright holders if permission to publish in this form has not been obtained. If any copyright material has not been acknowledged please write and let us know so we may rectify in any future reprint.

Except as permitted under U.S. Copyright Law, no part of this book may be reprinted, reproduced, transmitted, or utilized in any form by any electronic, mechanical, or other means, now known or hereafter invented, including photocopying, microfilming, and recording, or in any information storage or retrieval system, without written permission from the publishers.

For permission to photocopy or use material electronically from this work, please access www.copyright. com (http://www.copyright.com/) or contact the Copyright Clearance Center, Inc. (CCC), 222 Rosewood Drive, Danvers, MA 01923, 978-750-8400. CCC is a not-for-profit organization that provides licenses and registration for a variety of users. For organizations that have been granted a photocopy license by the CCC, a separate system of payment has been arranged.

Trademark Notice: Product or corporate names may be trademarks or registered trademarks, and are used only for identification and explanation without intent to infringe.

Library of Congress Cataloging-in-Publication Data

Marek, V. W. (V. Wiktor), 1943-
   Introduction to mathematics of satisfiability / Victor W. Marek.
      p. cm. -- (Chapman & Hall/CRC studies in informatics series)
   Includes bibliographical references and index.
   ISBN 978-1-4398-0167-3 (hardcover : alk. paper)
   1. Propositional calculus. 2. Logic, Symbolic and mathematical. I. Title. II. Series.

QA9.3.M37 2009
511.3--dc22
                                              2009016171

Visit the Taylor & Francis Web site at
http://www.taylorandfrancis.com

and the CRC Press Web site at
http://www.crcpress.com

Contents

Preface		**ix**
1	**Sets, lattices, and Boolean algebras**	**1**
	1.1 Sets and set-theoretic notation	1
	1.2 Posets, lattices, and Boolean algebras	3
	1.3 Well-orderings and ordinals	5
	1.4 The fixpoint theorem	6
	1.5 Exercises	9
2	**Introduction to propositional logic**	**11**
	2.1 Syntax of propositional logic	11
	2.2 Semantics of propositional logic	13
	2.3 Autarkies	23
	2.4 Tautologies and substitutions	28
	2.5 Lindenbaum algebra	32
	2.6 Permutations	34
	2.7 Duality	38
	2.8 Semantical consequence, operations *Mod* and *Th*	39
	2.9 Exercises	42
3	**Normal forms of formulas**	**45**
	3.1 Canonical negation-normal form	46
	3.2 Occurrences of variables and three-valued logic	48
	3.3 Canonical forms	50
	3.4 Reduced normal forms	54
	3.5 Complete normal forms	56
	3.6 Lindenbaum algebra revisited	58
	3.7 Other normal forms	59
	3.8 Exercises	60
4	**The Craig lemma**	**63**
	4.1 Craig lemma	63
	4.2 Strong Craig lemma	66
	4.3 Tying up loose ends	69
	4.4 Exercises	71

5 Complete sets of functors — 73
- 5.1 Beyond De Morgan functors — 74
- 5.2 Tables — 75
- 5.3 Field structure in *Bool* — 78
- 5.4 Incomplete sets of functors, Post classes — 83
- 5.5 Post criterion for completeness — 85
- 5.6 If-then-else functor — 88
- 5.7 Exercises — 90

6 Compactness theorem — 93
- 6.1 König lemma — 93
- 6.2 Compactness, denumerable case — 95
- 6.3 Continuity of the operator Cn — 99
- 6.4 Exercises — 100

7 Clausal logic and resolution — 101
- 7.1 Clausal logic — 102
- 7.2 Resolution rule — 107
- 7.3 Completeness results — 110
- 7.4 Query-answering with resolution — 113
- 7.5 Davis-Putnam lemma — 117
- 7.6 Semantic resolution — 119
- 7.7 Autark and lean sets — 124
- 7.8 Exercises — 132

8 Testing satisfiability — 133
- 8.1 Table method — 133
- 8.2 Hintikka sets — 135
- 8.3 Tableaux — 137
- 8.4 Davis-Putnam algorithm — 144
- 8.5 Boolean constraint propagation — 154
- 8.6 The DPLL algorithm — 158
- 8.7 Improvements to DPLL? — 161
- 8.8 Reduction of the search SAT to decision SAT — 162
- 8.9 Exercises — 163

9 Polynomial cases of SAT — 165
- 9.1 Positive and negative formulas — 165
- 9.2 Horn formulas — 167
- 9.3 Autarkies for Horn theories — 176
- 9.4 Dual Horn formulas — 181
- 9.5 Krom formulas and 2-SAT — 185
- 9.6 Renameable classes of formulas — 194
- 9.7 Affine formulas — 199
- 9.8 Exercises — 204

10 SAT, integer programming, and matrix algebra — 205
- 10.1 Representing clauses by inequalities — 206
- 10.2 Resolution and other rules of proof — 207
- 10.3 Pigeon-hole principle and the cutting plane rule — 209
- 10.4 Satisfiability and $\{-1, 1\}$-integer programming — 214
- 10.5 Embedding SAT into matrix algebra — 216
- 10.6 Exercises — 225

11 Coding runs of Turing machines, NP-completeness — 227
- 11.1 Turing machines — 228
- 11.2 The language — 231
- 11.3 Coding the runs — 232
- 11.4 Correctness of our coding — 233
- 11.5 Reduction to 3-clauses — 237
- 11.6 Coding formulas as clauses and circuits — 239
- 11.7 Decision problem for autarkies — 243
- 11.8 Search problem for autarkies — 245
- 11.9 Either-or CNFs — 247
- 11.10 Other cases — 249
- 11.11 Exercises — 252

12 Computational knowledge representation with SAT – getting started — 253
- 12.1 Encoding into SAT, DIMACS format — 254
- 12.2 Knowledge representation over finite domains — 261
- 12.3 Cardinality constraints, the language L^{cc} — 267
- 12.4 Weight constraints — 273
- 12.5 Monotone constraints — 276
- 12.6 Exercises — 283

13 Knowledge representation and constraint satisfaction — 285
- 13.1 Extensional and intentional relations, *CWA* — 285
- 13.2 Constraint satisfaction and SAT — 292
- 13.3 Satisfiability as constraint satisfaction — 297
- 13.4 Polynomial cases of Boolean CSP — 300
- 13.5 Schaefer dichotomy theorem — 305
- 13.6 Exercises — 317

14 Answer set programming — 321
- 14.1 Horn logic revisited — 321
- 14.2 Models of programs — 322
- 14.3 Supported models — 323
- 14.4 Stable models — 326
- 14.5 Answer set programming and SAT — 329
- 14.6 Knowledge representation and ASP — 333
    - 14.6.1 Three-coloring of graphs — 334

14.6.2 Hamiltonian cycles in ASP 335
14.7 Complexity issues for ASP 336
14.8 Exercises 337

15 Conclusions   **339**

References   **343**

Index   **347**

Preface

The subject of this book is *satisfiability* of theories consisting of propositional logic formulas, a topic with over 80 years of history. After an initial push, the subject was abandoned for some time by the mathematicians and analytical philosophers who contributed to its inception. Electrical and computer engineering, and also computer science, picked up the research of this area, starting with the work of Shannon in the 1930s. Several aspects of satisfiability are important for electronic design automation and have been pursued by computer engineers. Yet another reason to look at satisfiability has been a remarkable result of Levin and Cook - the fact that satisfiability expresses a huge class of problems of interest to computer science (we will present this fundamental result in Chapter 11). The fundamental algorithms for satisfiability testing (the simplest method, that of truth tables, was invented very early, in the 1920s, but is obviously very inefficient) were invented in the 1960s. While there was steady progress in understanding satisfiability, to transform it into useful technology the area had to mature and develop new techniques. Resolution proofs were studied in the decade of the 1970s, but it was not enough to make the area practically important (although the fans of the programming language PROLOG may be of a different opinion). Additional breakthroughs were needed and in the mid-1990s they occurred. The theoretical advances in satisfiability resulted in creation of a class of software, called *SAT solvers*. Solvers are used to find solutions to *search problems* encoded by propositional theories. Solvers found applications in a variety of areas: electronic design automation, hardware and software verification (but please do not conclude that all problems have been solved with SAT solvers!), combinatorial optimization and other areas.

As I mentioned above, the subject has a long history, with many contributors. Thus it is not possible to present everything that has been done (even if I knew it, which is not the case...). I had to select the topics for this book. Various important topics have been omitted, especially in the optimization of various normal forms (a topic taught extensively in the computer engineering curriculum and treated extensively in their monographs, e.g., [DM94]).

Now, about the contents of the book. Those who have patience to read the 300+ pages to follow will learn a variety of topics. In principle, the book is almost self-contained. By this I mean that very little of mathematics that is not explained in the book is needed to follow the material. Please do not take this as saying that there is no mathematics in the book; just the opposite. Most information that is needed beyond undergraduate computer science/mathematics education is provided. There are exceptions (for instance we ask the reader to accept König theorem in Chapter 7 and Geiger theorem in Chapter 13), but aside from such few pretty exotic topics almost everything is explained, often from first principles. One exception is a limited introduction to NP-completeness (Chapter 11).

So now, to the content. As is common in such texts, the first technical chapter, Chapter 1, prepares the reader for the material of further chapters and introduces basic means such as the Knaster-Tarski fixpoint theorem, as well as introduces the reader to the style of the book. The book is divided into three parts. Part One deals with logic fundamentals: syntax of propositional logic, complete sets of functors, normal forms, Craig lemma and compactness. Specifically, Chapter 2 introduces the reader to propositional logic, its syntax and semantics. The point of view accepted in this book stresses both the two-valued logic but also the three-valued logic of Kleene (there are several three-valued logics; the one we study here is Kleene logic). In addition to fundamentals of logic we study permutations of literals and of variables and show how these interact with formulas and with satisfaction of formulas. We introduce the consequence operator and its basic properties. In this chapter we study only De Morgan functors $\neg, \wedge, \vee, \Rightarrow$, and $\equiv$.

In Chapter 3 we investigate some normal forms of propositional formulas including negation normal form, conjunctive normal form, and disjunctive normal form. Whenever a set of functors is complete, it induces some normal form, and we discuss some other, more exotic forms.

Craig lemma, which is an "Occam Razor" principle for propositional logic (but surprisingly has applications in so-called model checking, a topic we do not discuss in this book at all), is discussed in Chapter 4. We show the usual form of Craig lemma (existence of an interpolant) and show that if the input is in a suitable form then interpolant can be easily computed.

Complete sets of functors are discussed in Chapter 5. We prove the Post criterion for completeness of sets of functors. We also look at the field structure in the set *Bool* and discuss the properties of Boolean polynomials (often called Zhegalkin polynomials) including representability of Boolean functions as polynomials.

We prove the compactness of propositional logic in Chapter 6. There are more general and less general versions of compactness, but we prove the most restricted version, for denumerable sets of formulas. In fact we show compactness for denumerable sets of clauses (but it implies compactness for arbitrary denumerable sets of formulas). We use compactness to show the continuity of consequence operator (but for denumerable sets of theories only).

In Part Two, we study clauses, their proof theory and semantics, and the algorithms for satisfiability testing.

The clausal logic, the most important topic of this book from the point of applications, is studied in Chapter 7. We show the completeness of the proof system based on rules of resolution and subsumption, and another based on the resolution refutation. We investigate autark sets and lean sets, proving along the way the relationships of resolution to these sets. There are many variations of resolution refutation proof systems. We look at one such variation: semantic resolution.

Algorithms for testing satisfiability of collections of propositional formulas are discussed in Chapter 8. We discuss four algorithms: table method, tableaux, variable elimination resolution, and backtracking search algorithm (usually called the DPLL algorithm). The first two apply to arbitrary finite sets of propositional formulas; the other two require the input in clausal form. We also discuss combinational circuits and their description by means of clauses.

It turns out that there are classes of theories with a simpler satisfiability problem. In Chapter 9 we discuss the classes of positive theories, negative theories, Horn theories, dual Horn theories, renameable-Horn theories, Krom theories (also known as 2SAT), and affine theories. For each of these classes we find the polynomial-time algorithm that tests satisfiability.

As we will see later on, many problems can be expressed as satisfiability problems for suitably chosen theories. But in Chapter 10 we do something else; we reduce satisfiability of clausal theories to classical problems of integer programming and also in the same chapter, to linear algebra problems. The idea is that scientific knowledge does not progress uniformly, and various areas of mathematics create their own problem-solving techniques. Those can be useful and, sometimes, may simplify solving for specific classes of problems. It should be clearly stated that both areas into which we delve in this chapter - integer programming and linear algebra - are huge, well-developed subjects in their own right. We selected results that appear to us very closely related to satisfiability. Hopefully readers will look for a deeper understanding of these areas on their own.

Part Three of the book is devoted to knowledge representation.

We code finite runs (of length bound by a polynomial in the length of the input) of Turing machines in Chapter 11. As a corollary we find that satisfiability of clausal theories is an NP-complete problem and then, since we can encode satisfiability of clausal theories by satisfiability of theories consisting of clauses of size at most three (additional variables are used), we find that the satisfiability problem for theories consisting of three-clauses is also NP-complete. We also reduce the satisfiability problem for arbitrary propositional theories to satisfiability of clausal theories. We also show that with few exceptions (listed in that chapter) "mix-and-match" of formulas between two classes leads to an NP-complete satisfiability problem.

Chapters 12 and 13 deal with theoretical, but also practical aspects of satisfiability. Chapter 12 studies encodings into SAT. The idea is that given a problem (it must be a search problem $\mathcal{P}$ in the class NP) the programmer translates the problem $\mathcal{P}$ into a propositional theory $T_\mathcal{P}$ so that there is a one-to-one correspondence between satisfying valuations for $T_\mathcal{P}$ and solutions to $\mathcal{P}$. Now, the process of translating may be tedious and the theory $T_\mathcal{P}$ may be large. The issue is what kind of shortcuts are available to such programmers. We discuss several such possible shortcuts. First we discuss a variation of predicate logic where the semantics is limited to Herbrand models. In some restricted cases formulas of predicate calculus (with semantics limited to finite models) can be used to represent problems. Yet another possibility is to use cardinality constraints (and more generally weight constraints). We show that the mathematics that forms a basis of the DPLL algorithm generalizes to such cases (of cardinality constraints and more generally weight constraints). We discuss monotone constraints showing that they are representable by positive formulas built out of cardinality constraints (thus also weight constraints). Chapter 13 drills the issue of knowledge representation. We show how constraint satisfaction systems over finite domains (and sometimes even infinite domains) can be solved by satisfiability solvers. We also discuss the so-called closed world reasoning. Then just to make the reader more miserable we prove the celebrated Schaefer theorem on the complexity of Boolean constraint satisfaction problems. In Chapter 14 we outline the foundations of a variation of the satisfiability called *Answer Set Programming*, in particular how this formalism can be used for Knowledge Representation.

This short discussion should show the reader that we cover quite a lot of material, but there is plenty that we do *not* cover. In what can be termed current satisfiability research we do not cover the issues discussed in Section 8.7, that is improvements to the DPLL algorithm. Those are, in our opinion, still a matter of research. We do not discuss the representation of formulas by means of binary decision diagrams. Traditionally this topic is treated by electrical engineers, not logicians. In fact electrical engineers know quite a lot about logical formulas (under different disguises) and it is not our intention to compete with them.

Acknowledgments: A number of colleagues contributed one way or another to this book. Special thanks go to Marc Denecker, Mike Dransfield, Rafi Finkel, John Franco, Joanna Golińska–Pilarek, Andy Klapper, Oliver Kullmann, Bill Legato, Anil Nerode, Jeff Remmel, John Schlipf, Marian Srebrny, Mateusz Srebrny, Mirek Truszczynski, Mark Vanfleet, Sean Weaver and Sarah Weissman. The audience of my Spring 2005 University of Kentucky course on satisfiability helped eliminate some errors, too. Likewise, the audience of my Spring 2007 University of Kentucky course on Boolean functions assisted me in tracing some errors. I am grateful to Marc Denecker for stressing the importance (in a different context, but close enough) of Kleene three-valued logic. My concern with autarkies, and more generally with partial valuations, comes from looking at satisfiability from the vantage point of Kleene three-valued logic. In particular I felt that the results on autarkies are important, and I put a number of those (mostly due to Kullmann and also to Truszczynski) in various places in this book. While Kullmann's approach to autarkies is algebraical, both Truszczynski and I were trying to put a set-theoretic spin on this area. I also stressed Knowledge Representation aspects of Propositional Logic and its extensions, presenting various constructs that are grounded in the experience of Answer Set Programming (see Chapter 12).

Of all the people mentioned above, Jeff Remmel and Mirek Truszczynski were my closest companions in the research of computational logic. I was blessed by their help and patience in explaining various subtle aspects of the areas covered in this book. Science is a social activity and collaboration often begets friendship. It certainly is the case with Jeff and Mirek.

All the colleagues mentioned above contributed to this work in a positive way. Any errors that remain are mine alone.

As I said, the material in this book consists mainly of topics that are well-known. Like many mathematicians, I would look at a proof of some result, did not like what I saw, and then provide my own proof (often worse than the original). An example in point is my own proof of Cook's theorem. The argument given in this book was heavily influenced by Jeff Remmel, but may be not enough.

I am grateful to the editing team of Taylor & Francis: Michele Dimont, David Grubbs, and Marsha Pronin for their help and encouragement. Chris Andreasen helped with the English. Shashi Kumar was very knowledgeable about LaTeX matters.

Finally, my family, and in particular my wife Elizabeth, was supportive during the years I worked on this book. Thanks!

For a number of years, various versions of this book were published on the Internet. The current version is much changed (due to the efforts of many individuals listed above and the reviewers). It is my intention to support the book through Web-errata. If the reader finds typos or errors he or she is welcome to send these to the author at marek@cs.uky.edu. Every message will be acknowledged, and corrections will be published on the errata page. Please look at the accompanying page http://www.cs.uky.edu/~marek/corr.html of additions, corrections, and improvements.

Cover image. In 1928, the Dutch mathematician, the great algebraist B.L. van der Waerden, proved a remarkable theorem. Given the positive integers k, l, there is a large enough m so that whenever a segment of integers of length $\geq m$ is partitioned in k disjoint blocks, then at least one of these blocks includes at least one arithmetic progression of length l. Thus there is a least m with this property, and such m is denoted in the literature by $W(k, l)$. There is no known closed form for the function $W(\cdot, \cdot)$, and, in fact only a small number of values of $W(k, l)$ are known. To see that given number n is smaller than $W(k, l)$, one needs a *certificate*, that is a partition of $[1..n]$ into k disjoint blocks, so that no block contains an arithmetic progression of length l. My collaborators and I used SAT solvers and techniques for Knowledge Representation discussed in Chapters 13 and 14 to compute certificates [DLMT04]. Subsequently, much stronger results were obtained by others, in particular by Marijn Heule and his coauthors [HHLM07]. As a side-effect of this research, Marijn invented a technique for visualizing certificates. The front cover of this book shows Marijn's visualization of the certificate showing $W(3, 5) > 170$. The back cover shows his visualization of our own certificate showing that $W(3, 5) > 125$. I find the first visualization, with its five-fold symmetries, esthetically appealing, and hope that the reader gets the same satisfaction looking at the cover of this book. Many thanks, Marijn.

Lexington, KY

Chapter 1

Sets, lattices, and Boolean algebras

1.1	Sets and set-theoretic notation	1
1.2	Posets, lattices, and Boolean algebras	3
1.3	Well-orderings and ordinals	5
1.4	The fixpoint theorem	6
1.5	Exercises	9

In this chapter we introduce the basics of set notation and fundamental notions that serve as technical support for our presentation: posets, lattices, and Boolean algebras. We will also state and prove the Knaster-Tarski fixpoint theorem, one of the fundamental results very often used in computer science. One caveat: this section is not meant to be the first contact for the reader with sets, posets, lattices and related concepts. Rather, it serves the purpose of setting up the terminology and "common language."

1.1 Sets and set-theoretic notation

Sets are collections of objects. We write $x \in X$ to denote that the object x belongs to the set X. Sets themselves may be elements of other sets. Often we will deal with subsets of some fixed set X. If this is the case, the family of all subsets of X is denoted by $\mathcal{P}(X)$. We assume that $\mathcal{P}(X)$ exists for every set X. Sets are often treated axiomatically, commonly using Zermelo-Fraenkel set theory (ZFC). We will not adopt the axiomatic approach to sets. It will be enough to realize that such an approach is possible and is developed in all of the classical texts of set theory. Given a fixed set X, and a formula φ, $\{x \in X : \varphi(x)\}$ is a subset of X consisting of those elements $x \in X$ which possess property φ. Such a set exists and is uniquely defined. Often we will not explicitly state that x belongs to X.

Generally, two sets are considered equal if they have precisely the same elements. This means that we can describe sets in a variety of ways, but what matters is what elements they have, not the form of their definition. We say that a set X is included in a set Y (in symbols $X \subseteq Y$) if every element of X is an element of Y. Thus for two sets X, Y

$$X = Y \text{ if and only if } X \subseteq Y \wedge Y \subseteq X.$$

There is a unique set without elements. It is called the *empty* set and is denoted by $\emptyset$. The empty set is included in every set. There are a couple of basic operations on sets. These include intersection, $X \cap Y$, which is $\{x : x \in X \land x \in Y\}$; union, $X \cup Y$, which is $\{x : x \in X \lor x \in Y\}$; and difference, $X \setminus Y$, $\{x : x \in X \land x \notin Y\}$. Two sets X, Y are *disjoint* if $X \cap Y = \emptyset$. When we fix a set X and consider only sets included in X we also have the complement operation: $-Y = \{x : x \notin Y\}$.

A *family* or *collection* of sets is a set consisting of sets. Given a collection $\mathcal{X}$, we define the union of $\mathcal{X}$ as $\{x : \exists_{Y \in \mathcal{X}}(x \in Y)\}$. Similarly, the intersection of $\mathcal{X}$, $\bigcap \mathcal{X}$, is $\{x : \forall_{Y \in \mathcal{X}}(x \in Y)\}$.

Given an object x, the set $\{x\}$ is one that has as an element x and nothing else. Likewise, $\{x, y\}$ is an *unordered pair* of elements of x and y. This set contains x, y and nothing else. There is also an *ordered* pair of x and y. This concept can be defined using just the membership relation and an unordered pair. We will denote the ordered pair of x and y by $\langle x, y \rangle$. The basic property of ordered pairs is that $\langle x, y \rangle = \langle z, t \rangle$ is equivalent to $x = z$ and $y = t$. The Cartesian product of sets X and Y, $X \times Y$, is $\{\langle x, y \rangle : x \in X \land y \in Y\}$. The operation $\times$ is not, in general, commutative. A *relation* is a subset of the Cartesian product of sets. A relation $R \subseteq X \times X$ is *reflexive* if for all $x \in X$, $\langle x, x \rangle \in R$. A relation $R \subseteq X \times X$ is *symmetric* if for all x, y in X, $\langle x, y \rangle \in R$ implies $\langle y, x \rangle \in R$. A relation $R \subseteq X \times X$ is *transitive*, if for all x, y, z in X, $\langle x, y \rangle \in R$ and $\langle y, z \rangle \in R$ imply $\langle x, z \rangle \in R$. A relation $R \subseteq X \times X$ is *antisymmetric* if for all $x, y \in X$, $\langle x, y \rangle \in R$ and $\langle y, x \rangle \in R$ implies $x = y$. A relation $R \subseteq X \times X$ is *connected* if for all $x, y \in X$, $\langle x, y \rangle \in R \lor \langle y, x \rangle \in R$. We will often write xRy instead of $\langle x, y \rangle \in R$.

With the ontology of relations defined above, we can further define various classes of relations. Here are a few important ones. A relation $R \subseteq X \times X$ is an *equivalence relation* if R is reflexive, symmetric, and transitive. Given an equivalence relation R in X, R defines a partition of X into its *cosets*, $[x] = \{y : yRx\}$. Conversely, given a partition $\mathcal{X}$ of a set X into (nonempty) blocks, we can find an equivalence relation R with cosets prescribed by $\mathcal{X}$. A *(partial) ordering* in the set X is a relation R such that R is reflexive, antisymmetric and transitive in X. We will use suggestive symbols such as $\leq, \preceq$, or $\sqsubseteq$ to denote orderings. When $\leq$ is an ordering of X, we call $\langle X, \leq \rangle$ a partially ordered set or *poset*. A linear ordering is a poset where $\leq$ is connected.

A function is a relation $R \subseteq X \times Y$ such that whenever $\langle x, y \rangle$ and $\langle x, z \rangle$ belong to R, $y = z$. We use letters f, g and the like to denote functions. We write $f(x) = y$ instead of $\langle x, y \rangle \in f$. The domain of a function f is the set $\{x : \exists_y f(x) = y\}$. This set, $dom(f)$, may be smaller than X. In such a case we talk about a *partial* function.

In our presentation we will often refer to specific sets. We will use N to denote the set of non-negative integers. We may sometimes use ω to denote the same set. We will use Z for the set of all integers, and R for the set of all reals.

1.2 Posets, lattices, and Boolean algebras

We recall that a partially ordered set (*poset*, for short) is a pair $\langle X, \leq \rangle$ where X is a set, and $\leq$ is an ordering of X, that is $\leq$ is reflexive, antisymmetric, and transitive in X. A classic example of a poset is $\langle \mathcal{P}(X), \subseteq \rangle$ where $\subseteq$ is the inclusion relation restricted to subsets of X. Of course, the definition of a poset is very general, and posets are abundant in all areas of mathematics and computer science. Given a poset $\langle X, \leq \rangle$ and an element $x \in X$ we say that x is *maximal* in $\langle X, \leq \rangle$ if $\forall_{y \in X}(x \leq y \Rightarrow x = y)$. Similarly we define *minimal* elements in the poset. The *largest* element of the poset $\langle X, \leq \rangle$ is an element x such that for all $y \in X$, $y \leq x$. The largest element (if one exists) is a unique maximal element. The least element of a poset is introduced similarly.

A *chain* in a poset $\langle X, \leq \rangle$ is a subset $Y \subseteq X$ such that for every $x, y \in Y$, $x \leq y \lor y \leq x$. If $\langle X, \leq \rangle$ is a poset and $Y \subseteq X$, we say that x is an upper bound for Y (we also say *of* Y) if for every element $y \in Y$, $y \leq x$. An upper bound for a set Y may or may not exist. A subset $Y \subseteq Y$ is *bounded from above* (in short *bounded*) if for some $x \in X$, for all $y \in Y, y \leq x$. The *least upper bound* of Y is an upper bound for Y which is $\leq$-smaller or equal to any other upper bound for Y. A least upper bound, if it exists, is unique. We denote such a least upper bound of Y by $\bigvee Y$. Analogously, we define lower bound and greatest lower bound, denoted $\bigwedge Y$. Existence of least upper bounds is a very strong property, and is, as we are going to see, very useful. We say that a poset $\langle X, \leq \rangle$ is *chain-complete* if every chain in $\langle X, \leq \rangle$ possesses a least upper bound.

Existence of bounds (not necessarily of least upper bounds, but bounds in general) entails existence of maximal elements. This is the extent of the so-called Zorn's lemma and generalizes a very useful property of finite posets.

PROPOSITION 1.1 (Zorn lemma)
If every chain in a poset $\langle X, \leq \rangle$ is bounded, then X possesses maximal elements. In fact, for every $x \in X$ there is a maximal element y such that $x \leq y$.

A version of Zorn's lemma, called *Hausdorff's maximal principle*, is a special case of the situation where X is a family of sets, and $\leq$ is inclusion relation. It says that any family of sets where every $\subseteq$-chain is bounded must contain $\subseteq$-maximal elements.

Whenever $\langle X, \leq \rangle$ is a poset, $\langle X, \geq \rangle$ is also a poset and elements maximal in one are minimal in the other and *vice-versa*. Thus a dual form of Zorn's lemma (with chains bound from below, and minimal elements) holds as well. Moreover, weakening of assumptions (assuming existence of bounds for well-ordered chains) does not change the conclusion. This last form will be used in one of our arguments.

A *lattice* is a poset where every pair of elements possesses a least upper bound and a greatest lower bound. Assuming we deal with a lattice, we denote the greatest lower bound of x and y by $x \wedge y$ and the least upper bound of x and y by $x \vee y$. The existence of bounds means that we gain an algebraic perspective on X, for we now have an algebra, $\langle X, \wedge, \vee \rangle$.

The set $\mathcal{P}(X)$ with operations of intersection ($\cap$), and union ($\cup$) forms a lattice. There are many interesting lattices, and an active area of research in universal algebra, called lattice theory, studies lattices.

A lattice $\langle L, \wedge, \vee \rangle$ is *complete* if every subset of L possesses a least upper bound and a greatest lower bound. The lattice $\langle \mathcal{P}(X), \cap, \cup \rangle$ is always complete, but there are many other complete lattices. In a complete lattice there is always a unique least element (denoted $\bot$) and a largest element ($\top$).

We defined the lattice operations from the ordering $\leq$ via bounds. We could go the other way around: given a lattice $\langle L, \wedge, \vee \rangle$ we can define an ordering $\leq$ in L by

$$x \leq y \text{ if } x \vee y = y.$$

With this ordering, $\langle L, \leq \rangle$ becomes a poset; this poset defines (via bounds) the operations $\wedge$ and $\vee$.

Lattices can be treated purely algebraically. Here is a set of postulates that a structure $\langle L, \wedge, \vee \rangle$ must satisfy to be a lattice.

L1	$x \wedge x = x$	$x \vee x = x$
L2	$x \wedge y = y \wedge x$	$x \vee y = y \vee x$
L3	$x \wedge (y \wedge z) = (x \wedge y) \wedge z$	$x \vee (y \vee z) = (x \vee y) \vee z$
L4	$x \wedge (x \vee y) = x$	$x \vee (x \wedge y) = x$

If a lattice possesses a largest ($\top$) and least ($\bot$) element, we list them in the presentation of the lattice, writing $\langle L, \wedge, \vee, \bot, \top \rangle$. Finite lattices always possess largest and smallest elements. An important class of lattices consists of *distributive lattices*. Those are lattices that, in addition to conditions (L1)–(L4), satisfy the following conditions.

L5	$x \wedge (y \vee z) = (x \wedge y) \vee (x \wedge z)$	$x \vee (y \wedge z) = (x \vee y) \wedge (x \vee z)$

A *Boolean algebra* is a distributive lattice $\langle B, \wedge, \vee, \bot, \top \rangle$ with an additional operation $-$, called a *complement*, subject to the additional conditions (B6), (B7), and (B8), below.

B6	$x \wedge -x = \bot$	$x \vee -x = \top$
B7	$-(x \wedge y) = -x \vee -y$	$-(x \vee y) = -x \wedge -y$
B8	$-(-x) = x$	

We do not claim that the axioms (L1)–(L5),(B6)–(B8) are independent (i.e., each cannot be proved from the others). In fact they are not independent, but we will not concern ourselves with this problem here.

Let us observe that for every set X the algebra $\langle \mathcal{P}(X), \cap, \cup, -, \emptyset, X \rangle$ is a Boolean algebra. In fact, $\langle \mathcal{P}(X), \cap, \cup, -, \emptyset, X \rangle$ is a complete Boolean algebra, that is it is complete as a lattice. The classic theorem of Stone states that every complete Boolean algebra is isomorphic to a field of sets (i.e., a family of sets closed under intersections, unions, and complements) but not necessarily of the form $\langle \mathcal{P}(X), \cap, \cup, -, \emptyset, X \rangle$.

In the case of finite B one gets a stronger result. We say that an element a of B is an *atom* in $\langle B, \wedge, \vee, -, \bot, \top \rangle$ if for all $x \in B$, $x \wedge a = \bot$ or $x \wedge a = a$. In other words, a cannot be split into proper smaller parts. It is easy to see that every finite Boolean algebra is atomic, i.e., every element must be bigger than or equal to an atom. Moreover, every finite Boolean algebra is complete. But it is almost immediate that a complete atomic Boolean algebra must be isomorphic to $\langle \mathcal{P}(A), \cap, \cup, -, \emptyset, A \rangle$, where A is the set of its atoms. Thus finite Boolean algebras are always (up to isomorphism) of the form $\langle \mathcal{P}(X), \cap, \cup, -, \emptyset, X \rangle$. Since $|\mathcal{P}(X)| = 2^n$, where $n = |X|$, we find that finite Boolean algebras have to have size 2^n for a suitably chosen n.

1.3 Well-orderings and ordinals

A relation $R \subseteq X \times X$ is *well-founded* if every non-empty subset Y of X possesses a least element with respect to R. That is, there is $y \in Y$ so that for all $x \in Y$, yRx. A well-founded poset is always linearly ordered, and in fact, being well-founded is a very strong property. It implies that there is no infinite strictly descending R-chain, that is, there is no sequence $\langle x_n \rangle_{n \in N}$ such that for all $i < j$, $x_j R x_i$ and $x_j \neq x_i$. A well-founded poset is often called a *well-ordered set*. A non-empty well-ordered set always has the least element, and the order-induction holds for such sets. By order-induction we mean the following property: once the least element of X possesses a property P and we establish that from the fact that all elements preceding a given element x have property P it follows that x itself has the property P, we can safely claim that all elements of X have property P. Well-ordered sets have many appealing properties and their properties are presented in every reasonable book on set theory (books without such presentation must be considered very incomplete). Here we will use a few facts about well-orderings (in fact already in the next section, when we prove the fixpoint theorem). The key fact here is that one can define so-called *ordinals* (also called von Neumann ordinals). These objects themselves are well-ordered by the membership relation $\in$. Moreover, for a von Neumann ordinal α, $\alpha = \{\beta : \beta \text{ is an ordinal and } \beta \in \alpha\}$. The ordinals are either *successors* (i.e., immediate successors of another ordinal) or *limits*. In the latter case a limit ordinal λ is bigger than all smaller ordinals, but there is no largest ordinal smaller than λ. The least infinite limit ordinal is ω, the order type of natural numbers.

There are various types of proof by transfinite induction. One is the order induction mentioned above. A version commonly used considers cases of successor and limit ordinals. We will use letters like α, β, etc. for ordinals (successors or otherwise), and λ for limit ordinals.

1.4 The fixpoint theorem

In this section we will formulate a theorem on existence of fixpoints of monotone operators (i.e., functions) in complete lattices. This result, known as the Knaster-Tarski theorem, is one of the fundamental tools of computer science. In fact we will be using the Knaster-Tarski theorem quite often in this book.

Recall that we had two distinct, though equivalent, perspectives of lattices: algebraic and order-theoretic. These approaches were equivalent and one allowed for defining the other. Here we will be more concerned with the order-theoretic aspects of lattices. Let $\langle L, \preceq \rangle$ be a complete lattice. We say that a function $f : L \to L$ is *monotone* if for all $x, y \in L$

$$x \preceq y \Rightarrow f(x) \preceq f(y).$$

Often such functions are called monotone *operators*. A *fixpoint* of the function f is any x such that $f(x) = x$. We will now prove the Knaster-Tarski fixpoint theorem for monotone operators. There are at least two different ways to prove this theorem; the argument below is the original Tarski argument. We refer the reader to the Exercise section below, for an alternative way to prove the fixpoint theorem.

We say that an element x of the complete lattice $\langle L, \preceq \rangle$ is a *prefixpoint* of a function $f : L \to L$ if $f(x) \preceq x$. Likewise, we say that an element x of the complete lattice $\langle L, \preceq \rangle$ is a *postfixpoint* of a function $f : L \to L$ if $x \preceq f(x)$.

LEMMA 1.1
If $f : L \to L$ is a function and $\langle L, \preceq \rangle$ is a complete lattice then f possesses both a prefixpoint and a postfixpoint.

Proof: Clearly, the largest element $\top$ of L is a prefixpoint, and the least element $\bot$ of L a postfixpoint. □

We will now denote by Pre_f and by $Post_f$ the sets of prefixpoints of the operator f and of postfixpoints of f, respectively. Lemma 1.1 says that both these sets are nonempty.

PROPOSITION 1.2 (Knaster and Tarski fixpoint theorem)
Let $f : L \to L$ be a monotone operator in a complete lattice $\langle L, \preceq \rangle$. Then:

(a) f possesses a least prefixpoint l, and l is, in fact, the least fixpoint of f.

(b) f possesses a largest postfixpoint m, and m is, in fact, the largest fixpoint of f.

Proof: (a). From Lemma 1.1 we know that $Pre_f \neq \emptyset$. Let $l = \bigwedge Pre_f$. That is, l is the greatest lower bound of Pre_f. We claim that l itself is a prefixpoint of f. To this end, let us consider any element x of Pre_f. Then, by definition, $l \preceq x$. Since $x \in Pre_f$, $f(x) \preceq x$. By monotonicity of f, since $l \preceq x$, $f(l) \preceq f(x)$. By transitivity of $\preceq$
$$f(l) \preceq x.$$
Since x was an arbitrary element of Pre_f, $f(l) \preceq \bigwedge Pre_f$, that is $f(l) \preceq l$. Thus l itself is a prefixpoint, and thus l is the least prefixpoint.

We now show that l is, in fact, a fixpoint. To this end let us observe that since l is a prefixpoint, i.e., $f(l) \preceq l$, $f(f(l)) \preceq f(l)$ thus $f(l)$ is also a prefixpoint of f. But then $l \preceq f(l)$, since l is the least prefixpoint. Thus $l = f(l)$, as desired.

All we need to do now is to show that l is a least fixpoint. To this end, let l' be any fixpoint of f. Then l' is a prefixpoint of f and so $l \preceq l'$, as desired.

(b) By looking at m, the supremum of postfixpoints of f, we easily adopt the argument of (a) to the task at hand. □

The proof of Proposition 1.2 should leave us pretty unhappy, for it gives no explicit way to compute the least fixpoint of a monotone operator. Fortunately we can find an explicit definition of the least fixpoint. This construction is, actually, due to Kleene. To this end let us define the sequence $\langle t_\alpha \rangle_{\alpha \in Ord}$ as follows:

1. $t_0 = \bot$
2. $t_{\beta+1} = f(t_\beta)$
3. $t_\lambda = \bigvee_{\beta < \lambda} t_\beta$ for λ limit.

As we will see in an exercise below, the least fixpoint of f is t_α where α is the least ordinal such that $t_\alpha = t_{\alpha+1}$. Right now, we will limit our attention to a class of monotone operators that we will encounter often below. There is a useful condition which, if satisfied, gives us a good handle on the place where the fixpoint is reached. Let us call a function $f : L \to L$ continuous, if for every $\preceq$-increasing sequence $\langle x_n \rangle_{n \in N}$

$$f\left(\bigvee_{n \in N} x_n\right) = \bigvee_{n \in N} f(x_n).$$

Intuitively (but it is only an intuition), continuity means that $f(\bigvee_{n \in N} x_n)$ has nothing but objects from $\bigvee_{n \in N} f(x_n)$. Let us observe that if f is a monotone function then for every $\preceq$-increasing sequence $\langle x_n \rangle_{n \in N}$, $\bigvee_{n \in N} f(x_n) \preceq f(\bigvee_{n \in N} x_n)$. We now have the following fact.

PROPOSITION 1.3
If $f : L \to L$ is continuous and monotone function in a complete lattice $\langle L, \preceq \rangle$, then f possesses a least fixpoint. This least fixpoint is the object t_ω

defined above, where ω is the least infinite ordinal.

Proof: We first need to show that the sequence $\langle t_n \rangle_{n \in N}$ is increasing. Clearly, $t_0 \preceq t_1$. Assuming that
$$t_0 \preceq t_1 \preceq \ldots \preceq t_n \preceq t_{n+1}$$
we have $t_n \preceq t_{n+1}$ and by monotonicity
$$f(t_n) \preceq f(t_{n+1}),$$
that is, $t_{n+1} \preceq t_{n+2}$. But then
$$t_0 \preceq t_1 \preceq \ldots \preceq t_n \preceq t_{n+1} \preceq t_{n+2}.$$
So now we know that the sequence $\langle t_n \rangle_{n \in N}$ is increasing and so by continuity
$$t_\omega = \bigvee_{n=0}^{\infty} t_n = \bigvee_{n=1}^{\infty} t_n = \bigvee_{n=0}^{\infty} f(t_n) = f\left(\bigvee_{n=0}^{\infty} t_n\right) = f(t_\omega).$$
Thus t_ω is a fixpoint of f. Let x be any fixpoint of f. Then obviously $t_0 \preceq x$. Assuming $t_n \preceq x$ we have
$$t_{n+1} = f(t_n) \preceq f(x) = x.$$
Thus for all n, $t_n \preceq x$. Therefore x is an upper bound of the sequence $\langle t_n \rangle_{n \in N}$. But then t_ω which is the least upper bound of $\langle t_n \rangle_{n \in N}$, must be $\preceq$-smaller than or equal to x, as desired. □.

Let us observe that Proposition 1.3 does not say that the fixpoint is *reached* at ω, but that the fixpoint is reached *at latest* at ω – it may be reached earlier, as will always happen in the case when L is finite.

On the other hand, let us observe that the continuity *does not* tell us that the largest fixpoint is reached in at most ω steps. In fact it is possible to construct examples of continuous operators where more than ω steps will be needed to reach the largest fixpoint.

Propositions 1.2 and 1.3 have been generalized in a variety of ways, in various circumstances where all the assumptions may not have been satisfied. We list a number of generalizations; they go beyond the scope of our work here. The arguments are variations on the proof of Proposition 1.2 or the alternative argument; see Exercises.

1. The poset $\langle L, \preceq \rangle$ does not have to be a lattice. It is enough that it is chain-complete, that is, that every $\preceq$-chain has a least upper bound.

2. The poset does not have to be even chain-complete, if we assume continuity. In such circumstances all we need is existence of a least upper bound for increasing sequences of length ω.

3. Instead of fixpoints, we may be interested in *prefixpoints* of f, i.e., elements x such that $f(x) \preceq x$. It so happens that the argument we gave shows existence of prefixpoints. Moreover, there is a least prefixpoint and it coincides with the least fixpoint.

4. In the general case of complete lattices and monotone operators, the collection of all fixpoints of f (under the ordering $\preceq$) forms a complete lattice.

5. Finally, we may have more than one monotone operator and be interested in common fixpoints. Each of cases (1)-(4) generalizes to this situation.

1.5 Exercises

1. Show that the sequence $\langle t_\alpha \rangle_{\alpha \in Ord}$ is increasing, that is,
$$\alpha \leq \beta \Rightarrow t_\alpha \preceq t_\beta.$$

2. Show that the sequence $\langle t_\alpha \rangle_{\alpha \in Ord}$ must have at least one pair (α, β) such that $\alpha \neq \beta$ and $t_\alpha = t_\beta$.

3. Show that if $\alpha < \beta$ and $t_\alpha = t_\beta$ then $t_\alpha = t_{\alpha+1}$ and so t_α is a fixpoint.

4. Show that the least fixpoint l of f is t_α for the least ordinal α such that t_α is a fixpoint.

5. Define the sequence $\langle k_\alpha \rangle_{\alpha \in Ord}$ as follows:

    (a) $k_0 = \top$
    (b) $k_{\beta+1} = f(k_\beta)$
    (c) $k_\lambda = \bigwedge_{\beta < \lambda}$, for λ limit.

    Prove the analogues of statements of problems (1)–(4) for this sequence $\langle k_\alpha \rangle_{\alpha \in Ord}$. Show that $\langle k_\alpha \rangle_{\alpha \in Ord}$ converges to the largest fixpoint of f.

6. (Harder problem.) Construct an example of a complete lattice $\langle L, \preceq \rangle$ and a continuous monotone operator f for which the largest fixpoint is reached at some $\alpha > \omega$.

7. Let f be a monotone operator in a complete lattice $\langle L, \preceq \rangle$. Let Fix_f be the set of fixpoints of f. Prove that $\langle Fix_f, \preceq \rangle$ is a complete lattice (although the $g.l.b.$ and $l.u.b.$ in both lattices are different, in general.)

8. An operator f in a poset L is *inflationary* if for all x, $x \preceq f(x)$. Observe that, in general, an inflationary operator does not have to be monotone. Nevertheless, prove that an inflationary operator in a chain-complete lattice with a least element $\bot$ must possess a fixpoint.

9. Assume f and g are two monotone operators in a complete lattice L. Show that f and g have common fixpoints. That is, there exists an element $x \in L$ such that $f(x) = x$ and $g(x) = x$. In fact there exists a least such common fixpoint.

Chapter 2

Introduction to propositional logic

2.1	Syntax of propositional logic	11
2.2	Semantics of propositional logic	13
2.3	Autarkies	23
2.4	Tautologies and substitutions	28
2.5	Lindenbaum algebra	32
2.6	Permutations	34
2.7	Duality	38
2.8	Semantical consequence, operations *Mod* and *Th*	39
2.9	Exercises	42

In this chapter we investigate basic properties of propositional logic. Actually, we do not expect that this book will be the first contact of the reader with logic. However, all the concepts will be rigorously introduced and a number of propositions and theorems proved.

2.1 Syntax of propositional logic

A propositional *variable* (variable for short) is a basic building block of propositional logic. From the point of view of propositional logic a variable has no inner structure. This, of course, is not necessarily true, as the applications of logic show. See the last part of this book, where we discuss various aspects of Knowledge Representation. In fact, we will see that we sometimes want variables to have some kind of structure. The alphabet of logic (over the given set of variables Var) consists of a symbol a for each variable $a \in Var$; two special symbols: $\bot$ and $\top$ (to describe formulas that are always false and always true); the unary symbol $\neg$; and four binary symbols: $\wedge$ (conjunction), $\vee$ (disjunction), $\Rightarrow$ (implication), and $\equiv$ (equivalence). These are called *functors*. Technically, one needs to be a bit careful (for we did not say what can be a variable), but we will not follow these concerns. This given, the set of formulas of propositional logic over the set of variables Var is defined as the least set of strings $Form_{Var}$ satisfying the following conditions:

1. $\bot \in Form_{Var}$
2. $\top \in Form_{Var}$

3. If $a \in Var$, then $a \in Form_{Var}$
4. If $\varphi_1, \varphi_2 \in Form_{Var}$, then $\neg \varphi_1 \in Form_{Var}$, $\varphi_1 \wedge \varphi_2 \in Form_{Var}$, $\varphi_1 \vee \varphi_2 \in Form_{Var}$, $\varphi_1 \Rightarrow \varphi_2 \in Form_{Var}$, and $\varphi_1 \equiv \varphi_2 \in Form_{Var}$.

It is easy to see that there exists such a least set of strings (for instance using the fixpoint theorem), it is the intersection of all sets of strings satisfying the conditions (1)–(4) above.

To eliminate ambiguity we will use brackets: (and). We will try to be as informal as possible in dealing with our syntax, though.

The representation of formulas as strings defined above is not the only one possible. We could use so-called Polish notation, or even reverse Polish notation. The reason is that, in reality, formulas are just ordered labeled binary trees. The leaves of those trees are labeled with special symbols $\top$ and $\bot$ or with variables. The internal nodes are labeled with functors: $\neg, \wedge, \vee, \Rightarrow$, and $\equiv$, with an obvious limitation that a node with one child must be labeled with $\neg$, whereas the node with two children cannot be labeled with $\neg$. Since we expect that the reader has had some rudiments of logic before, we will not discuss more details of these trees.

The representation of formulas as trees has, however, an obvious advantage. Namely, the nodes in a tree have a natural *rank*. Specifically, the leaves have rank 0, and the rank of an internal node is defined as the maximum of ranks of its children, incremented by 1. The *rank* of the formula φ is the rank of its root in the tree of φ. The availability of the rank function makes it possible to prove various properties of formulas by induction.

While every set of variables Var determines its own set of formulas $Form_{Var}$, we will see that, in reality, there is a very strong relationship between various sets $Form_{Var}$. First, let us observe that whenever $Var_1 \subseteq Var_2$, then $Form_{Var_1} \subseteq Form_{Var_2}$. In other words, more variables, more formulas. The rank of formulas does not change when moving between sets of variables. This, of course, needs a proof, which we leave to the reader. Moreover, for every formula φ there is a least set of variables Var so that $\varphi \in Form_{Var}$. We will denote that set by Var_φ. We may also use the symbol $Var(\varphi)$ for that set. Likewise, when T is a set of formulas, then Var_T (we may also write it $Var(T)$) is the set of all propositional variables occurring in formulas of T. Finally, when $Var_1 \subseteq Var_2$ and φ is a formula in $Form_{Var_2}$ which uses only symbols from the alphabet of Var_1, then $\varphi \in Form_{Var_1}$. These syntactic properties of formulas, as well as related semantic properties discussed below, imply that we do not have to be exceedingly formal when discussing the formulas of propositional logic.

We will write $Form$ instead of $Form_{Var}$ whenever we do not need to stress a particular set of variables.

In this book we will often talk about clauses. A *clause* is a formula of the form $l_1 \vee \ldots \vee l_k$, where each l_j, $1 \leq j \leq k$ is a *literal*, i.e., a variable or negated variable. We will see in Chapter 3 that sets of clauses express all sets of formulas. Specifically, for each set of formulas F there is a set of clauses G such that F are semantically equivalent. When we will limit our syntax to clauses, we will talk about *clausal logic*.

2.2 Semantics of propositional logic

We will now assign semantics to formulas. We first need to know what entities will provide the semantics. In this section we will consider two types of such entities. They will be *valuations of variables* and *partial valuations of variables*. Valuations will correspond to two-valued (i.e., Boolean) logic. By contrast, partial valuations are related to three-valued logic, *Kleene logic*, see below. We shall now discuss these two types of logic briefly.

Our first task is to discuss two-valued logic and its semantics of valuations. In two-valued logic *truth values* are 0 and 1 (although values in any Boolean algebra would do). We consider a relational structure

$$Bool = \langle \{0,1\}, \wedge, \vee, \neg, \Rightarrow, \equiv, 0, 1 \rangle.$$

Being naturally careless we often write *Bool* instead of $\{0,1\}$, that is, we do not distinguish between the algebra *Bool* and its universe $\{0,1\}$. The important fact here is that *Bool* forms a Boolean algebra (with additional operations $\Rightarrow$ and $\equiv$, but this is not crucial, for these operations are definable in terms of $\neg, \wedge,$ and $\vee$). Let us observe that we are using the same symbols for operations in the algebra *Bool* as we use in the language of logic. It may be a source of confusion and we will have to be a bit careful at times. The operations in the algebra *Bool* are given by their *truth tables* (Tables 2.1 and 2.2). The arguments in these tables always range over the set of Boolean elements, i.e., $\{0,1\}$.

TABLE 2.1: Truth table for negation

p	$\neg p$
0	1
1	0

TABLE 2.2: Truth table for binary functors

p	q	$p \wedge q$	$p \vee q$	$p \Rightarrow q$	$p \equiv q$	$p + q$
0	0	0	0	1	1	0
0	1	0	1	1	0	1
1	0	0	1	0	0	1
1	1	1	1	1	1	0

We have not discussed until now the binary functor $+$, shown in the last column of Table 2.2. We will need it in Chapter 5.

It should be clear that the structure *Bool* is a Boolean algebra, i.e., it satisfies the axioms for Boolean algebra listed in Chapter 1.

We now define a *valuation* as a function v defined on the set of variables *Var* and with values in $\{0, 1\}$. Thus a valuation assigns a truth value from $\{0, 1\}$ to each variable $x \in Var$. We will also call valuations *variable assignments*.

Given a valuation v of the set of variables *Var*, we extend the function v so it is defined on the entire set of formulas $Form_{Var}$. Here we use the fact that every formula has its rank, and we define the function $\tilde{v}$ by induction on the rank of formulas. Specifically we define:

1. $\tilde{v}(\bot) = 0$, $\tilde{v}(\top) = 1$
2. $\tilde{v}(p) = v(p)$, whenever $p \in Var$
3. $\tilde{v}(\neg \varphi) = \neg \tilde{v}(\varphi)$
4. $\tilde{v}(\varphi \wedge \psi) = \tilde{v}(\varphi) \wedge \tilde{v}(\psi)$
5. $\tilde{v}(\varphi \vee \psi) = \tilde{v}(\varphi) \vee \tilde{v}(\psi)$
6. $\tilde{v}(\varphi \Rightarrow \psi) = \tilde{v}(\varphi) \Rightarrow \tilde{v}(\psi)$
7. $\tilde{v}(\varphi \equiv \psi) = \tilde{v}(\varphi) \equiv \tilde{v}(\psi)$

We need to recognize that although we use the same symbols for operations on both sides of definitions, they have different meanings. On the left-hand side, the symbols $\neg, \wedge, \vee$, etc. are linguistic symbols from the alphabet of logic. On the right-hand side, they denote the names of operations in the structure *Bool* and are evaluated in that structure.

Example 2.1
Here the set of variables *Var* is $\{p, q, r\}$. The valuation v is defined by: $v(p) = v(q) = 0$, $v(r) = 1$. The formula φ is $\neg(\bot \wedge (\neg p \vee q))$. Let us compute $\tilde{v}(\varphi)$. We get, inductively: $\tilde{v}(\neg p) = 1$, $\tilde{v}(\neg p \vee q) = 1$. Then, $\tilde{v}(\bot) = 0$, $\tilde{v}(\bot \wedge (\neg p \vee q)) = 0$, and finally $\tilde{v}(\varphi) = 1$. □

Certainly we want to avoid superfluous terminology, and in particular writing $\tilde{v}$ each time we evaluate a formula with respect to the valuation v. Fortunately, the following is easy to see.

PROPOSITION 2.1
Every valuation v of variables Var uniquely extends to the function $\tilde{v}$ evaluating all formulas in $Form_{Var}$ and satisfying conditions (1)–(7) above.

The proof of Proposition 2.1 uses induction on the rank of formulas. We just show that any function w satisfying (1)–(7) must coincide with $\tilde{v}$. We leave it to the reader as an exercise.

The side effect of Proposition 2.1 is that we can drop the tilde symbol in $\tilde{v}$, and we will do this.

Traditionally, we write $v \models \varphi$ when $v(\varphi) = 1$. We also say that v *satisfies* φ. We will use both conventions; in the literature both are used. Let us observe that we can give an inductive definition of satisfaction relation $v \models \varphi$, as usual by induction on rank of formulas. Let us observe that in this definition we use negative information about $\models$ as well. The symbol $v \not\models \varphi$ means that v does not satisfy φ.

1. $v \models \top, v \not\models \bot$
2. $v \models p$ if p is a variable and $v(p) = 1$
3. $v \models \neg \varphi$ if $v \not\models \varphi$
4. $v \models \varphi \wedge \psi$ if $v \models \varphi$ and $v \models \psi$
5. $v \models \varphi \vee \psi$ if $v \models \varphi$ or $v \models \psi$
6. $v \models \varphi \Rightarrow \psi$ if $v \not\models \varphi$ or $v \models \psi$
7. $v \models \varphi \equiv \psi$ if $(v \models \varphi$ and $v \models \psi)$ or $(v \not\models \varphi$ and $v \not\models \psi)$

Let T be a set of formulas. We write $v \models T$ if for every $\varphi \in T$, $v \models \varphi$. Given a finite set of formulas T, by $\bigwedge T$ we mean the formula

$$\bigwedge \{\varphi : \varphi \in T\}.$$

For instance, when $T = \{\varphi_1, \varphi_2, \varphi_3\}$, then $\bigwedge T$ is $\varphi_1 \wedge \varphi_2 \wedge \varphi_3$. We then have the following fact (following by an easy induction on the size of T, i.e., the number of formulas in T, from the definition of satisfaction, point (4)).

PROPOSITION 2.2
Let T be a finite set of formulas. Then for every valuation v, $v \models T$ if and only if $v \models \bigwedge T$.

It turns out that the value of formula φ under the valuation v (and thus its satisfaction) depends *only* on the values assigned by v to those variables that actually occur in φ. We first need a bit of notation. Let $Var_1 \subseteq Var_2$ be two sets of variables, and let v be a valuation of the set Var_2. By $v\mid_{Var_1}$ we mean a function v' with the domain Var_1 and such that for all $p \in Var_1$, $v'(p) = v(p)$.
Here is a simple but fundamental fact (as usual proved by induction on the rank of formulas).

PROPOSITION 2.3 (Localization theorem)
If $\varphi \in Form_{Var_1}$, $Var_1 \subseteq Var_2$, and v is a valuation of the set Var_2, then $v \models \varphi$ if and only if $v\mid_{Var_1} \models \varphi$.

Here comes the fundamental definition of the main topic of this book. A *satisfiable* set of formulas is a set of formulas F such that there is a valuation v so that $v \models F$. Often, the word *consistent* is used instead of *satisfiable*. We will use both terms interchangeably.

The satisfaction relation $\models$ relates valuations and formulas, and also valuations and sets of formulas. The same relation allows us to relate sets of formulas and formulas. Here is how. Let T be a set of formulas (finite, or not). Let φ be a formula. We say that T entails φ, in symbols: $T \models \varphi$, if for every valuation v of $Var_T \cup Var_\varphi$, whenever $v \models T$ then also $v \models \varphi$. Now, it should be clear that we use the symbol $\models$ in two distinct meanings, but it should be clear from the context which one we mean. We now discuss *joint consistency* of sets of formulas. Given two sets of formulas T_1 and T_2, the sets of variables actually occurring in T_1 and T_2 may be different. It turns out that there is a sufficient condition allowing for conclusion of the consistency of the union of $T_1 \cup T_2$. Let V_1, V_2 be sets of variables occurring in T_1, and T_2, respectively. That is $V_1 = Var_{T_1}$, and $V_2 = Var_{T_2}$. Let $V = V_1 \cap V_2$. Let T be an arbitrary theory (i.e., set of formulas) in the language $\mathcal{L}' \supseteq \mathcal{L}_V$. We say that T is *complete* for V if for every $\varphi \in \mathcal{L}_V$, $T \models \varphi$ or $T \models \neg\varphi$.
We now have the following fact.

LEMMA 2.1
Let T be a satisfiable set of formulas and let us assume that T is complete for $\mathcal{L}_V$. Then there is a unique valuation v of V such that for any valuation v' of Var_T, if $v' \models T$ then $v'|_V = v$.

Proof: Let $x \in V$. Then, by completeness of T with respect to V, either $T \models x$ or $T \models \neg x$. Thus, setting

$$v(x) = \begin{cases} 1 & \text{if } T \models x \\ 0 & \text{if } T \models \neg x \end{cases}$$

we get the desired valuation of V. □
We now get the following corollary.

COROLLARY 2.1 (Robinson theorem)
Let T_i, $i = 1, 2$ be two satisfiable sets of formulas. Let $V = Var_{T_1} \cap Var_{V_2}$. Next, let us assume that each T_i, $i = 1, 2$ is complete with respect to V and that the consequences of T_1 in $\mathcal{L}_V$ and of T_2 in $\mathcal{L}_V$ coincide. Then $T_1 \cup T_2$ is a satisfiable set of formulas.

Proof: Since T_i, $i = 1, 2$ are satisfiable, there are valuations v_1 and v_2 of Var_{T_1} and of Var_{T_2}, respectively, so that $v_i \models T_i$, $i = 1, 2$. But then, by Lemma 2.1 there are unique valuations w_1 and w_2 of V so that v_1 extends w_1 and v_2 extends w_2. Now, w_1 coincides with w_2 since both w_1 and w_2 satisfy the same complete set of formulas in $\mathcal{L}_V$. But then $v = v_1 \cup v_2$ is a valuation, and by Proposition 2.3 $v \models T_1 \cup T_2$, as desired. □

We shall now look at alternative representations of valuations. Let us recall that a *literal* is a variable or negated variable. We talk about the *underlying variable* of a literal, when needed. Sometimes we will denote that variable by $|l|$. The *sign* of a literal l, $sgn(l)$, also called the *polarity* of literal l is 1 (positive) if l is a variable and is 0 (negative) if l is a negated variable. A literal *dual* to variable x is $\neg x$, and the dual of $\neg x$ is x. The literal dual to l is often denoted $\bar{l}$. We denote the set of literals determined by the set of variables Var by Lit_{Var}. We will drop the subscript Var if it will be clear from the context. It should be clear that if $|Var| = n$, then $|Lit| = 2n$.

A *complete set of literals* is a set $S \subseteq Lit$ such that for every $p \in Var$ exactly one of $p, \neg p$ belongs to S. It is easy to see that there is a bijective correspondence between valuations and complete sets of literals. Here is the mapping from valuations to complete sets of literals. When v is a valuation, we define

$$S_v = \{l : l \in Lit \text{ and } v \models l\}.$$

Then, clearly, S_v is a complete set of literals. The mapping $v \mapsto S_v$ is "onto"; given a complete set of literals S, define v_S as follows:

$$v_S(p) = \begin{cases} 1 & \text{if } p \in S \\ 0 & \text{if } \neg p \in S \end{cases}$$

It is now clear (this needs a proof, but it is quite easy) that the mappings $v \mapsto S_v$, $S \mapsto v_S$ are inverse to each other,

It follows immediately that, if $|Var| = n$ then there are precisely 2^n complete sets of literals over set Var.

Boolean algebra *Bool* induces the ordering of truth values 0 and 1, namely $0 \leq 1$. We can think about valuations as elements of the Cartesian product $K = \prod_{p \in Var} \{0, 1\}$. The elements of this Cartesian product are functions from Var to $\{0, 1\}$, thus valuations.

The Cartesian product K may be ordered in a variety of ways (for instance, lexicographically, antilexicographically, etc. if an ordering of the set Var is given). But there is one (partial, i.e., not necessarily linear) ordering of K which will be useful in our considerations. This is the *product* ordering. It is an ordering $\leq$ of K, defined as follows:

$$v_1 \leq v_2 \quad \text{if} \quad \forall_{p \in Var} \ v_1(p) \leq v_2(p).$$

(Recall that every coordinate has its own ordering $\leq$ with $0 \leq 1$.)

We denoted the ordering of the product $\prod_{p \in Var} \{0, 1\}$ by the same symbol, $\leq$, that we used for ordering of *Bool*. The context should make it easy to recognize which of these orderings we are dealing with.

We will now identify the product ordering in the Cartesian product $\prod_{p \in Var} \{0, 1\}$ with the inclusion ordering of subsets of the set of propositional variables. In this fashion we will have yet another representation of valuations.

Given a set of variables M, we assign to M a valuation v_M defined by the *characteristic function* of M, i.e.,

$$v_M(p) = \begin{cases} 1 & p \in M \\ 0 & \text{otherwise.} \end{cases}$$

(Let us observe that often, and we may also use this notation below, the function v_M is denoted by χ_M.)

The assignment $M \mapsto v_M$ establishes a bijective correspondence between the subsets of Var and valuations of Var. This assignment transforms inclusion relation into the relation $\leq$ in the product discussed above. For this reason we will be able to think about valuations in yet another way, namely as subsets of Var. It is easy to formally define the satisfaction relation between *sets* of variables and formulas. Here it is.

1. $M \models \top, M \not\models \bot$
2. $M \models p$ if p is a variable and $p \in M$
3. $M \models \neg\varphi$ if $M \not\models \varphi$
4. $M \models \varphi \wedge \psi$ if $M \models \varphi$ and $M \models \psi$
5. $M \models \varphi \vee \psi$ if $M \models \varphi$ or $M \models \psi$
6. $M \models \varphi \Rightarrow \psi$ if $M \not\models \varphi$ or $M \models \psi$
7. $M \models \varphi \equiv \psi$ if $(M \models \varphi$ and $M \models \psi)$ or $(M \not\models \varphi$ and $M \not\models \psi)$

Summarizing developments above, let us observe that we established three, seemingly different, but really equivalent ways of handling semantics for propositional formulas. These are: valuations, complete sets of literals, and subsets of the set of the propositional variables. We will apply these formalisms where their use can give us an advantage in an easier representation of various properties.

Next, we discuss three-valued logic. There are various three-valued logics. The one we consider here is the so-called Kleene logic. The formulas are the same here, but we have three truth values, not two, and the semantics is provided by partial valuations.

When actually *computing* valuations satisfying a given input formula φ, we will have to deal with *partial valuations*. Those are functions that are defined on *subsets* of the set of variables Var. We will now develop tools to deal with such entities. Like above, our goal is to establish several representations of partial valuations (we will have three) so we can use them whenever convenient. To this end we can think about partial valuations as partial functions from Var to $\{0, 1\}$ or as total functions, but with values in a set consisting of three, not two, values, $\{0, 1, u\}$. Specifically, we assign to a partial function v from Var to $\{0, 1\}$ a total function $v^\star : Var \to \{0, 1, u\}$ as follows:

$$v^\star(p) = \begin{cases} 0 & \text{if } v(p) = 0 \\ 1 & \text{if } v(p) = 1 \\ u & \text{if } v(p) \text{ is undefined.} \end{cases}$$

Then, clearly, we have a bijective correspondence between the partial valuations (with two values) and the (total) valuations taking values in $\{0, 1, u\}$. For this reason we will drop the superscript $\star$, identifying $v^\star$ with v.

There are two natural orderings in the set $\{0, 1, u\}$. The first one is called Kleene (or knowledge) ordering, $\leq_k$. In this ordering u precedes both 0 and 1, whereas 0 and 1 are incomparable. In the ordering $\leq_k$, u describes our ignorance, meaning that later on we may gain the knowledge of the Boolean value of p and hence change the value from u to either 0 or 1. In the second ordering, called Post ordering and denoted $\leq_p$, $0 \leq_p u \leq_p 1$. In this ordering, again u means undefined, but it can either collapse to 0, or increase to 1, we just do not know which.

Complete valuations are, of course, partial valuations. What distinguishes them is that they do not take value u at all. Now, the set of all partial valuations can be thought of as the Cartesian product $\prod_{p \in Var}\{0, 1, u\}$. Obviously $\prod_{p \in Var}\{0, 1\} \subseteq \prod_{p \in Var}\{0, 1, u\}$.

We noticed that the Cartesian product $\prod_{p \in Var}\{0, 1, u\}$ can be ordered in a variety of ways. The natural ordering that we will consider here comes from the ordering $\leq_k$ of $\{0, 1, u\}$. It is the product ordering of $\prod_{p \in Var}\{0, 1, u\}$ where the ordering of $\{0, 1, u\}$ is $\leq_k$. That is, we define the ordering $\leq_k$ in $\prod_{p \in Var}\{0, 1, u\}$ (we use the same symbol for this ordering as the Kleene ordering in $\{0, 1, u\}$, but it should not lead to confusion) by setting

$$v_1 \leq_k v_2 \text{ if } \forall_{p \in Var} \, v_1(p) \leq_k v_2(p).$$

The following should be clear about the ordering $\leq_k$:

$$v_1 \leq_k v_2 \text{ if and only if } \forall_p \, (v_1(p) \neq u \Rightarrow v_1(p) = v_2(p)).$$

In other words, in the ordering $\leq_k$ our knowledge about valuations increases. But once we commit to a Boolean value, we have to stay with it. That is, the value of a variable may be revised from u to some Boolean value, but never the other direction. There is nothing that prevents us from considering the ordering $\leq_p$ in the product. But in this ordering $\leq_p$, $0 \leq_p u$, so changing the value of a valuation from 0 to u results in a valuation that is strictly larger in $\leq_p$. This is *not* allowed in $\leq_k$.

Example 2.2
Let $Var = \{p, q, r, s\}$ and $v'(p) = 0, v'(q) = 1, v'(r) = v'(s) = u$. Now let v'' be defined by $v''(p) = 0, v''(q) = 1, v''(r) = u$ and $v''(s) = 1$. Then $v' \leq_k v''$. □

Two-valued assignments can be characterized as the maximal (in the ordering $\leq_k$) partial valuations. Here is a general result (again without proof) on the structure of the product ordering $\leq_k$.

PROPOSITION 2.4

1. *The poset $\langle \prod_{p \in Var}\{0, 1, u\}, \leq_k \rangle$ has a least element; it is the partial valuation constantly equal to u.*
2. *If $|Var| > 0$ then there is no largest element of $\langle \prod_{p \in Var}\{0, 1, u\}, \leq_k \rangle$.*
3. *The maximal elements of $\langle \prod_{p \in Var}\{0, 1, u\}, \leq_k \rangle$ are precisely two-valued, complete valuations, i.e., elements of $\prod_{p \in Var}\{0, 1\}$.*

4. The poset $\langle \prod_{p \in Var}\{0, 1, u\}, \leq_k \rangle$ is chain-complete, and thus the Knaster-Tarski theorem applies to monotone functions in $\langle \prod_{p \in Var}\{0, 1, u\}, \leq_k \rangle$.

A corresponding theorem for Post ordering of the product will be discussed in the Section 2.9 Exercises.

It turns out that partial valuations can be characterized in terms of sets of literals, in a manner similar to our characterization of valuations by complete sets of literals. Specifically, we say that a set of literals, S, is *consistent* if for every variable p, at most one of $p, \neg p$ belongs to S. We then have a bijective correspondence between consistent sets of literals and partial valuations. Namely, given a partial valuation v let us define

$$S_v = \{p : v(p) = 1\} \cup \{\neg p : v(p) = 0\}.$$

Then clearly S_v is a consistent (but not necessarily complete) set of literals. Conversely, given a consistent set S of literals, let us define

$$v_S = \begin{cases} 1 & \text{if } p \in S \\ 0 & \text{if } \neg p \in S \\ u & \text{if } p \notin S \text{ and } \neg p \notin S \end{cases}$$

It should be clear that we established a bijective correspondence between consistent sets of literals and three-valued valuations of Var.

Let us observe that, in one of the basic algorithms for satisfiability testing, the DPLL algorithm (which we will discuss in Chapter 8 as we construct a valuation in expectation that it will satisfy a given input set of formulas F), we build a chain of *partial* valuations, extending each other, i.e., growing along the ordering $\leq_k$. In the process, we need to proceed "locally," making sure that the partial valuations satisfy some formulas from F. Two issues must be settled to make this process useful. First we need to be able to evaluate formulas – assign to them a truth value. Second, we need to be sure that once a value that is assigned to a formula is Boolean (i.e., 0 or 1) it will stay this way in the future stages of construction. Let us observe that we are perfectly happy if a formula that is currently evaluated as u changes value. But once that value is changed to 0 or to 1, it must stay this way. So it should be clear that we need to define some evaluation function which will assign the value to formulas. This value will come from the set $\{0, 1, u\}$. Then if we do this right, the evaluation function will have the following property: once this value is settled as a Boolean, it will not change as we build bigger (in the ordering $\leq_k$) partial valuation.

To realize this idea let us introduce three-valued truth tables for our connectives in Tables 2.3 and 2.4.

Now, having defined the three-valued truth tables it is very natural to define the three-valued truth function for all formulas in $Form_{Var}$. It is the same definition of truth function we had in the two-valued case, except that we evaluate formulas in the set $\{0, 1, u\}$ as prescribed by our tables. Let us observe that the operations in truth values correspond to Post ordering, not Kleene ordering (but be careful with the implication!). In particular the logical value 1 is $\leq_p$-bigger than the *undefined*, which

Introduction to propositional logic

TABLE 2.3: Three-valued truth table for negation

p	$\neg p$
0	1
1	0
u	u

TABLE 2.4: Three-valued truth tables for binary operations

p	q	$p \wedge q$	$p \vee q$	$p \Rightarrow q$	$p \equiv q$
0	0	0	0	1	1
0	u	0	u	1	u
0	1	0	1	1	0
u	0	0	u	u	u
u	u	u	u	u	u
u	1	u	1	1	u
1	0	0	1	0	0
1	u	u	1	u	u
1	1	1	1	1	1

in turn is $\leq_p$-bigger than the Boolean value 0. Let us observe that the conjunction and disjunction are, respectively, min and max functions, whereas $\neg$ is an involution mapping 0 to 1 but keeping u fixed. Let us observe that we want to extend the algebra *Bool* by adding the value u. The truth values 0 and 1 must be ordered as they were in *Bool*. This, together with the interpretation of Boolean values as commitments, brings us to the tables for the three-valued logic. Thus we have the following inductive definition of evaluation function (we write $v_3(\varphi)$ to stress the fact that we are dealing with the three-valued logic here).

1. $v_3(\bot) = 0, v_3(\top) = 1$
2. $v_3(p) = v(p)$ whenever $p \in Var$
3. $v_3(\neg \varphi) = \neg v_3(\varphi)$
4. $v_3(\varphi \wedge \psi) = v_3(\varphi) \wedge v_3(\psi)$
5. $v_3(\varphi \vee \psi) = v_3(\varphi) \vee v_3(\psi)$
6. $v_3(\varphi \Rightarrow \psi) = v_3(\varphi) \Rightarrow v_3(\psi)$
7. $v_3(\varphi \equiv \psi) = v_3(\varphi) \equiv v_3(\psi)$

The operations on the right-hand side of the assignment operator refer to those defined in the set $\{0, 1, u\}$ by our tables whereas the same symbols on the left-hand side refer to functors.

We have the following facts which will be useful in many places in this book.

PROPOSITION 2.5 (Kleene theorem)

1. If v is a complete valuation (two-valued, that is, taking only Boolean values) then for every formula φ, $v_3(\varphi) = v(\varphi)$. In particular, for complete (two-valued) valuations v, v_3 takes only values 0 and 1.

2. If v and w are partial valuations, $v \leq_k w$, and φ is an arbitrary formula then $v_3(\varphi) \leq_k w_3(\varphi)$.

3. If v is a partial valuation, and φ is a formula and w is the restriction of v to variables that belong both to Var_φ and to the domain of v, then $v_3(\varphi) = w_3(\varphi)$.

Proof: To see (1), let us observe that the tables for three-valued functors, restricted to Boolean values coincide with the corresponding two-valued tables. Then, by induction on the rank of subformulas ψ of the formula φ we show that only values 0 and 1 are used.
(2) Here we need to proceed by induction on the rank of formula φ. The base case is obvious due to the fact how we evaluate constants and variables. Then the inductive step is a bit tedious (there are five cases to check), but is not complicated at all.
(3) is again proved by an easy induction on the rank of formulas. □

Let us see what Proposition 2.5 says. Namely, the first part says that for complete valuations we do not really get anything new; the evaluation functions $v(\cdot)$ and $v_3(\cdot)$ in two-valued and in three-valued logics coincide for complete (i.e., two-valued) valuations v's. But an easy example can be given of a situation where a valuation which is partial already commits us to a Boolean value. Let us take $\varphi : p \wedge q$ and a partial valuation defined on a single variable p and assigning to p the Boolean value 0. In such case $v_3(p \wedge q) = 0$. What the second part of Proposition 2.5 says is that the value can go only up in the ordering $\leq_k$. It may change from u to 0 or to 1. But if we committed ourselves to a *Boolean* value as in the example above, this value will never be undone in extensions of the current partial valuation. In particular, if we test a set of formulas F for satisfiability, we construct a partial valuation v and for some $\varphi \in T$, we get the value $v_3(\varphi) = 0$, then for *no* extension of v to a complete valuation w we will have $w(\varphi) = 1$. Thus w will not satisfy F and so we must backtrack. If, on the other hand, we already have $v(\varphi) = 1$, then we do not have to check for extensions w of v if they satisfy φ. They definitely do.

The property (3) tells us that, like in the case of two-valued valuations, the three-valued truth function $v_3(\cdot)$ depends only on the variables of φ that occur in the domain of v (in the two-valued case this last condition was immaterial – a valuation was defined on all variables).

So we now have the means to deal with partial valuations. The tool for doing this is three-valued logic which has Boolean truth values and an additional logical value u.

Let us conclude this section by reiterating that like in the case of two-valued logic we now have three different ways of handling partial valuations: by means of partial functions of Var into $\{0,1\}$, total functions of Var into $\{0,1,u\}$ and consistent sets of literals.

2.3 Autarkies

We will now discuss partial valuations v that have a useful property with respect to a given set of formulas F. Namely, the instantiation of variables according to such valuations does not change satisfiability of F. Those partial valuations are known as *autarkies*.

Let v be a partial valuation. Let us think about v as a partial function on Var. We say that v *touches* a formula φ if some variable in the domain of v belongs to Var_φ, that is, occurs in φ. Then, we say that v is an *autarky* for F if for all $\varphi \in F$ such that v touches φ, $v_3(\varphi) = 1$. While autarkies are used mostly in the context of clausal logic, that is when the formulas $\varphi \in F$ are clauses (and in fact in this section we will prove several properties of autarkies for clausal theories), it is, of course, perfectly natural to study autarkies in a general context.

The following observation is entirely obvious.

PROPOSITION 2.6
The empty partial valuation is an autarky for any set of formulas F. Moreover, any satisfying valuation for F is an autarky for F.

In the context of clausal logic (that is, when we limit the formulas to clauses, i.e., disjunctions of literals, we call a literal l *pure* in a set of clauses F if $\bar{l}$ does not occur in any clause of F. Later on, when we have appropriate language, we will generalize this notion to the general case.

We will discuss the conjunctive normal forms (yes, there is more than one) in Chapter 3 below, and prove that clausal logic (sets of clauses) is as expressive as full propositional logic. Right now, however, we need to discuss autarkies and this will require looking at sets of clauses. We first need a simple characterization of those clauses that are tautologies. We recall that a literal is a propositional variable or its negation, that each literal has its sign, and that each literal has an underlying propositional variable. We also recall the concept of a dual literal. The following fact is simple but useful.

PROPOSITION 2.7

1. *Every nonempty clause is satisfiable*
2. *A clause $C = l_1 \vee \ldots \vee l_k$ is a tautology (i.e., it is satisfied by all valuations) if and only if for some i, j, $1 \leq i < j \leq k$, l_j is dual of l_i.*

Proof: (1) is obvious, just locate a literal in C and take any valuation making it true. (2) If C contains a pair of dual literals l_i, l_j then C is a tautology because every valuation v evaluates one of l_i, l_j as true. If C does not contain a pair of dual literals, then here is a valuation making C false:

$$v(p) = \begin{cases} 1 & \text{if } p \text{ occurs in } C \text{ negatively} \\ 0 & \text{if } p \text{ occurs in } C \text{ positively} \\ 0 & \text{if } p \text{ does not occur in } C \text{ at all.} \end{cases}$$

It should be clear that v evaluates C as 0. □

We will often say that a formula φ is *nontautological* if φ is not a tautology.

PROPOSITION 2.8
Let F be a collection of clauses, and l be a pure literal in F. Then $v = \{l\}$ is an autarky for F.

Proof: If v touches C then it must be the case that $\bar{l}$ does not occur in C, so l occurs in C and so $v_3(C) = 1$. □

Now, given any partial valuation v, and a set of formulas F, v determines a partition $F = F'_v \cup F''_v$ as follows: F''_v is the set of those formulas in F that are touched by v, F'_v is the rest of the formulas in F.

PROPOSITION 2.9
If v is an autarky for F, then F is satisfiable if and only if F'_v is satisfiable.

Proof: Since $F'_v \subseteq F$, if F is satisfiable, so is F'_v. Conversely, assume that w is a valuation satisfying F'_v. Define a new valuation w' as follows:

$$w'(l) = \begin{cases} v(l) & \text{if } v \text{ is defined on } l \\ w(l) & \text{otherwise.} \end{cases}$$

We claim that w' satisfies F. Indeed, let $\varphi \in F$. If $\varphi \in F''$, that is, if φ is touched by v then $(w')_3(\varphi) = 1$ because $v_3(\varphi) = 1$ and $v \prec_k w'$. As w' is a valuation, $w' \models \varphi$. If $vph \in F'_w$ then φ is not touched by v then $w_3(\varphi) = 1$ (as $\varphi \in F'_v$). Then no variable occurring in φ belongs to the domain of v. But then, on every variable x occurring in φ, $w(x) = w'(x)$. As $w \models \varphi$, $w' \models \varphi$, as desired. □

Later on in this section we prove a stronger property of autarkies, generalizing Proposition 2.9.

Before we prove our next fact, we need a simple definition. Let v and w be two consistent sets of literals. Define $v \oplus w$ to be the set of literals

$$v \oplus w := v \cup \{l \in w : \bar{l} \notin v\}.$$

We then have the following properties of autarkies.

PROPOSITION 2.10

1. If F is a set of formulas, v is a partial valuation, $F' \subseteq F$ and v is an autarky for F, then v is an autarky for F'.
2. If v is a partial valuation, and F_1, F_2 are two sets of formulas, and v is an autarky for F_i, $i = 1, 2$, then v is an autarky for $F_1 \cup F_2$.
3. If v_1, v_2 are autarkies for F, then $v_1 \oplus v_2$ is an autarky for F, and thus:
4. If v_1, v_2 are autarkies for F and the union $v_1 \cup v_2$ is consistent, then $v_1 \cup v_2$ is an autarky for F.

Proof: (1) and (2) are obvious.
For (3), consider $\varphi \in F$ and let us assume $v_1 \oplus v_2$ touches φ. If v_1 touches φ then $v_1 \models \varphi$, thus $(v_1 \oplus v_2)_3(\varphi) = 1$ because $v_1 \leq_k v_1 \oplus v_2$. So now let us assume that v_1 does not touch φ. Then v_2 touches φ. But since v_1 does not touch φ, and v_2 touches φ it must be the case that v_2 restricted to those variables that do not occur in v_1 satisfies φ (cf. Proposition 2.5(3)). Let us call this partial valuation w. Then $w_3(\varphi) = 1$, and since $w \leq_k v_1 \oplus v_2$, $(v_1 \oplus v_2)_3(\varphi) = 1$, as desired.
(4) follows from (3) since, if the union $v_1 \cup v_2$ is consistent, $v_1 \cup v_2 = v_1 \oplus v_2$. □

Let us call an autarky v for F *complete* if for every variable p, either $p \in F$ or $\neg p \in F$. Of course complete autarkies are just satisfying valuations. An *incomplete* autarky for F is an autarky for F which is *not* complete. Here is a characterization of incomplete autarkies.

PROPOSITION 2.11
Let v be a consistent set of literals. Let p be a variable, and let us assume that neither p nor $\neg p$ belongs to a set v. Then v is an autarky for F if and only if v is an autarky for $F \cup \{p, \neg p\}$.

Proof: Clearly, by our choice of v, v does not touch either p or $\neg p$. Thus all v touch in $F \cup \{p, \neg p\}$ are formulas in F which are satisfied by v since v is an autarky for F.
Conversely, since v is an autarky for $F \cup \{p, \neg p\}$, it is an autarky for its subset, F (Proposition 2.10(1)). □

One property that does *not* hold for autarkies is inheritance by subsets. Specifically, if v is an autarky for F and $w \subseteq v$, then w does not need to be an autarky for F. Here is an example. Let F consists of clauses $p \vee q$ and $\neg p \vee \neg q$. Then $\{p, \neg q\}$ is an autarky for F (because it satisfies F), but $\{p\}$ is not an autarky for F.

Our goal now is to characterize autarkies in terms of extensions of partial valuations. The characterizations we provide can be expressed in various terms, for instance in terms of two-element Boolean algebra. Here we will state those results in terms of partial valuations. We need some notation. We will think about partial valuations both in terms of partial functions and consistent sets of literals. Let us fix a set of clauses F. Given a partial valuation v, Var_v is the domain of v. By O_v we denote the set of other variables in the set $Var := Var_F$, that is, those which are not in Var_v, in other words, $Var \setminus Var_v$. By definition $O_v \cap Var_v = \emptyset$, and $Var_v \cup O_v$ is entire Var. Valuations of variables of F are of the form $v_1 \oplus v_2$ where the domain of v_1 is Var_v and the domain of v_2 is O_v.

We now have the following characterization of autarkies.

PROPOSITION 2.12

Let F be a set of nonempty, nontautological clauses. A nonempty partial valuation v is an autarky for F if and only if for every clause $C \in F$ and for every partial valuation w such that $Dom(w) = O_v$, the following are equivalent:

(a) There exists a partial valuation u such that $Dom(u) = Var_v$ and $w \oplus u \models C$.

(b) $w \oplus v \models C$.

Proof: First, let us assume that v is an autarky for F. Let C be an arbitrary clause in F. The implication (b)$\Rightarrow$(a) is obvious. For the other implication, let us assume that for some u such that $Dom(u) = Var_v$,

$$w \oplus u \models C.$$

If v touches C then we are done because $v_3(C) = 1$ and so $(v \oplus w)_3(C) = 1$ and since $v \oplus w$ is complete,

$$v \oplus w \models C.$$

If v does not touch C then $Var_C \subseteq O_v = Dom(w)$. If for some u with the domain of u equal to Var_v,

$$w \oplus u \models C$$

then $w \models C$ and so

$$v \oplus w \models C.$$

Conversely, let us assume the equivalence of (a) and (b). If v does not touch C there is nothing to prove. So let us assume that v touches C. Let us consider a valuation w with $Dom(w) = O_v$ which evaluates all literals l in C that have the underlying variable in O_v as 0. Since C is not tautological and so does not contain a pair of dual literals at least one such partial valuation exists. There may be many such partial valuations w; let us fix one. Since v touches C it cannot be the case that w evaluates all literals in C, for w is defined on O_v and nothing else. But we could take that literal in C that is not evaluated by w and make it 1. This means that we can define a partial valuation u on Var_v so that $u_3(C) = 1$, and therefore

$$w \oplus u \models C.$$

But then, since the equivalence of (a) and (b) is assumed,

$$v \oplus w \models C.$$

Now, $w \oplus v$ evaluates variables in O_v so that all literals in C with an underlying variable in O_C are evaluated as 0. But then some other literal l in C is evaluated as 1 and it must be the case that $v(l) = 1$, i.e., $v_3(C) = 1$, as desired. □

Proposition 2.12 allows us to obtain a useful property of autarkies.

PROPOSITION 2.13

Let F be a set of nonempty, nontautological clauses. If a nonempty partial valuation v is an autarky for F then for every partial valuation w such that $Dom(w) = Var_v$, the following are equivalent:
(a) There exists a partial valuation u such that $Dom(u) = Var_v$ and $w \oplus u \models \bigwedge F$.
(b) $w \oplus v \models \bigwedge F$.

Proof: Let us choose w as in assumptions. The implication $(b) \Rightarrow (a)$ is again obvious, for we can take v for u.

So let us assume that there exists a partial valuation u such that $Dom(u) = Var_v$ and $w \oplus u \models \bigwedge F$. Then, in particular, for every clause C of F there is a valuation u with the domain Var_v so that $w \oplus u \models C$. But then by Proposition 2.12, $w \oplus v \models C$. But then

$$w \oplus v \models \bigwedge F$$

as desired. □

Finally, we have another property of autarkies. Recall that F'_v is the set of clauses in F which are *not* touched by v.

PROPOSITION 2.14

Let F be a set of nonempty, nontautological clauses. If a nonempty partial valuation v is an autarky for F then for every partial valuation w such that $Dom(w) = Var_v$, the following are equivalent:
(a) There exists a partial valuation u such that $Dom(u) = Var_v$ and $w \oplus u \models \bigwedge F$
(b) $w \models \bigwedge F'_v$.

Proof: Let w be a partial valuation with the domain Var_v. First, let us assume that for some partial valuation u with the domain Var_v

$$w \oplus u \models \bigwedge F.$$

Since v is an autarky, by Proposition 2.13,

$$w \oplus v \models \bigwedge F.$$

Now, $F'_v \subseteq F$ and so

$$w \oplus v \models \bigwedge F'_v.$$

But by the choice of F'_v, all variables of F'_v are already evaluated by w, so

$$w \models \bigwedge F'_v.$$

Conversely, let us assume that w is defined on Var_w and

$$w \models \bigwedge F'_v.$$

Then, clearly,

$$w \oplus v \models \bigwedge F'_v.$$

Now, for clauses C in F''_v, that is, those touched by v, $v_3(C) = 1$. But then $(v \oplus w)_3(C) = 1$ and so

$$v \oplus w \models C.$$

Therefore $v \oplus w \models \bigwedge F$, and so there is u (namely v) making (a) true. □

2.4 Tautologies and substitutions to tautologies

A *tautology* is a formula $\varphi \in Form_{Var}$ such that for all valuations v of Var, $v \models \varphi$. The notion of tautology is closely related to that of satisfiable formula. Since for every formula φ and for every valuation v, either $v \models \varphi$, or $v \models \neg\varphi$ (but not both at once) holds, we have the following fact.

PROPOSITION 2.15
A formula φ is satisfiable if and only if $\neg\varphi$ is not a tautology.

Proposition 2.15 tells us that an engine capable of testing whether a formula is a tautology can be used to test whether a formula is satisfiable (by testing $\neg\varphi$ for being a tautology).

In principle, to test if a formula $\varphi \in Form_{Var}$, with $|Var| = n$ is a tautology requires testing if all 2^n valuations satisfy φ. We will later see that the satisfiability problem (i.e., the language consisting of satisfiable sets of formulas) is NP-complete. But if we are naïve, and all we know is the table method, then the order of $O(n \cdot 2^n)$ operations is needed for tautology testing.

In Table below we list a sample of tautologies (we will see below that there are infinitely many tautologies, even if the language has only finitely many variables).

All these tautologies are given traditional names. For instance the first of these tautologies is commonly called the "law of excluded middle." The fourth and fifth tautologies in Table 2.5 are called "idempotence laws." The twelfth and thirteenth tautologies are called "De Morgan laws." The tenth and eleventh tautologies are called "distributive laws."

TABLE 2.5: A sample of tautologies

$p \vee \neg p$
$\neg(p \wedge \neg p)$
$p \Rightarrow p$
$p \Rightarrow (q \Rightarrow p)$
$(p \wedge p) \equiv p$
$(p \vee p) \equiv p$
$p \wedge q \equiv q \wedge p$
$p \vee q \equiv q \vee p$
$p \wedge (q \wedge r) \equiv (p \wedge q) \wedge r$
$p \vee (q \vee r) \equiv (p \vee q) \vee r$
$p \wedge (q \vee r) \equiv (p \wedge q) \vee (p \wedge r)$
$p \vee (q \wedge r) \equiv (p \vee q) \wedge (p \vee r)$
$\neg(p \wedge q) \equiv (\neg p \vee \neg q)$
$\neg(p \vee q) \equiv (\neg p \wedge \neg q)$
$\neg\neg p \equiv p$

We will now show how these and other tautologies can be used to generate more tautologies. To this end, a *substitution* is a function s that assigns formulas to variables. When $p_1, \ldots, p_n$ are variables and $\psi_1, \ldots, \psi_n$ are formulas we write such substitution s as

$$\begin{pmatrix} p_1 & \cdots & p_n \\ \psi_1 & \cdots & \psi_n \end{pmatrix}$$

if $s(p_1) = \psi_1, s(p_2) = \psi_2$, etc. The formulas ψ_j, $1 \leq j \leq n$ may themselves involve variables from $p_1, \ldots, p_n$ but also additional variables. For instance $\begin{pmatrix} p & q \\ \neg q \vee r & \neg t \end{pmatrix}$ is a substitution s such that $s(p) = \neg q \vee r$ and $s(q) = \neg t$.

Substitutions act on formulas. We will denote the result of the action of the substitution s on a formula φ by φs. The result of the action of substitution s on the formula φ is a *simultaneous* substitution of ψ_1 for p_1, ψ_2 for p_2 etc. For instance if φ is $p \Rightarrow (q \Rightarrow p)$ and $s = \begin{pmatrix} p & q \\ \neg q \vee r & \neg t \end{pmatrix}$ then φs is the formula

$$(\neg q \vee r) \Rightarrow (\neg t \Rightarrow (\neg q \vee r)).$$

It is convenient (at least for the author, that is) to think about formulas as trees while dealing with substitutions. What happens here is that for each *leaf* labeled with p_j we substitute the entire tree of formula ψ_j (the label in that node changes from the variable p_j to the label of the root of ψ_j). This operation affects leaves, but not the internal nodes of the tree for φ. The result of the action of substitution s on a formula φ maintains the "top part" of the structure of φ, but allows us to make the overall structure more complex.

Given the substitution $\begin{pmatrix} p_1 & \cdots & p_n \\ \psi_1 & \cdots & \psi_n \end{pmatrix}$ and a valuation v of variables occurring in formulas $\psi_1, \ldots, \psi_n$, we can construct a new valuation v' defined on variables $p_1, \ldots p_n$

by setting
$$v'(p_j) = v(\psi_j).$$

Here is an example. Let us consider the substitution $\begin{pmatrix} p & q \\ \neg q \vee r & \neg t \end{pmatrix}$. The formulas $\neg q \vee r$ and $\neg t$ have three variables: q, r, and t. Let v be a valuation given by $v(q) = 1$, $v(r) = 0$ and $v(t) = 0$. Then the valuation v' is defined on variables p and q, and $v'(p) = v(\neg q \vee r) = 0$ whereas $v'(q) = v(\neg t) = 1$.

We will now establish a property that will allow us to generate new tautologies at will.

PROPOSITION 2.16 (Substitution lemma)

Let $\varphi \in Form_{\{p_1,\ldots,p_n\}}$ be a formula in propositional variables $p_1, \ldots, p_n$ and let $\langle \psi_1, \ldots, \psi_n \rangle$ be a sequence of propositional formulas. Let v be a valuation of all variables occurring in $\psi_1, \ldots, \psi_n$. Finally, let v' be a valuation of variables $p_1, \ldots, p_n$ defined by $v'(p_j) = v(\psi_j)$, $1 \leq j \leq n$. Then

$$v'(\varphi) = v\left(\varphi \begin{pmatrix} p_1 & \cdots & p_n \\ \psi_1 & \cdots & \psi_n \end{pmatrix}\right).$$

Proof: By induction of the rank of the formula φ. To begin induction, let us look at the formulas $\bot, \top$ and variables.

If $\varphi = \bot$ then $\varphi\left(\begin{pmatrix} p_1 & \cdots & p_n \\ \psi_1 & \cdots & \psi_n \end{pmatrix}\right)$ is also $\bot$, and $\bot$ is evaluated to 0 both by v and v', so the desired equality holds. The argument for $\top$ is very similar.

If $\varphi = p_i$ for a variable p_i, then $\varphi\left(\begin{pmatrix} p_1 & \cdots & p_n \\ \psi_1 & \cdots & \psi_n \end{pmatrix}\right) = \psi_i$. By the definition of v', $v'(p_i) = v(\psi_i)$. Now, we have:

$$v'(\varphi) = v'(p_i) = v(\psi_i) = v\left(p_i\left(\begin{pmatrix} p_1 & \cdots & p_n \\ \psi_1 & \cdots & \psi_n \end{pmatrix}\right)\right).$$

Thus $v'(\varphi) = v\left(\varphi \begin{pmatrix} p_1 & \cdots & p_n \\ \psi_1 & \cdots & \psi_n \end{pmatrix}\right)$ as desired.

In the inductive step there are five cases, corresponding to connectives $\neg, \wedge, \vee, \Rightarrow$, and $\equiv$. We shall discuss the case of $\Rightarrow$; the other cases are similar. So let $\varphi := \varphi' \Rightarrow \varphi''$. Clearly

$$(\varphi' \Rightarrow \varphi'')\begin{pmatrix} p_1 & \cdots & p_n \\ \psi_1 & \cdots & \psi_n \end{pmatrix} = \left(\varphi'\begin{pmatrix} p_1 & \cdots & p_n \\ \psi_1 & \cdots & \psi_n \end{pmatrix}\right) \Rightarrow \left(\varphi''\begin{pmatrix} p_1 & \cdots & p_n \\ \psi_1 & \cdots & \psi_n \end{pmatrix}\right).$$

The reason for this is obvious when we think about formulas as trees. The tree for φ has a root labeled with $\Rightarrow$, and the left subtree (which is the tree for φ') and the right subtree (the tree for φ''). The substitution works on the left subtree and on the right subtree separately.

Introduction to propositional logic

So now the use of inductive assumption for simpler formulas φ' and φ'' makes matters quite simple.

$$\begin{aligned}
v'(\varphi' \Rightarrow \varphi'') &= v'(\varphi') \Rightarrow v'(\varphi'') \\
&= v\left(\varphi'\begin{pmatrix} p_1 & \cdots & p_n \\ \psi_1 & \cdots & \psi_n \end{pmatrix}\right) \Rightarrow v\left(\varphi''\begin{pmatrix} p_1 & \cdots & p_n \\ \psi_1 & \cdots & \psi_n \end{pmatrix}\right) \\
&= v\left(\varphi'\begin{pmatrix} p_1 & \cdots & p_n \\ \psi_1 & \cdots & \psi_n \end{pmatrix} \Rightarrow \varphi''\begin{pmatrix} p_1 & \cdots & p_n \\ \psi_1 & \cdots & \psi_n \end{pmatrix}\right) \\
&= v\left(\varphi\begin{pmatrix} p_1 & \cdots & p_n \\ \psi_1 & \cdots & \psi_n \end{pmatrix}\right)
\end{aligned}$$

as sought. □

Now we get a desired corollary.

COROLLARY 2.2

Let φ be a tautology, with variables of φ among $p_1, \ldots, p_n$. Then for every choice of formulas $\langle \psi_1, \ldots, \psi_n \rangle$, the formula $\varphi\begin{pmatrix} p_1 & \cdots & p_n \\ \psi_1 & \cdots & \psi_n \end{pmatrix}$ is a tautology.

Proof: Let *Var* be the set of variables occurring in formulas $\psi_1, \ldots, \psi_n$. Only variables from *Var* occur in $\varphi\begin{pmatrix} p_1 & \cdots & p_n \\ \psi_1 & \cdots & \psi_n \end{pmatrix}$. Now let v be any valuation of variables from *Var*. Let v' be as in the substitution lemma (Proposition 2.16). Then

$$v'(\varphi) = v\left(\varphi\begin{pmatrix} p_1 & \cdots & p_n \\ \psi_1 & \cdots & \psi_n \end{pmatrix}\right).$$

But φ is a tautology. Therefore $v'(\varphi) = 1$. Thus $v\left(\varphi\begin{pmatrix} p_1 & \cdots & p_n \\ \psi_1 & \cdots & \psi_n \end{pmatrix}\right) = 1$ and since v was an arbitrary valuation of *Var*, $\varphi\begin{pmatrix} p_1 & \cdots & p_n \\ \psi_1 & \cdots & \psi_n \end{pmatrix}$ is a tautology. □

Coming back to our example we conclude that since $p \Rightarrow (q \Rightarrow p)$ is a tautology, the formula

$$(\neg q \vee r) \Rightarrow (\neg t \Rightarrow (\neg q \vee r))$$

is also a tautology.

Since there are infinitely many formulas, there are infinitely many substitutions. Thus from a single tautology (for instance any tautology from our list in Table 2.5) we can generate infinitely many tautologies.

2.5 Lindenbaum algebra

Let Var be a set of propositional variables. There is a natural equivalence relation in the set of all formulas $Form_{Var}$ induced by the notion of valuation. Namely, let us say that two formulas φ_1 and φ_2 in $Form_{Var}$ are *equivalent* if for every valuation v, $v(\varphi_1) = v(\varphi_2)$. We write $\varphi_1 \sim \varphi_2$ if φ_1 and φ_2 are equivalent. The following fact should be pretty obvious.

PROPOSITION 2.17
The relation $\sim$ is an equivalence relation, that is, it is reflexive, symmetric, and transitive.

It turns out that the relation $\sim$ is closely related to tautologies. Indeed, we have the following fact.

PROPOSITION 2.18
Let φ_1 and φ_2 be formulas in $Form_{Var}$. Then $\varphi_1 \sim \varphi_2$ if and only if the formula $\varphi_1 \equiv \varphi_2$ is a tautology.

Proof: If $\varphi_1 \sim \varphi_2$ then for all valuations v, $v(\varphi_1) = v(\varphi_2)$. But

$$v(\varphi_1 \equiv \varphi_2) = \begin{cases} 1 & \text{if } v(\varphi_1) = v(\varphi_2) \\ 0 & \text{otherwise.} \end{cases}$$

Thus if $\varphi_1 \sim \varphi_2$ then for all v, $v(\varphi_1 \equiv \varphi_2) = 1$, i.e., $\varphi_1 \equiv \varphi_2$ is a tautology. Converse reasoning is equally obvious. □

Let us denote the equivalence class of a formula φ (w.r.t. the relation $\sim$) by $[\varphi]$. The collection of all these equivalence classes is called *Lindenbaum algebra* and denoted $Form/\sim$. We shall introduce some operations in this set. Let us define:

1. $\top = [\top]$
2. $\bot = [\bot]$
3. $\neg [\varphi] = [\neg \varphi]$
4. $[\varphi_1] \wedge [\varphi_2] = [\varphi_1 \wedge \varphi_2]$
5. $[\varphi_1] \vee [\varphi_2] = [\varphi_1 \vee \varphi_2]$
6. $[\varphi_1] \Rightarrow [\varphi_2] = [\varphi_1 \Rightarrow \varphi_2]$
7. $[\varphi_1] \equiv [\varphi_2] = [\varphi_1 \equiv \varphi_2]$

Although the same symbols are used on the left-hand side and on the right-hand side, there is a significant difference. On the left-hand side we define a constant (first two cases) or an operation in Lindenbaum algebra. The corresponding symbols on the right-hand side are functors of the propositional language.

It is not even clear that the definition given above is proper, that is, that the result does not depend on the choice of representatives of the equivalence classes. But a bit of work shows that this is, indeed, the case.

PROPOSITION 2.19
The values of operations $\neg, \wedge, \vee, \Rightarrow,$ and $\equiv$ do not depend on the choice of representatives of equivalence classes.

Proof: There are five cases to prove, but we will deal with one, leaving the remaining four cases to the reader. Let us look at the operation $\Rightarrow$ (chosen randomly by our random number generator).

What we need to prove is that if $\varphi_1 \sim \psi_1$ and $\varphi_2 \sim \psi_2$ then $(\varphi_1 \Rightarrow \varphi_2) \sim (\psi_1 \Rightarrow \psi_2)$. To this end, let v be any valuation assigning values to all the variables occurring in all the formulas involved. We prove that

$$v(\varphi_1 \Rightarrow \varphi_2) = v(\psi_1 \Rightarrow \psi_2).$$

Case 1: $v(\varphi_1 \Rightarrow \varphi_2) = 1$. Again we need to look at two subcases.
Subcase 1.1: $v(\varphi_1) = 1$. Then $v(\varphi_2) = 1$, $v(\psi_2) = 1$ and so $v(\psi_1 \Rightarrow \psi_2) = 1$. Thus the desired equality holds.
Subcase 1.2: $v(\varphi_1) = 0$. Then $v(\psi_1) = 0$, and so $v(\psi_1 \Rightarrow \psi_2) = 1$, as desired.
Case 2: $v(\varphi_1 \Rightarrow \varphi_2) = 0$. Then it must be the case that $v(\varphi_1) = 1$ and $v(\varphi_2) = 0$. But then $v(\psi_1) = 1$ and $v(\psi_2) = 0$, i.e., $v(\psi_1 \Rightarrow \psi_2) = 0$, as desired. $\square$

Thus we proved that, indeed, $\neg, \wedge, \vee, \Rightarrow,$ and $\equiv$ are operations in Lindenbaum algebra.

So now we have a structure (in fact an algebra) $\langle Form/\sim, \neg, \wedge, \vee, \Rightarrow, \equiv, \bot, \top \rangle$. We need to ask a couple of questions. The first is: "What kind of algebra is it?", i.e., "Are there familiar axioms satisfied by this algebra?" Fortunately, this question has a simple answer. We have the following fact.

PROPOSITION 2.20
The Lindenbaum algebra satisfies the axioms of Boolean algebra.

Proof: We select one of the axioms of Boolean algebra and check that it is true in the Lindenbaum algebra. We hope the reader checks the remaining axioms of Boolean algebra. The one we selected is the second De Morgan law:

$$\neg(x \vee y) = (\neg x) \wedge (\neg y).$$

Thus we need to prove that for all formulas φ_1 and φ_2,

$$\neg([\varphi_1] \vee [\varphi_2]) = (\neg[\varphi_1]) \wedge (\neg[\varphi_2]).$$

That is:
$$[\neg(\varphi_1 \vee \varphi_2)] = ([\neg\varphi_1 \wedge \neg\varphi_2]).$$

But this is equivalent to the fact that the formula

$$\neg(\varphi_1 \vee \varphi_2)) \equiv (\neg\varphi_1 \wedge \neg\varphi_2)$$

is a tautology. This last fact is a substitution to second De Morgan law, thus a tautology. □

But Proposition 2.20 opens another collection of questions. While it is obvious that if Var is infinite then the Lindenbaum algebra is infinite (all variables are in different classes), we do not know yet if for finite set Var, the Lindenbaum algebra is finite (it is), or if every finite Boolean algebra can be characterized as a Lindenbaum algebra for a suitably chosen set of variables Var (it is not the case). These questions will be discussed in Chapter 3 once we get appropriate tools to analyze the situation and revisit the Lindenbaum algebra.

2.6 Satisfiability and permutations

Problems encoded by collections of formulas often exhibit symmetries. Let us look at an example.

Example 2.3
Assume that we want to compute partitions of integers from 1 to 10 into two nonempty blocks B_1 and B_2. Here is how we can encode this problem as a satisfiability problem. We consider the language with 20 variables $p_1, \ldots, p_{20}$ assuming that the variable $p_{(i-1)\cdot 10+j}$, ($1 \leq i \leq 2, 1 \leq j \leq 10$) denotes the fact that the number j belongs to the block B_i. Thus p_1 denotes the fact that 1 belongs to the block B_1 whereas p_{11} denotes the fact that 1 belongs to the block B_2. Likewise, p_2 denotes the fact that 2 belongs to the block B_1, and p_{12} denotes the fact that 2 belongs to B_2 etc. Then the formula that describes our problem (i.e., formula whose satisfying valuations are in one-to-one correspondence with partitions of $\{1, \ldots, 10\}$ into two blocks) consists of 22 clauses in four groups (A)–(D) below.

(A) $p_1 \vee p_{11}, \ldots, p_{10} \vee p_{20}$

(B) $\neg p_1 \vee \neg p_{11}, \ldots, \neg p_{10} \vee \neg p_{20}$

(C) $p_1 \vee \ldots \vee p_{10}$

(D) $p_{11} \vee \ldots \vee p_{20}$

The clauses of group (A) tell us that 1 belongs to one of two blocks, 2 belongs to one of the blocks, etc. The clauses of group (B) say that no number in (1..10) belongs to both blocks at once. Finally, (C) and (D) express the constraints that none of the blocks B_1 and B_2 is empty. Any assignment v satisfying all $(A..D)$ generates a partition, namely $B_i = \{j : v(p_{(i-1)\cdot 10+j}) = 1\}$.

The solutions to our problem exhibit a symmetry: if $\langle X, Y \rangle$ is a solution, then $\langle Y, X \rangle$ is also a solution. When we look at the satisfying valuations for the theory T consisting of groups (A)–(D), we see that whenever a valuation satisfies the theory T, then the valuation v' defined by

$$v'(p_{i\cdot 10+j}) = v(p_{(1-i)\cdot 10+j})$$

has v' is also a satisfying valuation for T. In fact v' is the valuation corresponding to the solution $\langle Y, X \rangle$, obtained from $\langle X, Y \rangle$ by symmetry. □

There is more to symmetries. Namely, we do not have to move a variable to a variable. Thinking about symmetries as renamings, we may move a variable p to a literal $\neg q$ instead of q. In such case we rename the variable *and* change polarity (i.e., sign) as well. But then it is only appropriate that $\neg p$ is renamed into q.

To formalize this intuition we recall that $\bar{l}$, for a literal l, is its opposite (also called *dual*): $\bar{p} = \neg p$, $\overline{\neg p} = p$. We will be permuting literals, but according to the above intuition we will impose a limitation on the permutations: if a permutation π of literals moves l to a literal m, then it also must move $\bar{l}$ to $\bar{m}$, that is, $\pi(\bar{l}) = \bar{m}$. We will call such permutations *consistent*. In other words, a permutation π is consistent if and only if for every literal l, $\pi(\bar{l}) = \overline{\pi(l)}$.

Now let us do some counting. There are, of course, $n!$ permutations of variables. There are precisely 2^n binary sequences of length n. We will use this information to count consistent permutations of literals. We list here a fact concerning consistent permutations of literals.

PROPOSITION 2.21

1. *Consistent permutations of literals based on Var form a group. That is: composition of consistent permutations is consistent, and the inverse of a consistent permutation is consistent.*

2. *A consistent permutation of literals based on Var $= \{p_1, \ldots, p_n\}$ is uniquely determined by a permutation γ of numbers $\{1, \ldots, n\}$ and a binary sequence $\langle i_1, \ldots, i_n \rangle$ as follows:*

$$\pi(p_j) = \begin{cases} p_{\gamma(j)} & \text{if } i_j = 0 \\ \neg p_{\gamma(j)} & \text{if } i_j = 1 \end{cases}$$

$$\pi(\neg p_j) = \begin{cases} \neg p_{\gamma(j)} & \text{if } i_j = 0 \\ p_{\gamma(j)} & \text{if } i_j = 1 \end{cases}$$

3. *Consequently, there are precisely $2^n \cdot n!$ consistent permutations of literals.*

Proof: (1) Let π_1, π_2 be two consistent permutations of literals. Let $\pi_3 = \pi_1 \circ \pi_2$. Of course π_3 is a permutation of literals. But why is it consistent? Let l be a literal. Then

$$\pi_3(\bar{l}) = (\pi_1 \circ \pi_2)(\bar{l}) = \pi_1(\pi_2(\bar{l})) = \pi_1(\overline{\pi_2(l)})$$
$$= \overline{\pi_1(\pi_2(l))} = \overline{(\pi_1 \circ \pi_2)(l)} = \overline{\pi_3(l)}.$$

as desired.

We will ask the reader to prove the second part of (1), which is equally demanding, in one of the exercises.

(2) We want to prove that a consistent permutation of literals is uniquely determined by the choice of a permutation of variables, and the choice of polarity for each variable. But let us observe that the once we know where p is moved and to what polarity, we also know where $\neg p$ is moved. So, all we need to prove is that the choice of permutation of variables and the choice of polarities uniquely determines a consistent permutation of literals. But if the underlying permutations of variables are different then clearly for any choice of polarities we get a different permutation of literals. And if the same permutation of variables is used, but sequences of polarities differ, then again resulting permutations of literals are different.

(3) This follows directly from (2). $\square$

It will be convenient to think about valuations of variables in terms of complete sets of literals. We have the basic fact tying complete sets of literals and consistent permutations.

PROPOSITION 2.22

1. If $S \subseteq Lit$ is a complete set of literals and π is a consistent permutation of literals, then the image of S under π, $\pi(S) = \{m : \exists_{l \in S}(\pi(l) = m)\}$, is a complete set of literals.

2. Moreover, for every pair of complete sets of literals, S' and S'', there exists a consistent permutation π such that $\pi(S') = S''$. In fact, there is $n!$ of such permutations.

Proof: (1) Let S be a complete set of literals, and π a consistent permutation of literals. We claim $\pi(S)$ is complete. Indeed, if $m \in \pi(S)$ then there is a unique literal $l \in S$ such that $\pi(l) = m$. Since π is a consistent permutation, $\pi(\bar{l}) = \bar{m}$. Now, S is complete, so $\bar{l} \notin S$, and since π is one-to-one, $\bar{m} \notin \pi(S)$, as desired.

(2) To see the second part, let us observe that, given S' and S'' the following permutation π brings S' to S''.

$$\pi(p_j) = \begin{cases} p_j & \text{if } (p_j \in S' \text{ and } p_j \in S'') \text{ or } (\neg p_j \in S' \text{ and } \neg p_j \in S'') \\ \neg p_j & \text{if } (p_j \in S' \text{ and } \neg p_j \in S'') \text{ or } (\neg p_j \in S' \text{ and } p_j \in S''). \end{cases}$$

$\square$

Introduction to propositional logic

Since complete sets of literals are nothing but an alternative way of representing valuations, we now have the concept of $\pi(v)$, where v is a valuation of variables and π is a consistent permutation. It can be done purely formally, without resorting to complete sets of literals; the set-of-literals representation just makes it more natural. Let us look at an example.

Example 2.4
Let v be a valuation of variables p, q, and r defined by $v(p) = v(q) = 1$ and $v(r) = 0$. Let π be a consistent permutation of literals determined by the values on variables (the remaining values determined uniquely by the consistency condition): $\pi(p) = \neg q$, $\pi(q) = r$, and $\pi(r) = \neg p$. Then the valuation $\pi(v)$ is the valuation with these values assigned to variables: $\pi(v)(p) = 1, \pi(v)(q) = 0$, and $\pi(v)(r) = 1$. Indeed, v is determined by the complete set of literals $S = \{p, q, \neg r\}$. The image of S under π is $\pi(S) = \{\neg q, r, p\} = \{p, \neg q, r\}$, which gives the assignment listed above. □

Proposition 2.22 tells us that we can extend the action of a consistent permutation π to valuations. We will now show how to extend the action of consistent permutations of literals to formulas. It will be convenient to think about formulas as trees (although it makes little difference in this case). What we do is we uniformly substitute $\pi(p)$ for every variable p in the leaves of the tree of the formula φ. Here is an example.

Example 2.5
Here $\varphi := \neg p \wedge q$. The permutation π is specified by $\pi(p) = \neg q$, $\pi(q) = r$, $\pi(r) = \neg p$. The formula $\pi(\varphi)$ is $(\neg\neg q) \wedge r$. □

Here is the fundamental result tying consistent permutations of literals and valuations to the result of action of permutations on formulas.

PROPOSITION 2.23 (Permutation lemma)
Let π be a consistent permutation of literals, let v be a valuation of variables, and let φ be a formula. Then $v \models \varphi$ if and only if $\pi(v) \models \pi(\varphi)$.

Proof: By induction on the rank of φ. The case of $\varphi = \bot$ or $\varphi = \top$ is obvious: these formulas are never (resp., always) satisfied. Let φ be a variable, say p. Let $v(p) = 1$ (the case of $v(p) = 0$ is similar). Thinking about v as a complete set of literals S_v, $p \in S_v$. Then $\pi(p) \in \pi(S_v)$. But $\pi(v)$ is a valuation whose set-of-literals representation is $\pi(S_v)$. Thus $\pi(v) \models \pi(p)$. The converse reasoning ($\pi(v) \models \pi(p)$ implying $v \models p$) follows from the fact that π is one-to-one.

There are five cases in the inductive step. We deal with a randomly selected one, the rest is left to the reader as an exercise: the proofs are very similar. Let φ be $\psi \wedge \vartheta$. Then $\pi(\varphi) = \pi(\psi) \wedge \pi(\vartheta)$ because the permutation affects the leaves but not internal nodes. Assuming $v \models \psi \wedge \vartheta$, $v \models \psi$ and $v \models \vartheta$. By inductive assumption

$\pi(v) \models \pi(\psi)$ and $\pi(v) \models \pi(\vartheta)$. Thus $\pi(v) \models \pi(\psi) \wedge \pi(\vartheta)$, i.e., $\pi(v) \models \pi(\varphi)$. Converse reasoning is similar. □

Before we move on, let us see how permutations of literals deal with the third representation of valuations; that is, one by means of sets of variables. Here is what happens. Let π be a consistent permutation. Let $M \subseteq \mathit{Var}$. Let us define:

$$\pi(M) = \{x \in \mathit{Var} : \exists_y (y \in M \wedge \pi(y) = x) \vee \exists_y (y \notin M \wedge \pi(y) = \neg x)\}.$$

In other words, a variable x belongs to $\pi(M)$ if it is the image of a variable in M or if it is the image of a negated variable which does not belong to M. It should now be clear that π acts as a permutation of $\mathcal{P}(\mathit{Var})$. As in the case of Proposition 2.22, we can transform any set M into any other N. We formalize this in the following fact.

PROPOSITION 2.24
If $M \subseteq \mathit{Var}$ and $M \models \varphi$, then $\pi(M) \models \pi(\varphi)$.

We will illustrate the technique of permutations within logic by looking at so-called *symmetric* formulas. We will say that a formula φ is *symmetric* if for every permutation of variables π, $\pi(\varphi) \equiv \varphi$. We stress the fact that we deal with permutations of variables, not only with a consistent permutation of literals. We will now give a characterization of symmetric formulas in terms of models. Given a valuation v, define $size(v) = |\{x : x \in \mathit{Var} \text{ and } v(x) = 1\}|$. In other words, in the representation of valuations as sets of variables we just count the set of atoms representing v. We limit our attention to a finite set Var because with an infinite set of variables there cannot be *any* symmetric formulas except tautologies and false formulas. Given two valuations v_1, v_2, we write $v_1 \sim v_2$ if $size(v_1) = size(v_2)$. Clearly, $\sim$ is an equivalence relation, and it splits valuations according to the cardinality of the set of atoms evaluated as 1. We now have the following fact.

PROPOSITION 2.25
A formula φ is symmetric if and only if for every equivalence class C of $\sim$, either all valuations in C satisfy φ, or none of them does.

2.7 Duality

The structure $\mathit{Bool} = \langle \{0,1\}, \wedge, \vee, 0, 1 \rangle$ is isomorphic to the structure $\langle \{0,1\}, \vee, \wedge, 1, 0 \rangle$. One can visualize this isomorphism as a "flip," 0 goes to 1, and conversely 1 turns into 0. We can use that isomorphism to get results on the closure of tautologies under some operations. For instance, it should be clear that if we have a formula φ that involves *only* the functors of conjunction and disjunction (and no constants), and we change in φ every conjunction to disjunction and conversely, and the resulting formula is ψ, then φ is a tautology if and only if ψ is a tautology.

With the presence of constants ($\top$ and $\bot$) we can further extend this result, except that now every occurrence of $\top$ must be changed to $\bot$, and conversely, every $\bot$ to $\top$. Again, the resulting formula is a tautology if the original formula was.

Finally, we have yet another duality result, this time in the presence of negation, conjunction, and disjunction. This time substitute every $\vee$ for $\wedge$, and $\wedge$ for $\vee$, $\bot$ for $\top$, and $\top$ for $\bot$. Moreover, we do two more things: in the leaves (recall we are thinking about formulas as ordered, binary, labeled trees) we substitute $\neg p$ for p for every variable p, and put the negation functor in front of the resulting formula. Then the original formula is a tautology if and only if the resulting formula is a tautology. We note in passing that all these kinds of duality are present in our list of tautologies.

2.8 Semantical consequence, operations *Mod* and *Th*

When a valuation satisfies a set of formulas S it may have to satisfy other formulas as well. For instance, if $v \models p \wedge \neg q$, then $v \models p$ and $p \models \neg q$. Thus we get a natural notion of *semantical consequence* or *entailment*. Formally, given a set F of formulas, let us define

$$Cn(F) = \{\varphi : \forall_v (v \models F \Rightarrow v \models \varphi)\}.$$

Operator Cn assigns to each set of formulas the collection of their consequences. Here are the basic properties of the operation Cn.

PROPOSITION 2.26

1. For every set of formulas F, $F \subseteq Cn(F)$.

2. For all sets of formulas F_1, F_2, if $F_1 \subseteq F_2$ then $Cn(F_1) \subseteq Cn(F_2)$.

3. For all sets of formulas F, $Cn(Cn(F)) = Cn(F)$.

4. $Cn(\emptyset)$ consists of all tautologies.

It should be observed at this point that the operator Cn is continuous. At this moment we have no means to prove it. We will do so in Chapter 6 as a corollary to the compactness theorem.

A common reasoning property, used so frequently that it is easy to forget its source, is the following property of the implication functor $\Rightarrow$.

PROPOSITION 2.27 (Deduction theorem)
Let F be a set of formulas, and let φ, ϑ be two formulas. Then, the following are equivalent:

1. $\varphi \Rightarrow \vartheta \in Cn(F)$.
2. $\vartheta \in Cn(F \cup \{\varphi\})$.

Proof: Let v be an arbitrary valuation satisfying F. If $v \models \varphi$, then $v \models \vartheta$, thus $v \models F \cup \{\varphi\}$ implies $v \models \vartheta$.
Conversely, if $\vartheta \in Cn(F)$ then either $v \not\models \varphi$ and then $v \models \varphi \Rightarrow \vartheta$, or $v \models \varphi$, then $v \models \vartheta$ and again $v \models \varphi \Rightarrow \vartheta$. □

There are natural mappings assigning collections of valuations to sets of formulas, and assigning sets of formulas to collections of valuations. Let us first define the collection of *models* of a set of formulas F, $Mod(F)$.

$$Mod(F) = \{v : v \text{ is a valuation and } v \models F\}.$$

Thus $Mod(F)$ is the set of all satisfying valuations for F. These satisfying valuations are often called *models* of F, hence the notation. We can also go in the opposite direction, assigning to a set of valuations V the set formulas that are satisfied by all valuations $v \in V$. Thus

$$Th(V) = \{\varphi : \forall_{v \in V}(v \models \varphi)\}.$$

Before we look at the connections between the operations Mod and Th, let us look at the case when $V = \{v\}$, that is, of a single valuation. In this case the set $Th(\{v\}) \cap Lit$ is nothing but the set-of-literals representation of v. It follows that, in this case, $Th(\{v\}) \cap Lit$ uniquely determines $Th(\{v\})$.
Here is the basic relationship between the operations Mod and Th.

PROPOSITION 2.28
Let v be a valuation and V a set of valuations. Then $v \in Mod(Th(V))$ if and only if for every finite set of variables A there is a valuation $w \in V$ such that $v \mid_A = w \mid_A$.

Proof: ($\Leftarrow$). We need to show that under the assumption of the right-hand side, $v \models Th(V)$. Let $\varphi \in Th(V)$. We need to show that $v \models \varphi$. Let A be the set of variables occurring in φ. Then there is a valuation $w \in V$ such that $w \mid_A = v \mid_A$. Since $\varphi \in Th(V)$, $w \models \varphi$. But then, by the localization theorem (Proposition 2.3), $w \mid_A \models \varphi$. But then $v \mid_A \models \varphi$ and again, by the localization theorem, $v \models \varphi$. As φ was an arbitrary formula in $Th(V)$, we are done.
($\Rightarrow$). Conversely, let us assume that the right-hand side is false. Then there is a set of variables A such that for every valuation $w \in V$, $w \mid_A \neq v \mid_A$. Let us consider the following formula φ.

$$\varphi := \bigwedge\{p : p \in A \text{ and } v \models p\} \cup \bigwedge\{\neg p : p \in A \text{ and } v \models \neg p\}.$$

Then clearly, $v \models \varphi$. On the other hand, the formula φ uniquely characterizes the behavior of any valuation on A. In other words, if $w \models \varphi$ then $w\mid_A = v\mid_A$. Now, since no $w \in V$ coincides with v on the set A, each such $w \in V$ must satisfy $\neg\varphi$. But then $\neg\varphi \in Th(V)$, while $w \models \varphi$. Thus $v \not\models Th(V)$, as desired. □

We get a corollary completely characterizing the case of finite sets of variables.

COROLLARY 2.3
Let Var be a finite set of propositional variables. Let V be a collection of valuations of set Var and v be a valuation of Var. Then $v \in Mod(Th(V))$ if and only if $v \in V$.

Thus in the case of finite set of variables Var, there is a formula characterizing each valuation (in fact one such formula was constructed in the proof of Proposition 2.28). But this is not necessarily the case for infinite sets of variables.

What about the inverse composition, that is, $Th(Mod(F))$? Here the situation is very simple; we have the equality $Th(Mod(F)) = Cn(F)$. To check this, observe that every formula which is a semantical consequence of F is true in every model of F, and thus belongs to $Th(Mod(F))$. The other inclusion is obvious.

What about more compositions? Let us observe that $Mod(F) = Mod(Cn(F))$. Indeed, since $F \subseteq Cn(F)$, $Mod(Cn(F)) \subseteq Mod(F)$. Conversely, if $v \models F$, then $v \models Cn(F)$ and so $v \in Mod(Cn(F))$.

Knowing that much we can easily see that

$$Mod(Th(Mod(F))) = Mod(Cn(F)) = Mod(F)$$

and

$$Th(Mod(Th(V))) = Th(V).$$

We were able to describe each valuation of a finite set of propositional variables by a suitably chosen formula. The same happens for a set of valuations. The reader recalls that we had several representations of valuations and one of those was by means of subsets of the set of atoms. It will be convenient to express the representability of the set of valuations by means of representability of collections of sets of atoms.

PROPOSITION 2.29
Let Var be a finite set of propositional variables, and let $\mathcal{M} \subseteq \mathcal{P}(Var)$. Then there exists a formula φ such that $\mathcal{M} = Mod(\varphi)$.

Proof: For each $M \in \mathcal{M}$, let φ_M be the formula constructed above in the proof of $\Rightarrow$ of Proposition 2.28. This formula φ_M has the property that M is the *only* model of φ_M. Now let us form φ as follows:

$$\varphi = \bigvee_{M \in \mathcal{M}} \varphi_M.$$

We claim that $\mathcal{M} = Mod(\varphi)$. Indeed, if $M \in \mathcal{M}$ then, since $M \models \varphi_M$, $M \models \varphi$. Conversely, if $N \models \varphi$, then for some $M \in \mathcal{M}$, $N \models \varphi_M$, and so $N = M$ for that M, i.e., $N \in \mathcal{M}$. □

2.9 Exercises

1. Let Var_1, Var_2 be two sets of propositional variables, $Var_1 \subseteq Var_2$. Let $\varphi \in \mathcal{L}_{Var_1}$. Let v be a valuation of Var_2. Prove that $v(\varphi) = v|_{Var_1}(\varphi)$ and in particular $v \models \varphi$ if and only if $v|_{Var_1} \models \varphi$. Show that the assumption $\varphi \in \mathcal{L}_{Var_1}$ is meaningful.

2. Investigate the properties of $\prod_{p \in Var}\{0, 1, u\}$. In particular, show that the function constantly equal to 0 is the least element, and the function constantly equal to 1 is the largest element in $\langle \prod_{p \in Var}\{0, 1, u\}, \leq_p \rangle$.

3. Prove that $\langle \prod_{p \in Var}\{0, 1, u\}, \leq_p \rangle$ is a complete lattice. Then, after you prove it, comment why it is obvious. Give a procedure for computation of the least upper bound in the lattice $\langle \prod_{p \in Var}\{0, 1, u\}, \leq_p \rangle$.

4. Why is the poset $\langle \prod_{p \in Var}\{0, 1, u\}, \leq_k \rangle$ only chain-complete, and not a complete lattice (that is, if $Var \neq \emptyset$)?

5. This is a completely misleading problem. Maybe the reader does not want to solve it at all. Prove that the poset $\langle \{f : f \leq_p \mathbf{u}\}, \leq_p \rangle$ is order-isomorphic to $\langle \prod_{p \in Var}\{0, 1\}, \leq_p \rangle$. Here $\mathbf{u}$ is the function constantly equal to u.

6. (Almost obvious) Show that whenever φ is a formula, and v is a partial valuation, and $v_3(\varphi) = 1$, then φ is satisfiable.

7. Let us assume that a formula ψ is a tautology, that is, for all (complete two-valued) valuations v, $v \models \varphi$. Can we deduce that for all partial valuations v, $v_3(\varphi) = 1$? If not, provide a suitable counterexample, and explain what it means. Your explanation should use the informal notion of "commitment" discussed above.

8. Prove Proposition 2.6. It is so obvious that proofs requiring more than four sentences are too long.

9. Let F be this set of clauses:

$$\{p \vee q,\ p \vee \neg q,\ \neg p \vee r\}.$$

   (a) Is the partial assignment $\{\neg p, \neg r\}$ an autarky for F?

   (b) Is the partial assignment $\{p, \neg r\}$ an autarky for F?

10. Prove that the operation $\oplus$ discussed in Section 2.3 is associative, and that the algebra $\langle PVal, \oplus, \emptyset \rangle$ is a monoid. If you do not know what "monoid" is, check Wikipedia. $PVal$ is the set of all partial valuations.

11. Is the operation $\oplus$ commutative?
12. Prove that for every set F of formulas the algebra $\langle Auk_F, \oplus, \emptyset \rangle$ is a monoid. Here Auk_G is the set of all autarkies for F.
13. Select at random one of the functors from the set $\{\neg, \wedge, \vee, \equiv\}$ and prove that it is an operation in the Lindenbaum algebra.
14. Add a symbol f for any operation $f : Bool^n \rightarrow Bool$, define the satisfaction relation for the functor f, define the Lindenbaum algebra of the extended language, and prove that f is an operation in this algebra.
15. Complete the argument of Proposition 2.21(2).
16. Complete the argument of Proposition 2.22(2). The author proved that there is *one* consistent permutation of literals transforming a given valuation v to another given valuation w. Adapt this argument to the more general case.
17. Prove the proposition from Section 2.7. At least three statements await your proof. Comment on the relationship to the following fact: if we change the lattice ordering of $\langle L, \leq \rangle$ to $\leq^{-1}$ we also get a lattice ordering. If $\langle L, \leq \rangle$ is not only a lattice, but also a Boolean algebra, we get an isomorphic Boolean algebra.
18. Prove Proposition 2.26. There is nothing deep about it, but some effort is needed.

Chapter 3

Normal forms of formulas

3.1	Canonical negation-normal form	46
3.2	Occurrences of variables and three-valued logic	48
3.3	Canonical forms	50
3.4	Reduced normal forms	54
3.5	Complete normal forms	56
3.6	Lindenbaum algebra revisited	58
3.7	Other normal forms	59
3.8	Exercises	60

We will now study normal forms. Generally, by a *normal form* formula we mean a formula with some limitation on its syntactic form. The reader recalls that we introduced formulas as textual representations of some labeled trees. One, then, can think about formulas as *trees*. This was very useful, because trees have rank function, and so various arguments were easier. We can also think about formulas as representations of tables. Many formulas may (and in fact will) represent a single table, and we may want to use some specific representations. These representations are limited by the syntactic form we allow. It turns out that it makes a lot of difference *how* we represent formulas. If we think about formulas as representing constraints on valuations (namely that a valuation needs to satisfy the formula) then the specific syntactic form we use to represent constraints may make a lot of difference. In fact, having a specific syntactic form of a formula may make the task of testing some properties entirely trivial. For instance, if we have a formula φ in disjunctive normal form then testing its satisfiability is very easy. Likewise, testing a conjunctive normal form for being a tautology is completely trivial. The only problem with this approach is that finding the disjunctive (resp. conjunctive) normal form of a formula is an expensive task in itself. But there are often advantages. For instance, if we have a formula φ in conjunctive normal form and we are testing satisfiability, then there is an additional mechanism (called Boolean constraint propagation) that can be used to speed up the DPLL process by cutting portions of the search tree. Moreover, in the same case of conjunctive normal form, we have a general processing technique, called resolution, that can be used to test satisfiability.

First, we will discuss a simpler normal form, called *negation normal form*. This normal form preserves *polarity* and will be useful when we discuss the Craig lemma. We will also discuss "complete forms" of disjunctive and conjunctive normal forms, and connection between conjunctive and disjunctive normal forms. This allows us to get a better handle on the Lindenbaum algebra of the language.

The completeness results for sets of functors always produce normal form (in fact completeness of a set of functors already gives *some* normal form). In this spirit we will discuss implicational normal form, this one involving the fact that the functor $\Rightarrow$ forms by itself a complete set of functors (in a weak sense, that is in the presence of constants) and the *ite*-normal form, again the consequence of the fact that *ite* is (weakly) complete. Finally, we notice that there exist normal forms involving *NAND* and *NOR*. These are of practical importance, but we will not discuss them here.

3.1 Canonical negation-normal form

In this section we will limit ourselves to formulas built of constants, negation, conjunction and disjunction. With an additional effort we could extend the negation-normal form to all formulas, but we will not do so.

Formulas can be complex and, in particular in the inductive definition of formulas we did not put any restriction on the occurrences of negation in a formula. Now we will do so. We say that a formula φ is in *negation-normal form* if all negation symbols occur only in front of variables. For instance, the formula $p \wedge (q \vee \neg r)$ is in negation normal form, while the formula $p \wedge \neg(\neg q \wedge r)$ is not in negation normal form. It turns out that, for a given formula φ we can find a formula ψ in negation normal form such that φ and ψ are satisfied by exactly the same valuations. This transformation (from φ to ψ) is based on the following three tautologies, which in our context work as rewriting rules that "push negation inward":

1. (Double negation elimination) $\neg\neg\varphi \equiv \varphi$
2. (De Morgan I) $\neg(\varphi \wedge \psi) \equiv \neg\varphi \vee \neg\psi$
3. (De Morgan II) $\neg(\varphi \vee \psi) \equiv \neg\varphi \wedge \neg\psi$

Repeatedly scanning the formula from left to right and applying these rewrite rules (i.e. substituting right hand side for the left hand side) we decrease the rank of subformulas where the negation symbol is a top symbol, until the negation is applied only to variables. Since all three rewriting rules preserve the satisfying valuations, it follows that after our algorithm terminates the resulting (output) formula is equivalent to the original input formula and that the output formula is in the negation normal form. This was, really, a top-down procedure, but we can also go bottom-up. We can think about negation-normal form as pushing negations as deeply as they can be pushed, that is, right above propositional variables.

We define the formula $\mathrm{cNN}(\varphi)$ inductively as follows:

1. $\mathrm{cNN}(a) = a$ if a is a variable
2. $\mathrm{cNN}(\neg a) = \neg a$ if a is a variable
3. $\mathrm{cNN}(\varphi_1 \wedge \varphi_2) = \mathrm{cNN}(\varphi_1) \wedge \mathrm{cNN}(\varphi_2)$
4. $\mathrm{cNN}(\varphi_1 \vee \varphi_2) = \mathrm{cNN}(\varphi_1) \vee \mathrm{cNN}(\varphi_2)$

5. $\mathrm{cNN}(\neg\varphi) = \begin{cases} \mathrm{cNN}(\psi) & \text{if } \varphi = \neg\psi \\ \mathrm{cNN}(\neg\psi) \vee \mathrm{cNN}(\neg\vartheta) & \text{if } \varphi = \psi \wedge \vartheta \\ \mathrm{cNN}(\neg\psi) \wedge \mathrm{cNN}(\neg\vartheta) & \text{if } \varphi = \psi \vee \vartheta. \end{cases}$

Now, it should be clear that we have the following important property of the canonical negation normal form.

PROPOSITION 3.1
The canonical negation normal form of φ, $\mathrm{cNN}(\varphi)$, can be computed in linear time.

We define occurrences of variables in formulas. We do this inductively on the complexity of formulas.

1. A variable a occurs in the formula a positively.
2. A variable a occurs in the formula $\neg a$ negatively.
3. A variable a occurs in a formula $\neg\varphi$ positively (negatively) if a occurs in the formula φ negatively (positively).
4. A variable a occurs in the formula $\varphi \wedge \psi$ positively if a occurs positively in φ or a occurs positively in ψ.
5. A variable a occurs in the formula $\varphi \vee \psi$ negatively if a occurs negatively in φ or a occurs negatively in ψ.

In particular a variable may occur both positively and negatively in a formula. For instance, p occurs both positively and negatively in the formula $p \vee \neg(p \vee \neg q)$. We now have the following useful fact.

THEOREM 3.1
A variable p occurs positively (negatively) within a formula φ if and only if p occurs positively (negatively) within the formula $\mathrm{cNN}(\varphi)$.

Proof: By induction on the complexity of the formula φ. The base cases are obvious since in this case $\mathrm{cNN}(\varphi) = \varphi$. If the top connective (the label of the root) of φ is either $\vee$ or $\wedge$ then the result is obvious since both $\vee$ and $\wedge$ preserve both positive and negative occurrences of variables. Consequently we only need to look at the case when the top connective of φ is $\neg$. Let us make the following observation: a variable p occurs positively (negatively) in the formula φ if and only if p occurs negatively (positively) in the formula $\neg\varphi$. So let us now assume that $\varphi = \neg\psi$. We need to consider three cases. First, assume that the top connective of ψ is $\neg$. Then $\psi = \neg\vartheta$. Now, if p occurs positively in φ then p occurs positively in ϑ and thus p occurs positively in $\mathrm{cNN}(\varphi) = \mathrm{cNN}(\vartheta)$. If p occurs positively in $\mathrm{cNN}(\varphi)$, i.e., in $\mathrm{cNN}(\vartheta)$, then by inductive assumption (ϑ is simpler than φ) the variable p occurs positively in ϑ, thus in φ, as desired. The case of negative occurrences is similar.

The second case is when the top connective of ψ is $\wedge$, i.e., $\psi = \vartheta_1 \wedge \vartheta_2$. Now, in this case,

$$\mathrm{cNN}(\neg\psi) = \mathrm{cNN}(\neg\vartheta_1) \vee \mathrm{cNN}(\neg\vartheta_2).$$

Assume p occurs in φ positively. Then p occurs negatively in ψ. Without loss of generality we can assume that p occurs negatively in ϑ. By the inductive assumption, p occurs negatively in $\mathrm{cNN}(\vartheta_1)$. Thus p occurs positively in $\mathrm{cNN}(\neg\vartheta_1)$, and thus p occurs positively in $\mathrm{cNN}(\neg\vartheta_1) \vee \mathrm{cNN}(\neg\vartheta_2)$, that is in $\mathrm{cNN}(\varphi)$, as desired. Conversely, assume p occurs positively in $\mathrm{cNN}(\varphi)$. That is, p occurs positively in $\mathrm{cNN}(\neg\vartheta_1) \vee \mathrm{cNN}(\neg\vartheta_2)$. Without the loss of generality we can assume that p occurs positively in $\mathrm{cNN}(\neg\vartheta_1)$. Then, by inductive assumption, p occurs positively in $\neg\vartheta_1$. This means that p occurs negatively in ϑ_1. This, in turn, implies that p occurs negatively in $\vartheta_1 \wedge \vartheta_2$. Finally, we conclude that p occurs positively in $\neg(\vartheta_1 \wedge \vartheta_2)$, that is, p occurs positively in φ, as desired.

The case of negative occurrence of p in φ is similar. Likewise, the case when $\psi = \vartheta_1 \vee \vartheta_2$ is again similar. □

Next, let us observe that the canonical negation-normal form of a formula φ has length at most twice that of φ (and this is a pessimistic estimate). The reason for that is that, as we transform the formula φ into its negation normal form, the total number of internal nodes labeled with $\vee$ or with $\wedge$ does not change (of course the labels of nodes change, but not the total number of nodes labeled with $\vee$ or $\wedge$, as can easily be checked by induction). But then, we can lengthen the formula only by putting $\neg$ in front of leaves of the original formula. Thus the length of $\mathrm{cNN}(\varphi)$ can be estimated by the length of φ incremented by the number of leaves of the tree for φ, thus twice the length of φ. Since every node is visited once, we established that there is an inexpensive way to check occurrences of a variable (say p) in a formula φ. All we need to do is to compute the canonical negation-normal form of φ, and then to scan that normal form $\mathrm{cNN}(\varphi)$ for the positive and negative occurrences of the variable p.

3.2 Occurrences of variables and three-valued logic

We will now prove an interesting property of formulas with respect to three-valued logic. Let us assume that $V \subseteq Var$ is a set of propositional variables. Let φ be a propositional formula. The formula φ_V arises from φ by substitution: we substitute every positive occurrence of every variable $x \in V$ by $\bot$ and every negative occurrence of every variable $y \in V$ by $\top$. Here is one example. Let φ be $(\neg(x \vee \neg z) \wedge x) \vee y$. Let $V = \{x\}$. The variable x has two occurrences in our formula. The first one is negative, the second one is positive. When we substitute $\top$ for the first one, and $\bot$ for the second one (according to our definition) we get the formula $(\neg(\top \vee \neg z) \wedge \bot) \vee y$. Let us observe that the variables of V do *not* occur in

φ_V at all. One can interpret the operation φ_V (the elimination of variables from V) as the "worst-case scenario": we make positively occurring variables from V false and negatively occurring variables from V true.

Now, we have the following important property of the three-valued evaluation function. We will use this property when we discuss autarkies for Horn formulas.

PROPOSITION 3.2

Let φ be a propositional formula. Let v be a nonempty partial assignment of some or all propositional variables in φ. Let $V \subseteq Var \setminus Var(v)$. Then $v_3(\varphi) = 1$ if and only if $v_3(\varphi_V)) = 1$, and $v_3(\varphi) = 0$ if and only if $v_3(\varphi_V) = 0$.

Proof: We proceed by simultaneous induction on both parts of the assertion on the complexity of the formula φ.
Base case: There is nothing to prove if φ is either $\bot$ or $\top$. When $\varphi = p$, where p is a propositional variable, then since v is nonempty, $p_V = p$ and both parts of the assertion are obvious.
Inductive step: There will be three cases, each corresponding to the main connective of the formula φ.
(a): $\varphi = \neg\psi$. Since the operation $\varphi \mapsto \varphi_V$ affects only the leaves but not the internal nodes of the tree representation of φ, $(\neg\varphi)_V = \neg\varphi_V$. Let us assume that $v_3(\varphi) = 1$. Then $v_3(\psi) = 0$, $v_3(\psi_V) = 0$ (inductive assumption, ψ is simpler), $v_3(\neg\psi_V) = 1$, $v_3(\varphi_V) = 1$. It is easy to see that all implications in this chain are, in fact, equivalences, and so the other implication also follows. If $v_3(\varphi) = 0$, the argument is very similar.
(b): $\varphi = \psi_1 \wedge \psi_2$. Then $\varphi_V = (\psi_1)_V \wedge (\psi_2)_V$ because the operation $\varphi \mapsto \varphi_V$ affects leaves, and not internal nodes. Assuming $v_3(\varphi) = 1$ we have: both $v_3(\psi_1) = 1$ and $v_3(\psi_2) = 1$. Then by inductive assumption $v_3((\psi_1)_V) = 1$ and $v_3((\psi_2)_V) = 1$. But then $v_3((\psi_1)_V \wedge (\psi_2)_V) = 1$. Thus $v_3(\varphi_V) = 1$. All the implications were equivalences, thus the converse implication holds as well.
Next we have to consider the case when $v_3(\varphi) = 0$. The argument is similar, except that now we use the fact that $v_3(\varphi) = 0$ if and only if at least one of $v_3(\psi_1)$ and $v_3(\psi_2)$ is 0.
(c): $\varphi = \psi_1 \vee \psi_2$. Then $\varphi_V = (\psi_1)_V \vee (\psi_1)_V$ because the operation $\varphi \mapsto \varphi_V$ affects leaves, and not internal nodes. Now, we reason as in case (b), except that two subcases are reversed. That is $v_3(\psi_1 \vee \psi_2) = 1$ if and only if at least one of $v_3(\psi_1), v_3(\psi_2)$ is 1, $v_3(\psi_1 \vee \psi_2) = 0$ if and only if both $v_3(\psi_1), v_3(\psi_2)$ are 0. □

3.3 Canonical disjunctive and conjunctive normal forms

We found that it is easy to "push negation inward" and to make sure that the negation functor applies only to variables. Now, we will see that we can push inward disjunctions and conjunctions. But unlike the case of negation, the cost associated with such transformation may be significant. Within this section we will assume that formulas are always in negation normal form. In other words the procedure outlined above in Section 3.1 allows us to make sure that the formulas under consideration are cNN-formulas.

The following four tautologies are important for building disjunctive normal forms and conjunctive normal forms.

1. $\varphi \wedge (\psi \vee \vartheta) \equiv (\varphi \wedge \psi) \vee (\varphi \wedge \vartheta)$
2. $(\psi \vee \vartheta) \wedge \varphi \equiv (\psi \wedge \varphi) \vee (\vartheta \wedge \varphi)$
3. $\varphi \vee (\psi \wedge \vartheta) \equiv (\varphi \vee \psi) \wedge (\varphi \vee \vartheta)$
4. $(\psi \wedge \vartheta) \vee \varphi \equiv (\psi \vee \varphi) \wedge (\vartheta \vee \varphi)$

We will now use the formulas (1) and (2) when we want to "push conjunction inward," while (3) and (4) will be used to "push disjunction inward." Formulas where all conjunctions were pushed inward will be called disjunctive normal form formulas (DNFs). Likewise, formulas where all disjunctions have been pushed inward will be called conjunctive normal form formulas (CNFs).

Thus assume our formula φ is in negation normal form. We will compute a disjunctive normal form of φ. A conjunctive normal form can be computed in an analogous manner (except that instead of tautologies (1) and (2) we would use (3) and (4)).

Before we do this in detail, let us observe that both conjunction and disjunction are associative, in other words:

5. $\varphi \wedge (\psi \wedge \vartheta) \equiv (\varphi \wedge \psi) \wedge \vartheta$
6. $\varphi \vee (\psi \vee \vartheta) \equiv (\varphi \vee \psi) \vee \vartheta$

This means that we do not have to be concerned with the order of brackets when the formula consists of conjunctions only. We define a *term* as a formula of the form $l_1 \wedge l_2 \wedge \ldots \wedge l_k$ where each l_j is a literal, that is, a variable or negated variable. A formula φ is in *disjunctive normal form* (DNF) if $\varphi = D_1 \vee D_2 \vee \ldots \vee D_m$ where each D_j, $1 \leq j \leq m$, is a term.

We define the formula $\mathrm{cDNF}(\varphi)$ by induction on the complexity of formulas. Recall that we assume our formulas to be in negation-normal forms. We define:

1. $\mathrm{cDNF}(a) = a$ if a is a variable.
2. $\mathrm{cDNF}(\neg a) = \neg a$ if a is a variable.

3. Assuming $D_1^1 \vee D_2^1 \vee \ldots \vee D_{m_1}^1 = \mathrm{cDNF}(\psi_1)$ and $D_1^2 \vee D_2^2 \vee \ldots \vee D_{m_2}^2 = \mathrm{cDNF}(\psi_2)$, define

   (a) $\mathrm{cDNF}(\psi_1 \wedge \psi_2) = \bigvee_{1 \leq i \leq m_1, 1 \leq j \leq m_2} D_i^1 \wedge D_j^2$.
   (b) $\mathrm{cDNF}(\psi_1 \vee \psi_2) = \mathrm{cDNF}(\psi_1) \vee \mathrm{cDNF}(\psi_2)$.

Since the distribution is applied only to terms we see that starting with a formula in a negation-normal form we end up with a formula in a negation-normal form as well.

PROPOSITION 3.3

For every formula φ in negation-normal form, the formula $\mathrm{cDNF}(\varphi)$ is in disjunctive normal form. Moreover, $\varphi \equiv \mathrm{cDNF}(\varphi)$ is a tautology.

Proof: Proceeding by induction on the complexity of formulas, while assuming the input formula is in negation-normal form, the output is also in negation-normal form. Moreover the positive and negative occurrences of variables do not change in the process of computation of $\mathrm{cDNF}(\varphi)$. Clearly, the operations utilized in (3.a) and (3.b) produce the outputs in disjunctive normal form.

Thus all we need to do is to show that for every valuation v, $v \models \varphi$ if and only if $v \models \mathrm{cDNF}(\varphi)$. We proceed by induction. The base cases are obvious. All we need to take care of are the top functor symbols being $\vee$ and the top function symbol being $\wedge$. The first case is obvious. Turning our attention to the case when the top functor of φ is $\wedge$, we observe that if $v \models \psi_1$ and $v \models \psi_2$, then for some i, $1 \leq i \leq m_1$, and some j, $1 \leq j \leq m_2$, $v \models D_i^1$ and $v \models D_j^2$, so $v \models D_i^1 \wedge D_j^2$, thus $v \models \mathrm{cDNF}(\psi_1 \wedge \psi_2)$, as desired. Conversely, if $v \models \mathrm{cDNF}(\psi_1 \wedge \psi_2)$, then for some i, j, $1 \leq i \leq m_1, 1 \leq j \leq m_2$,

$$v \models D_i^1 \wedge D_j^2.$$

Therefore $v \models D_i^1$ and $v \models D_j^2$. Hence $v \models \mathrm{cDNF}(\psi_1)$ and $v \models \mathrm{cDNF}(\psi_2)$. By inductive assumption, $v \models \psi_1$ and $v \models \psi_2$ and so $v \models \psi_1 \wedge \psi_2$ as desired. □

One consequence of the existence of disjunctive normal form (but we could use *ite* normal form instead, too) is the following fact, closely related to the fact we established when showing completeness of the De Morgan set of functors (Proposition 5.2).

PROPOSITION 3.4

For every formula φ and a variable $p \in \mathrm{Var}_\varphi$ there exist two formulas ψ_i, $i = 1, 2$ such that

1. $p \notin \mathrm{Var}_{\psi_i}$, $i = 1, 2$
2. *The formula $\varphi \equiv ((p \wedge \psi_1) \vee (\neg p \wedge \psi_2))$ is a tautology.*

Proof: Let us consider a canonical disjunctive normal form for φ. We can assume that there is no contradictory conjunction in that normal form. Now, the disjunction splits in three parts: χ_1 consisting of terms containing p, χ_2 consisting of terms containing $\neg p$, and χ_3 consisting of remaining terms. Then χ_1 can be easily transformed (by taking p out of terms of χ_1) to the formula $p \wedge \vartheta_1$, $p \notin Var_{\vartheta_1}$. Similarly we have χ_2 equivalent to $\neg p \wedge \vartheta_2$ with $p \notin Var_{\vartheta_2}$. Then we form ψ_1 equal to $\vartheta_1 \vee \chi_3$ and ψ_2 equal to $\vartheta_2 \vee \chi_3$. We leave it to the reader to check that we computed desired formulas. □

The substitution lemma (Proposition 2.16) implies that substitutions of equivalent formulas for variables result in equivalent formulas. Likewise, substituting same formulas into equivalent formulas results in equivalent formulas. We will show how these observations are of use.

PROPOSITION 3.5 (Tarski propositional fixpoint theorem)

Let φ be a propositional formula, and let $p \in Var_\varphi$. Let $\psi = \varphi\left(\frac{p}{\top}\right)$. Then, for every valuation v, if $v \models \varphi$ then $v \models \varphi\left(\frac{p}{\psi}\right)$.

Proof: Without loss of generality we can assume that $\varphi \equiv ((p \wedge \psi_1) \vee (\neg p \wedge \psi_2))$ is a tautology. We first claim that $\psi \equiv \psi_1$ is a tautology. Indeed, let v be any valuation. Then by the substitution lemma, $v(\psi) = v'(\varphi)$ where $v'(p) = v(\top) = 1$ and $v'(x) = v(x)$ for all variables x different from p. But

$$v'(\varphi) = v'\left((p \wedge \psi_1) \vee (\neg p \wedge \psi_2)\right) = v'(\psi_1)$$

because $v'(p) = 1$. Now, $v(\psi_1) = v'(\psi_1)$ because ψ_1 does not contain any occurrence of p. Thus $v(\psi) = v(\psi_1)$, as desired.

Now, since $\psi \equiv \psi_1$ is a tautology, $\varphi\left(\frac{p}{\psi}\right) \equiv \varphi\left(\frac{p}{\psi_1}\right)$ is a tautology. But $\varphi\left(\frac{p}{\psi_1}\right)$ is nothing else but

$$(\psi_1 \wedge \psi_1) \vee (\neg \psi_1 \wedge \psi_2)$$

which in turn is equivalent to $\psi_1 \vee (\neg \psi_1 \wedge \psi_2)$.

Thus all we need to do is to show that $v \models \psi_1 \vee (\neg \psi_1 \wedge \psi_2)$.

Two cases are possible:

Case 1: $v \models \psi_1$. Then, clearly $v \models \psi_1 \vee (\neg \psi_1 \wedge \psi_2)$.
Case 2: $v \models \neg \psi_1$. Then, since $v \models \varphi$, it must be the case that $v \models \neg p \wedge \psi_2$, in particular $v \models \psi_2$. But then $v \models \neg \psi_1 \wedge \psi_2$, and so $v \models \psi_1 \vee (\neg \psi_1 \wedge \psi_2)$, as desired. □

What about the polarity of variables? A close inspection shows that application of distributive laws preserves polarity. In particular, the CNF computed out of cNNF(φ) (canonical negation-normal form of φ) has exactly the same occurrences of variables. This fact will be important in our proof of the Craig lemma below.

Complexitywise, computation of cDNF(φ) is expensive. The reason for this is that the size of the result of the application of transformation 3.a above is proportional to the product of the inputs. This results in the exponential size increase.

Example 3.1
Let $\varphi_n = (a_1^1 \vee a_2^1) \wedge (a_1^2 \vee a_2^2) \wedge \ldots \wedge (a_1^n \vee a_2^n)$, where $a_1^1, a_2^1, \ldots, a_1^n, a_2^n$ are $2n$ distinct propositional variables.
Then the formula φ_n consists of n clauses, each of size 2. But the formula cDNF(φ_n) consists of 2^n terms, each of size n. Thus the exponential growth in the computation of disjunctive normal form cannot be avoided. □

The result analogous to Proposition 3.3 can be easily proved for the conjunctive normal form. We just use the other two distributive laws. As before we can either push the disjunction inward or define inductively conjunctive normal form formulas as we did before.

Assuming that we are dealing with negation-normal form formulas, we assign the canonical conjunctive-normal form (cCNF) to such formula as follows:

1. $\text{cCNF}(a) = a$ if a is a variable
2. $\text{cCNF}(\neg a) = \neg a$ if a is a variable
3. Assuming $C_1^1 \wedge C_2^1 \wedge \ldots \wedge C_{m_1}^1 = \text{cCNF}(\psi_1)$ and $C_1^2 \wedge C_2^2 \wedge \ldots \wedge C_{m_2}^2 = \text{cCNF}(\psi_2)$, define
    (a) $\text{cCNF}(\psi_1 \vee \psi_2) = \bigwedge_{1 \leq i \leq m_1, 1 \leq j \leq m_2} C_i^1 \vee C_j^2$
    (b) $\text{cCNF}(\psi_1 \wedge \psi_2) = \text{cCNF}(\psi_1) \wedge \text{cCNF}(\psi_2)$

No wonder, we get:

PROPOSITION 3.6

For every formula φ in negation-normal form, the formula $\text{cCNF}(\varphi)$ is in conjunctive-normal form. Moreover, the formula $\varphi \equiv \text{cCNF}(\varphi)$ is a tautology.

Given a theory (i.e., a set of formulas) T we can compute for each $\varphi \in T$ its conjunctive normal form $\text{cCNF}(\varphi)$. Because satisfaction of a formula of the form $\bigwedge_{i=1}^n \varphi_i$ means just the satisfaction of all φ_i, $1 \leq i \leq n$, we can drop all conjunctions, and in this way we compute a set of clauses. We will call that set $\text{cCNF}(T)$. With a bit of impreciseness (identifying $\text{cCNF}(\varphi)$ with a set of clauses) we just have:

$$\text{cCNF}(T) = \bigcup_{\varphi \in T} \text{cCNF}(\varphi).$$

We then have the following fact.

PROPOSITION 3.7
The set of clauses $\text{cCNF}(T)$ has exactly the same models as T. Consequently, for every propositional theory T there is a set of clauses F such that $\text{Mod}(F) = \text{Mod}(T)$.

It would be tempting to think that if a variable p has only positive (resp. negative) occurrences in all formulas φ in a set of formulas F, then $\{p\}$ is an autarky for F. This is not the case, as witnessed by the following example: $F = \{(p \wedge q) \vee r\}$. Here p has only positive occurrences in the only formula φ of F, $v = \{p\}$ touches that formula, but $v_3(\varphi) = u$. Nevertheless, a weaker (but still desirable) property of such sets of formulas holds. Namely, we have the following.

PROPOSITION 3.8
Let F be a set of formulas in which a variable p has only positive (resp. negative) occurrences. Then F is satisfiable if and only if $F \cup \{p\}$ (resp. $F \cup \{\neg p\}$) is satisfiable. In other words, F is satisfiable if and only if it is satisfiable by a valuation v such that $v(p) = 1$ (resp. $v(p) = 0$).

Proof: We prove it for positive occurrences of p only; the reader should have no problem with the other case. We need to prove that if $v \models F$ and w is defined by

$$w(x) = \begin{cases} v(x) & \text{if } x \neq p \\ 1 & \text{if } x = p \end{cases}$$

then $v \models \varphi$ for all $\varphi \in F$. Various arguments are possible. Here is one that uses the canonical conjunctive normal form. Since p has only positive occurrences in φ, p has only positive occurrences in cCNF(φ). But then p is a free literal in the set of clauses G defined as $\bigcup_{\varphi \in F} \text{cCNF}(\varphi)$. But then $\{p\}$ is an autarky for G and so the result follows. □

One consequence of Proposition 3.8 is that the concept of autarky is *not* preserved by equivalence of formulas.

3.4 Reduced normal forms

If we computed a cDNF(φ) or cCNF(φ) we should check if all the conjunctions (resp. clauses) are really needed there. We will now discuss the case of conjunctive normal form (the case of DNF is very similar).

There are two distinct reduction rules allowing for elimination of clauses from CNF. The first one is a simple fact that characterizes clauses that are tautologies. The second one is a characterization of subsumption between clauses.

PROPOSITION 3.9

1. *A clause $C = p_1 \vee \ldots \vee p_k \vee \neg q_1 \vee \ldots \vee \neg q_l$ is a tautology if and only if for some i, $1 \leq i \leq k$ and j, $1 \leq j \leq l$, $p_i = q_j$.*

2. Given two clauses C_1 and C_2, $C_1 \models C_2$ if and only if all literals occurring in C_1 occur in C_2, that is, C_1 subsumes C_2.

Proposition 3.9 tells us that, at a cost, we can reduce the cCNF by a repeated application of two rewrite rules: first, given a clause C, if C contains a pair of complementary literals, then C can be eliminated. Second, if a clause C_1 is subsumed by a clause C_2, then C_1 can be eliminated. Assuming that the formula is in conjunctive normal form and is sorted in the sense that shorter clauses are listed earlier, this task can be done in time proportional to the square of the size of the formula. The resulting formula is *reduced*; there are no unnecessary clauses (tautologies), and there are no subsumptions. There is a cost involved in such preprocessing, but it may pay, eventually. Moreover, as before, we have the following fact.

PROPOSITION 3.10
For every formula φ there is a reduced conjunctive-normal form formula ψ such that $\varphi \equiv \psi$ is a tautology.

There are the corresponding results for disjunctive-normal form. We will state them now.

PROPOSITION 3.11

1. A term $D = p_1 \wedge \ldots \wedge p_k \wedge \neg q_1 \wedge \ldots \wedge \neg q_l$ is evaluated false (i.e., 0) by all valuations if and only if for some i, $1 \leq i \leq k$ and j, $1 \leq j \leq l$, $p_i = q_j$.
2. Given two terms D_1 and D_2, $D_1 \models D_2$ if and only if all literals occurring in D_2 occur in D_1.

Proposition 3.11 entails a reduction procedure and the concept of a reduced DNF. An algorithm for reduction is similar, except that in the case of second reduction, the inclusion is reversed, and so we need to sort the formulas in the opposite order. We still get an analogous result.

PROPOSITION 3.12
For every formula φ there is a reduced disjunctive normal form formula ψ such that $\varphi \equiv \psi$ is a tautology.

Testing of a formula in a reduced disjunctive normal form for satisfiability is entirely trivial. As long as the reduced normal form is nonempty, the formula is satisfiable. The reader should have a nagging suspicion that there is something similar in the normal forms of related formulas, and that there should be some connection.
Let us see what happens if we take a formula φ already in (say) disjunctive normal form, put the negation functor in front, and push the negation inward (as we did in the

computation of canonical negation-normal form). We get a conjunctive normal form formula. This formula is a conjunctive normal form for $\neg\varphi$. The procedure itself is very efficient: every disjunction is rewritten to a conjunction and vice versa, and at the literal level, every literal is changed to its dual literal. We certainly can perform this computation in linear time. It is pretty obvious that we can apply exactly the same algorithm to a conjunctive normal form of φ, getting a disjunctive normal form of $\neg\varphi$. It takes little effort to see that the reduced DNF becomes a reduced CNF and vice versa. Thus we get the following fact.

PROPOSITION 3.13
The procedure outlined above converts the canonical conjunctive normal form of φ, $\mathrm{cCNF}(\varphi)$, to the canonical disjunctive normal form $\mathrm{cDNF}(\neg\varphi)$ of $\neg\varphi$. If the input DNF has been reduced, the output formula is a reduced CNF. The same conversion procedure transforms $\mathrm{cCNF}(\varphi)$ into $\mathrm{cDNF}(\neg\varphi)$.

Proposition 3.13 tells us that if we ever get a device that efficiently computes one of the normal forms, then such device could be used, together with postprocessing described above, to compute the other normal form.

3.5 Complete normal forms

There is even more to disjunctive and conjunctive normal forms. We will first focus on the disjunctive normal forms, and then use the transformation of DNF of $\neg\varphi$ into the CNF for φ to get similar results for CNFs.

The limitation we impose in this section is that we assume the set *Var* to be finite, $Var = \{p_1, \ldots, p_n\}$. Yet another piece of terminology needed here is that of a *minterm*. A minterm is a term which contains exactly one literal from each pair $\{p, \neg p\}$. It should be clear that a minterm is a minimal possible term, in the sense that all the other terms that are entailed by it, are included in it. Here is an example. Let $Var = \{p, q, r\}$. Then $p \wedge \neg q \wedge \neg r$ is a minterm. Another minterm is $p \wedge q \wedge r$. Clearly there is a bijection between minterms and valuations. Namely, a minterm is a conjunction of a valuation, if we think about a valuation as a complete set of literals. The following fact may remind the reader of atoms in Boolean algebra.

LEMMA 3.1
For every formula φ and every minterm t, either the formula $(\varphi \wedge t) \equiv t$ is a tautology, or $\varphi \wedge t$ is false.

Proof: As observed above, the term t determines a valuation v_t, a unique valuation satisfying t. Then we have two cases.

Normal forms of formulas

Case 1: $v_t \models \varphi$. Then v_t is a unique valuation satisfying $\varphi \wedge t$. Since v_t is a unique valuation satisfying t, the formula $(\varphi \wedge t) \equiv t$ is a tautology.
Case 2: $v_t \not\models \varphi$. Then there is no valuation satisfying $\varphi \wedge t$, and so it is false. □

But now, we show another disjunctive normal form of a given formula φ. Let us form the set $S = Mod(\{\varphi\})$, of all valuations satisfying φ. For each such valuation v, let us form a minterm t_v, where t_v is the conjunction of literals evaluated by v as 1. Then let us form the formula $d_\varphi := \bigvee_{v \in S} t_v$. We then show the following fact.

PROPOSITION 3.14
For every formula φ, $\varphi \equiv d_\varphi$ is a tautology. Up to the order of variables, and the listing of minterms the representation $\varphi \mapsto d_\varphi$ is unique.

Proof: Formulas φ and d_φ are satisfied by precisely the same valuations. □

Assuming that the set Var is finite, a minterm can be identified with a valuation. Namely, given a minterm t let us define the function v_t as follows:

$$v_t(x) = \begin{cases} 1 & \text{if } x \text{ occurs in } t \text{ positively} \\ 0 & \text{if } x \text{ occurs in } t \text{ negatively.} \end{cases}$$

Since t is a minterm, v_t evaluates the entire set Var. The assignment $t \mapsto v_t$ is bijective. In particular every valuation v is of the form v_t. Let V be a set of valuations. Since Var is finite, V is finite. Let us form the following formula φ_V:

$$\bigvee \{t : v_t \in V\}.$$

We then have the following property.

PROPOSITION 3.15
$V = Mod(\varphi_V)$. *In other words, the models of φ_V are precisely the valuations from V.*

Proof: If $v \in V$ then for minterm t such that $v = v_t$, $v \models t$. Therefore $v \models \varphi_V$. Since v is an arbitrary valuation in V, inclusion $\subseteq$ follows.
Conversely, if $v \models \varphi_V$, then for some t such that $v_t \in V$, $v \models t$. But then, by Lemma 3.1, $v = v_t$ so $v \in V$. Thus the inclusion $\supseteq$ follows, too. □

The formula φ_V constructed above is a DNF formula. We can find its conjunctive-normal form; let us call it ψ_V for the lack of better name. Let F_V be the set of all clauses of ψ_V. Then, clearly, $Mod(F_V) = Mod(\psi_V) = Mod(\varphi_V) = V$. Consequently we get the following fact.

PROPOSITION 3.16
Let Var be a finite set of propositional variables. Then for every set V of valuations of Var there is a set of clauses S_V such that $V = Mod(S_V)$.

The formula d_φ is another DNF for φ. This form is very inefficient, and amounts to just another representation of the table for φ, except that we list only those rows of the table where the value is 1. On average a formula in $Form^{Var}$ where $|Var| = n$ has a representation as 2^{n-1} minterms, an unpleasant fact of life. It should also be clear that if a formula φ is in a complete disjunctive normal form, then testing for tautology is trivial and reduces to counting. Such a formula is a tautology if and only if it contains precisely 2^n minterms. Likewise, counting the number of satisfying assignments of a formula φ is trivial if we have d_φ – we just count the number of minterms in d_φ.

We can assign to minterm t its *size* by counting positive literals occurring in t. Then here is one technique to construct a symmetric formulas (cf. Section 2.6). Let $|Var| = n$. Let us select any increasing sequence $1 \le i_1 < \ldots < i_k \le n$. For each j, $1 \le j \le k$, we form $\psi_j := \bigvee \{t : t \text{ is a minterm and } size(t) = j\}$. Then let us form $\psi_{j_1,\ldots,j_k} := \bigvee_{1 \le j \le k} \psi_{i_j}$. Then $\psi_{j_1,\ldots,j_k}$ is a symmetric formula. All symmetric formulas are equivalent to formulas of the form $\psi_{j_1,\ldots,j_k}$ for a suitably chosen sequence $\langle j_1, \ldots, j_k \rangle$.

Let us observe that we can count symmetric formulas (up to equivalence); there are as many of them as the strictly increasing sequences $\langle j_1, \ldots, j_k \rangle$ that is 2^n. There are 2^{2^n} elements of the Lindenbaum algebra, but we just found that only 2^n of them are symmetric. However we observe that symmetric formulas form a nice subclass of the Lindenbaum algebra. They are closed under negation, conjunction, and disjunction.

There are analogous results about *maxclauses*, that is, non-tautological clauses of size precisely n. Every formula φ is equivalent to the conjunction c_φ of such clauses. All we need to do is to compute $d_{\neg\varphi}$, then transform it using the transformation pushing the DNF of $\neg\varphi$ to CNF of φ. In this manner we get the following fact.

PROPOSITION 3.17
For every formula φ, $\varphi \equiv c_\varphi$ is a tautology. Up to the order of variables and the listing of maxclauses the representation $\varphi \mapsto c_\varphi$ is unique.

3.6 Lindenbaum algebra revisited

The complete disjunctive normal form discussed in Section 3.5 allows us to characterize completely the Lindenbaum algebra of propositional logic based on n variables. First, Lemma 3.1 tells us that the coset of every minterm is an atom in the Lindenbaum algebra. There are 2^n such atoms. Our unique representation result (Proposition 3.14) was just restating a well-known result of the representation of complete atomic Boolean algebras by the power set of the set of atoms. Moreover, we also know how many objects altogether there are in the Lindenbaum algebra; since there are 2^n atoms, there are altogether 2^{2^n} cosets altogether. Moreover, the

Normal forms of formulas 59

number of atoms determines the algebra up to isomorphism. Altogether this gives us the following fact.

PROPOSITION 3.18
If the set Var is finite then the corresponding Lindenbaum algebra is finite.

Our estimation of the number of atoms in the Lindenbaum algebra tells us that not every finite Boolean algebra is isomorphic to the Lindenbaum algebra of a propositional logic. (Let us recall that different sets *Var* determine different Lindenbaum algebras.) However, by the same token, every finite Boolean algebra is isomorphic to a subalgebra of a suitably chosen Lindenbaum algebra, namely that of symmetric formulas. Further results can be obtained in this direction, but we will not pursue this matter here.

3.7 Other normal forms

Every functor can be represented in terms of $\Rightarrow$ and $\bot$ (we will study completeness of sets of functors in Chapter 5). This leads to another normal form, called the implication-normal form. That is, for every formula φ there is a formula $\text{INF}(\varphi)$ built only of variables, constants, and the functor $\Rightarrow$ so that $\varphi \equiv \text{INF}_\varphi$ is a tautology. We define this formula inductively. We will assume that the formula φ does not involve the functor $\equiv$ but it is of course a minor limitation. We could impose other limitations (e.g., require that the formula be in negation normal form), but it is not necessary.

1. $\text{INF}(a) = a$ if a is a variable
2. $\text{INF}(\neg\varphi) = (\text{INF}(\varphi) \Rightarrow \bot)$
3. Proceeding by induction on the rank of formulas (recall that all formulas are in negation-normal form) let us define

    (a) $\text{INF}(\varphi \wedge \psi) = (\text{INF}(\varphi) \Rightarrow (\text{INF}(\psi) \Rightarrow \bot)) \Rightarrow \bot$
    (b) $\text{INF}(\varphi \vee \psi) = (\text{INF}(\varphi) \Rightarrow \bot) \Rightarrow \text{INF}(\psi)$
    (c) $\text{INF}(\varphi \Rightarrow \psi) = \text{INF}(\varphi) \Rightarrow \text{INF}(\psi)$

PROPOSITION 3.19
For every formula φ, $\varphi \equiv \text{INF}(\varphi)$ is a tautology.

The implication-normal form has been well-known since the 1920s. Computation of INF forms a first step of the so-called Stålmarck's algorithm for testing formulas for tautology.[1]

Finally, let us turn our attention to the *ite* ternary functor. We will look at it again in Section 5.6. In terms of De Morgan functors, $ite(p, q, r)$ is $(p \wedge q) \vee (\neg p \wedge r)$. The functor *ite* is complete, providing constants can be used. We will now, as we did in the case of implicational-normal form, define the ITE-normal form of formulas, $\text{ITE}(\varphi)$.

1. $\text{ITE}(a) = a$ if a is a variable
2. $\text{ITE}(\neg\varphi) = ite(\text{ITE}(\varphi), \bot, \top)$
3. $\text{ITE}(\varphi \wedge \psi) = ite(\text{ITE}(\varphi), \text{ITE}(\psi), \bot)$
4. $\text{ITE}(\varphi \vee \psi) = ite(\text{ITE}(\varphi), \top, \text{ITE}(\psi))$
5. $\text{ITE}(\varphi \Rightarrow \psi) = ite(\text{ITE}(\varphi, \text{ITE}(\psi), \top)$

The following result shows that we can represent formulas in ITE-normal forms.

PROPOSITION 3.20
For every formula φ, $\varphi \equiv \text{ITE}(\varphi)$ is a tautology.

The ITE-normal form is useful when we represent formulas as so-called binary decision diagrams, BDDs. These expressions are representations of Boolean functions as graphs. The specific form of BDD, reduced ordered BDD, offers a particularly efficient representation of Boolean functions. ROBDD is based on efficient reuse of components of Boolean functions. While a fascinating topic, BDD are mostly beyond the scope of this book. In Exercises to Chapter 11 we discuss, however, the construction of ROBDD and the fundamental result of the theory of ROBDDs, Bryant theorem.

It should be quite clear that *disjunctive* normal form allows for a fast computation of the ITE form. The reason for it is that once we have φ in a disjunctive normal form, we can group terms containing a given variable p into a single conjunction, take p out of it, forming a formula $p \wedge \psi_1$, and do the same thing for $\neg p$, forming $(\neg p) \wedge \psi_2$. There is the third part, ψ_3, not involving p at all. Then the original formula φ is equivalent to $ite(p, \psi_1 \vee \psi_3, \psi_2 \vee \psi_3)$. This forms a basis for an entire theory useful in electronic design automation.

3.8 Exercises

1. Find a canonical negation normal form for this formula:

$$\neg(\neg(p \wedge q) \vee (p \wedge \neg q))$$

---

[1] We will not discuss the Stålmarck algorithm here since it is *patented*.

Normal forms of formulas

2. List positive and negative occurrences of variables in the formula of Problem (1).

3. Investigate the relationship between positive and negative occurrences of variables and the number of negations on the path from the root of the tree to the occurrence of the variable.

4. In Section 3.1 we defined the negation-normal form for formulas based on the functors $\{\neg, \wedge, \vee\}$. Can we do this for all formulas built from De Morgan functors? How do we need to define occurrences of variables in formulas of the form $\varphi \Rightarrow \psi$ and in formulas of the form $\varphi \equiv \psi$ to preserve the validity of Theorem 3.1?

5. In the Tarski fixpoint theorem (Proposition 3.5) we eliminated a variable p, substituting for it a formula arising from substitution of variable p by $\top$. We could substitute p with $\bot$, instead. What will happen?

6. In Proposition 3.7 we established that for every theory T there is a set of clauses F such that $Mod(T) = Mod(F)$. First, complete the argument (which was only outlined there). Second, the set F is not unique, even if there are no subsumptions and no tautological clauses. Give an appropriate example.

7. In Proposition 3.6 we established that if the set Var is finite, and of size n, then the corresponding Lindenbaum algebra has 2^n atoms and hence that algebra is finite. Why is it so? Specifically, do a small excursion into Boolean algebra and prove the following fact: *If $k \in N$, then there is exactly one (up to isomorphism) Boolean algebra with exactly k atoms. That algebra is finite, and has exactly 2^k elements.*

Chapter 4

The Craig lemma

4.1 Craig lemma 63
4.2 Strong Craig lemma 66
4.3 Tying up loose ends 69
4.4 Exercises 71

Our goal now is to prove the Craig lemma, a kind of "Occam Razor" principle for logic. The idea here is that when an implication $\psi \Rightarrow \varphi$ is a tautology, then this phenomenon means that there is a constraint on the variables that occur both in ψ and φ. This constraint, ϑ must be entailed by the constraint expressed by ψ and in itself must entail φ. This limitation on the variables that are mentioned in ϑ is the reason why we mentioned "Occam Razor." It turns out that, actually, the requirements can be more stringent, and we can find even better intermediate constraints (they are called interpolants), with additional limitations on the polarity of variables involved.

4.1 Syntactic transformations and the Craig lemma

The Craig lemma, in its simplest form, states that whenever $\psi \Rightarrow \varphi$ is a tautology, then there exists an *interpolant*, a formula ϑ, involving only the variables occurring both in ψ and in φ, such that both $\psi \Rightarrow \vartheta$ and $\vartheta \Rightarrow \varphi$ are tautologies.

Our proof will be constructive (a nonconstructive proof can be provided using "consistency properties"). This is not to say that our proof will not involve work. The work involved in *computing* an interpolant will require (in our presentation) to compute either the disjunctive normal form of ψ or the conjunctive normal form of φ. If any of these is available, the amount of work is linear in the size of $Var_\psi \cup Var_\varphi$.

Let us observe that if φ itself is a tautology, or ψ is false, then the Craig lemma is obvious. Indeed, in the first case we can select $\top$ as an interpolant, and in the second case $\bot$ serves as an interpolant. Thus the only interesting case is when φ is not a tautology, and ψ is not false.

In preparation for our proof of the Craig lemma we have a lemma that allows us to simplify checking that some formulas of the form $\psi \Rightarrow \varphi$ are tautologies.

LEMMA 4.1

Let ψ be a DNF formula, $\psi = D_1 \vee \ldots \vee D_k$, and let φ be an arbitrary formula. Then the formula $\psi \Rightarrow \varphi$ is a tautology if and only if for all i, $1 \leq i \leq k$, $D_i \Rightarrow \varphi$ is a tautology.

Proof: It is easy to check that the following formula,

$$((D_1 \vee \ldots \vee D_k) \Rightarrow \varphi) \equiv \bigwedge_{j=1}^{k} (D_j \Rightarrow \varphi),$$

is a tautology.

So now, $D_1 \vee \ldots \vee D_k \Rightarrow \varphi$ is a tautology if and only if $\bigwedge_{j=1}^{k} (D_j \Rightarrow \varphi)$ is a tautology. But it is easy to see that a conjunction of formulas is a tautology if and only if all the conjuncts are tautologies. $\square$

Next, let us assume that a formula $\psi \Rightarrow \varphi$ is a tautology. Then let us consider the canonical disjunctive normal form $D_1 \vee \ldots \vee D_k$ of ψ. Without the loss of generality we can assume that ψ is $D_1 \vee \ldots \vee D_k$. We can also assume that no term D_i, $1 \leq i \leq k$ is false. That is, D_i, $1 \leq i \leq k$ does not contain a pair of contradictory literals. Given a term (elementary conjunction) $D = l_1 \wedge \ldots \wedge l_m$, and a set of variables $X \subseteq Var$, define $D \mid_X$ as

$$\bigwedge \{l : l \text{ occurs in } D \text{ and } l \in \mathcal{L}_X\}.$$

Thus $D \mid_X$ eliminates from D all literals with underlying variables *not* in X. Let L be a non-contradictory set of literals, that is, one that does not contain a pair of complementary literals. We define v_L as the valuation making all literals in L true and defined on those variables that actually appear in L (that is the representation of L as a partial valuation). We now have the following lemma.

LEMMA 4.2

Let D be a non-contradictory conjunction of literals, and φ an arbitrary formula. Then the implication $D \Rightarrow \varphi$ is a tautology if and only if the implication $D \mid_{Var(\varphi)} \Rightarrow \varphi$ is a tautology.

Proof: First, assume $D \mid_{Var(\varphi)} \Rightarrow \varphi$ is a tautology. Then it is easy to see that the formula $D \Rightarrow \varphi$ is a tautology. The reason is that D has more literals than $D \mid_{Var(\varphi)}$ and so the formula $D \Rightarrow D \mid_{Var(\varphi)}$ is a tautology. Then, using transitivity of $\Rightarrow$ we get the validity of our implication $D \Rightarrow \varphi$.

Conversely, let us assume that $D \Rightarrow \varphi$ is a tautology, but $D \mid_{Var(\varphi)} \Rightarrow \varphi$ is not a tautology. Then there must exist a valuation v such that $v(D \mid_{Var(\varphi)}) = 1$ but $v(\varphi) = 0$. Now, all variables occurring in $D \mid_{Var(\varphi)} \Rightarrow \varphi$ occur in φ. Since $D \Rightarrow \varphi$ is a tautology, it must be the case that $D \neq D \mid_{Var(\varphi)}$. Define now the following set of literals: $L = \{l : l \text{ occurs in } D \text{ and } l \notin \mathcal{L} \mid_{Var(\varphi)}\}$. Then L is non-contradictory, and v_L is well-defined. Recall that v_L assigns 1 to all literals in L. We now define a new valuation w, defined on all variables occurring in D or in φ, as follows:

$$w(p) = \begin{cases} v(p) & \text{if } p \in Var(\varphi) \\ v_L(p) & \text{if } p \text{ appears in } L. \end{cases}$$

Now, let us observe that for $p \in Var(\varphi)$, $w(p) = v(p)$, thus $w(\varphi) = 0$. Next, let us look at D. Without a loss of generality

$$D = l_1 \wedge \ldots \wedge l_k \wedge l_{k+1} \wedge \ldots \wedge l_m$$

where $l_1, \ldots, l_k \in \mathcal{L}\,|_{Var(\varphi)}$, and $\{l_{k+1}, \ldots, l_n\} = L$. Thus for all j, $1 \leq j \leq m$, $w(l_j) = 1$. The reason for this is that $v(D\,|_{Var}\,(\varphi)) = 1$, and so $v(l_j) = 1$ for $1 \leq j \leq k$, thus $w(l_j) = 1$. For $k+1 \leq j \leq m$, $v_L(l_j) = 1$ and so, again, $w(l_j) = 1$.
Thus $w(D) = 1$. But $w(\varphi) = 0$, so $v(\varphi) = 0$ as w and v coincide on Var_φ. This is a desired contradiction. □
The argument given above entails the following fact of independent interest.

COROLLARY 4.1
If D is a term that is not false, and φ is a formula such that φ is not a tautology, and if $D \Rightarrow \varphi$ is a tautology, then D and φ must share at least one variable.

COROLLARY 4.2
Let $\psi = D_1 \vee \ldots \vee D_k$ be a DNF formula, and let us assume that $\psi \Rightarrow \varphi$ is a tautology, and that none of D_i, $1 \leq i \leq k$, is false. Then the formula

$$D_1\,|_{Var(\varphi)} \vee \ldots \vee D_k\,|_{Var(\varphi)} \Rightarrow \varphi$$

is a tautology.

Proof: By Lemma 4.1 the formulas $D_j \Rightarrow \varphi$, $1 \leq j \leq m$, are all tautologies. Thus for each j, $1 \leq j \leq m$, the formula $D_j\,|_{Var(\varphi)} \Rightarrow \varphi$ is a tautology. Applying Lemma 4.1 again we find that the formula $D_1\,|_{Var(\varphi)} \vee \ldots \vee D_k\,|_{Var(\varphi)} \Rightarrow \varphi$ is a tautology. □

We now have the following fact.

LEMMA 4.3
For every formula φ, the formula

$$D_1 \vee \ldots \vee D_k \Rightarrow D_1\,|_{Var(\varphi)} \vee \ldots \vee D_k\,|_{Var(\varphi)}$$

is a tautology.

Proof: Clearly for each j, $1 \leq j \leq k$, $D_j \Rightarrow D_j\,|_{Var(\varphi)}$ is a tautology. Thus $D_1 \vee \ldots \vee D_k \Rightarrow D_1\,|_{Var(\varphi)} \vee \ldots \vee D_k\,|_{Var(\varphi)}$ is a tautology. □

But now, if we combine Corollary 4.2 and Lemma 4.3 we get the desired interpolation result.

PROPOSITION 4.1 (Craig lemma)
If $\psi \Rightarrow \varphi$ is a tautology, then there exists a formula ϑ such that $Var(\vartheta) \subseteq Var(\psi) \cap Var(\varphi)$ and such that both $\psi \Rightarrow \vartheta$ and $\vartheta \Rightarrow \varphi$ are tautologies.

Proof: Let $D_1 \vee \ldots \vee D_k$ be a disjunctive normal form for ψ. We can assume that the variables of $D_1 \vee \ldots \vee D_k$ all occur in ψ. By Lemma 4.3 the formula

$$D_1 \vee \ldots \vee D_k \Rightarrow D_1 \,|_{Var(\varphi)} \vee \ldots \vee D_k \,|_{Var(\varphi)}$$

is a tautology. By Corollary 4.2 the formula

$$D_1 \,|_{Var(\varphi)} \vee \ldots \vee D_k \,|_{Var(\varphi)} \Rightarrow \varphi$$

is a tautology. By the construction all the variables of $D_1 \,|_{Var(\varphi)} \vee \ldots \vee D_k \,|_{Var(\varphi)}$ occur both in ψ and in φ. Thus $\vartheta : D_1 \,|_{Var(\varphi)} \vee \ldots \vee D_k \,|_{Var(\varphi)}$ is a desired interpolation formula. □

4.2 Strong Craig lemma

Clearly, the Craig interpolation lemma (Proposition 4.1) can be, at least sometimes, improved. To see this, let us look at the following example.

Example 4.1
Assume that the formula ψ in disjunctive normal form has as one of its conjuncts the elementary conjunction $\neg p \wedge q \wedge s$, whereas the formula φ is $p \vee q$. Then the procedure outlined above in Lemma 4.2 will compute the interpolant $\neg p \wedge q$, whereas it is easy to see that the formula $\vartheta : q$ is an interpolant. In other words, at least sometimes, the procedure outlined in Lemma 4.2 is too weak; a stronger interpolant should be computed. □

The problem encountered in Example 4.1 was that we did not take into account the polarity of variables. We will now show a stronger version of the Craig lemma (Proposition 4.1). First we will prove a lemma.

LEMMA 4.4
Let us assume φ is a formula in which a variable p occurs positively but not negatively. Then there exist two formulas ψ_1 and ψ_2 without occurrences of p at all, such that
$$\varphi \equiv ((p \wedge \psi_1) \vee \psi_2)$$

is a tautology.

Proof: Without the loss of generality we can assume that φ is in negation normal form (because the transformation to canonical negation normal form does not alter polarity of occurrences of variables). Now proceed by induction the complexity of formula φ. The base case is when φ is p. Then $\psi_1 = \top$, whereas $\psi_2 = \bot$.

Since the formula φ is in negation normal form, in the inductive step we need to consider only two cases: first, when the main connective of φ is $\wedge$, and second, when the main connective of φ is $\vee$.

Case 1. The main connective of φ is $\wedge$. Then $\varphi : \varphi_1 \wedge \varphi_2$. By the definition, the variable p must occur positively in at least one of the formulas ψ_1 and ψ_2 but it cannot occur negatively in either of them.

Subcase 1.1. The variable p occurs positively in both φ_1 and φ_2. Then there are four formulas $\psi_{1,1}, \psi_{1,2}, \psi_{2,1}$, and $\psi_{2,2}$ such that $\varphi_1 \equiv (p \wedge \psi_{1,1}) \vee \psi_{1,2}$ is a tautology, and $\varphi_2 \equiv (p \wedge \psi_{2,1}) \vee \psi_{2,2}$ is a tautology. But then it is easy to see that the formula

$$\psi_1 \vee \psi_2 \equiv (p \wedge ((\psi_{1,1} \wedge \psi_{2,2}) \vee (\psi_{1,1} \wedge \psi_{2,1}) \vee (\psi_{1,1} \wedge \psi_{2,1}))) \vee (\psi_{1,2} \wedge \psi_{2,2})$$

is a tautology. Thus we can take

$$\psi_1 : (\psi_{1,1} \wedge \psi_{2,2}) \vee (\psi_{1,1} \wedge \psi_{2,1}) \vee (\psi_{1,1} \wedge \psi_{2,1})$$

and $\psi_2 : \psi_{1,2} \wedge \psi_{2,2}$.

Subcase 1.2. The variable p does not occur in one of the formulas φ_1 or φ_2. In such case we can assume without the loss of generality that p does not occur in φ_2. By inductive assumption we have formulas ψ_1 and ψ_2, none of those with an occurrence of p such that $\varphi_1 \equiv (p \wedge \psi_{1,1}) \vee \psi_{1,2}$. But then, since φ_2 has no occurrence of p at all, it is easy to see that

$$\varphi_1 \wedge \varphi_2 \equiv (p \wedge (\psi_{1,1} \wedge \varphi_2)) \vee (\psi_{1,2} \wedge \varphi_2),$$

so we can take $\psi_1 : \psi_{1,1} \wedge \varphi_2$ and $\psi_2 : \psi_{1,2} \wedge \varphi_2$.

Case 2. The main connective of φ is $\vee$. Again two cases need to be considered (in analogy with the Case 1), but the argument is even simpler this time and we leave it to the reader. □

We now state without a proof a lemma similar to Lemma 4.4, and proved in a similar fashion.

LEMMA 4.5

Assume φ is a formula in which a variable p occurs negatively but not positively. Then there exist two formulas ψ_1 and ψ_2 without occurrences of p at all, such that:

$$\varphi \equiv ((\neg p \wedge \psi_1) \vee \psi_2)$$

is a tautology.

We now prove a lemma allowing for elimination from the interpolant the literals with opposite polarities in the antecedent and the consequent of the implication. The specific case we prove is in the special situation when the antecedent is a conjunction of literals. Recall that we are dealing with NNF formulas. For that reason we will now say that $\neg p$ has an occurrence in a formula χ if the tree for $\neg p$ is a subtree of the tree for χ.

LEMMA 4.6
Let $l_1 \wedge \ldots \wedge l_k \Rightarrow \varphi$ be a tautology, with $\{l_1, \ldots, l_k\}$ non-contradictory set of literals, and assume that the literal l_1 does not occur in φ at all. Then the formula $l_2 \wedge \ldots \wedge l_k \Rightarrow \varphi$ is a tautology.

Proof: Without the loss of generality we can assume that $l = p$ for some variable p. Our assumption means that either p does not occur in φ at all, or if it does, it occurs in φ only negatively, that is, as a part of the literal $\neg p$.

Case 1. The variable p does not occur in φ at all. Define $D : l_1 \wedge \ldots \wedge l_k$. Then by Lemma 4.2 the conjunction $D\,|_{Var(\varphi)}$ of those literals in the antecedent for which the underlying variable has an occurrence in φ is an interpolant. But this conjunction does not include p. Thus we have two tautologies: $l_2 \wedge \ldots \wedge l_k \Rightarrow D\,|_{Var(\varphi)}$ and $D\,|_{Var(\varphi)} \Rightarrow \varphi$. But then $l_2 \wedge \ldots \wedge l_k \Rightarrow \varphi$ is a tautology, as desired.

Case 2. The variable p occurs in φ negatively. That is, $\neg p$ occurs in φ (but p does not occur in φ). By Lemma 4.5, there are two formulas, none involving any occurrence of p, ψ_1, and ψ_2 such that

$$\varphi \equiv ((\neg p \wedge \psi_1) \vee \psi_2)$$

is a tautology. But then the implication

$$p \wedge l_2 \wedge \ldots \wedge l_k \Rightarrow ((\neg p \wedge \psi_1) \vee \psi_2)$$

is a tautology. We claim that the implication

$$p \wedge l_2 \wedge \ldots \wedge l_k \Rightarrow \psi_2$$

is a tautology. Indeed, let v be any valuation. If $v(D) = 0$ then v evaluates our implication as 1. If $v(D) = 1$, then it must be the case that $v(p) = 1$. But then $v(\neg p) = 0$, and $v((\neg p \wedge \psi_1)) = 0$. But $v(D \Rightarrow \varphi) = 1$. Since $v(D) = 1$, $v(\varphi) = 1$. Thus it is the case that $v((\neg p \wedge \psi_1) \vee \psi_2) = 1$. But then $v(\psi_2)$ is 1, as desired.

So now we have established that the formula $D \Rightarrow \psi_2$ is a tautology. But ψ_2 does not involve p at all. We now apply Lemma 4.2 to the implication $D \Rightarrow \psi_2$. Then we get an interpolant D' for the implication $D \Rightarrow \psi_2$, involving only variables occurring in ψ_2, in particular one that does not involve p at all. But the formula $\psi_2 \Rightarrow \varphi$ is a tautology. Therefore D' is an interpolant for $D \Rightarrow \varphi$. Indeed, the formulas $D' \Rightarrow \psi_2$, and $\psi_2 \Rightarrow \varphi$ are all tautologies. Thus $D' \Rightarrow \varphi$ is a tautology. Since $l_2 \wedge \ldots \wedge l_k \Rightarrow D'$, the assertion follows. □

In order to get a stronger form of the Craig lemma, we need a stronger form of Lemma 4.2. Here it is.

LEMMA 4.7
Let D be a non-contradictory elementary conjunction, and assume that $D \Rightarrow \varphi$ is a tautology. Let D' be the conjunction of those literals in D which have occurrences in φ. Then $D' \Rightarrow \varphi$ is a tautology.

Proof: By repeated application of Lemma 4.6, those literals in D for which the underlying variable does not occur in φ or it does, but only with the opposite polarity, will be eliminated. The resulting D' is an interpolant. □

Let us observe that the difference between Lemma 4.7 and Lemma 4.2 is that now we look at occurrences of literals, not only variables, that is, we take the polarity into account.

Thus the interpolant computed in Lemma 4.7 has the property that whenever a variable p occurs in the interpolant D', it must occur in both D and in φ with the same polarity (otherwise it is eliminated). We call such interpolant a *strong interpolant*. Now let us apply Lemma 4.3. We then get the following.

COROLLARY 4.3
Let ψ and ϕ be two formulas such that $\psi \Rightarrow \varphi$ is a tautology. Then there exists a formula ϑ such that

1. Both $\psi \Rightarrow \vartheta$ and $\vartheta \Rightarrow \varphi$ are tautologies
2. Whenever a variable p occurs in ϑ then p must occur with the same polarities in ψ, ϑ, and φ.

4.3 Tying up loose ends

The procedure we established for the computation of an interpolant (Lemma 4.7) can be easily used for the situation when φ is given in conjunctive-normal form, that is, as a conjunction of clauses. The reason for this is simple; the formula $\psi \Rightarrow \varphi$ is a tautology if and only if the formula $\neg \varphi \Rightarrow \neg \psi$ is a tautology. Now, assume φ is in conjunctive normal form. Then with the work linear in the size of formulas φ and ψ we can convert the implication $\neg \varphi \Rightarrow \neg \psi$ into an implication $\psi' \Rightarrow \varphi'$ where ψ' is in disjunctive normal form, φ' is in negation normal form. The reason for this amount of work is that the conversion of negation of a CNF to a DNF requires the following simple actions: First conversion of every literal l to its dual $\bar{l}$, and then simultaneous substitution of every $\wedge$ for $\vee$ and $\vee$ for $\wedge$. That is, it is inexpensive to compute a DNF of $\neg \varphi$, *provided* that φ is in CNF. Likewise, given a formula φ in DNF, a CNF for $\neg \varphi$ can be computed in linear time.

Next, we compute the interpolant for $\psi' \Rightarrow \varphi'$, say χ. The formula χ will be a DNF. Then we can easily convert $\neg \chi$ to a CNF ϑ which is an interpolant for $\psi \Rightarrow \varphi$.

But, of course, those two transformations (first $\neg\varphi \mapsto \psi'$ and then $\neg\chi \mapsto \vartheta$) are entirely superfluous. A momentary reflection shows that both transformations can be combined into a single procedure that does not require computation of DNF of $\neg\varphi$ at all. Namely, it should be clear that the composition of the two procedures outlined above results in shortening of clauses constituting φ (in a way similar to shortening elementary conjunctions constituting ψ), provided ψ is a DNF. In each of these clauses we eliminate those literals which do not occur in the antecedent with the same polarity or their underlying variables do not occur in ψ at all.

We could, of course, prove analogues of all the lemmas we proved for DNFs, for the CNF φ (instead of DNF ψ). The starting point would be the tautology

$$(\psi \Rightarrow \bigwedge_{1 \leq j \leq k} \varphi_i) \equiv \bigwedge_{1 \leq j \leq k} (\psi \Rightarrow \varphi_i)$$

instead of the one used in Lemma 4.1. Then we could prove lemmas that, instead of pruning conjunctions in the antecedent, would prune disjunctions in the consequent. The second "loose end" involves the interpolant. If we look at the procedure outlined in Lemma 4.7, then it is clear that regardless of the fact whether $\psi \Rightarrow \varphi$ is a tautology or not, the implication $\psi \Rightarrow \vartheta$ is a tautology. If the formula $\psi \Rightarrow \varphi$ is a tautology then, as we see in Lemma 4.7, the formula $\vartheta \Rightarrow \varphi$ is a tautology. Conversely, if the formula $\vartheta \Rightarrow \varphi$ is a tautology, then since $\psi \Rightarrow \vartheta$ is a tautology, $\psi \Rightarrow \varphi$ is a tautology. Thus we have proved another useful form of the Craig lemma.

THEOREM 4.1
If ψ and φ are two formulas, then there is a formula ϑ such that ϑ involves only variables occurring in both ψ and in φ such that the following are equivalent:

1. *$\psi \Rightarrow \varphi$ is a tautology*

2. *$\vartheta \Rightarrow \varphi$ is a tautology*

Moreover, we can assume that any variable occurring in ϑ must occur in both ψ and φ with the same polarities as it occurs in ϑ.

There are other questions that can be investigated. For instance, if the formula ψ is given in the ITE format (i.e., defined using only the *ite* operation, and with a fixed order of variables), can we compute a similar form of the interpolant in polynomial time? This is certainly possible *if* the formula ψ is an elementary conjunction.

What happens if the formula $\psi \Rightarrow \varphi$ is a tautology, but ψ and φ have no variables in common? The interpolant takes a particularly simple form: it has to be either $\bot$ or $\top$. If the interpolant is $\bot$, then, because $\psi \Rightarrow \bot$ is a tautology, it must be the case that the negation of ψ is a tautology (the only case possible). If the interpolant is $\top$, then φ must be a tautology (again the only case possible). Thus we get the following corollary.

COROLLARY 4.4
If $\psi \Rightarrow \varphi$ is a tautology, but formulas ψ and φ do not share variables, then either $\neg \psi$ is a tautology, or φ is a tautology.

For the enthusiasts of Boolean algebras, and Lindenbaum algebras in particular, we will now provide an interpretation of the Craig lemma in topological terms. Let ψ and φ be two formulas and let $V = Var_\psi \cap Var_\varphi$ and let $W = Var_\psi \cup Var_\varphi$. Then $V \subseteq W$. The corresponding propositional languages are related: $\mathcal{L}_V \subseteq \mathcal{L}_W$. Each of these languages has its own Lindenbaum algebra $\mathcal{A}_V$ and $\mathcal{A}_W$, resp. Given $\vartheta \in \mathcal{L}_V$ it has two equivalence classes: one, $[\vartheta]_1$, in $\mathcal{A}_V$ and another $[\vartheta]_2$, in $\mathcal{A}_W$. So it is natural to define

$$e([\vartheta]_1) = [\vartheta]_2$$

and see what happens. Clearly the function e is a monomorphism. This follows from the fact that satisfaction is preserved if the set of variables grows. Next, keeping V and W fixed, we can assign to every element of the algebra $\mathcal{A}_2$ its *closure* in $\mathcal{A}_1$, namely,

$$\bar{a} = \bigwedge \{b : b \in \mathcal{A}_1 \text{ and } a \leq b\}.$$

Likewise we can define an interior of an element a:

$$\underline{a} = \bigvee \{b : b \in \mathcal{A}_1 \text{ and } b \leq a\}.$$

The operations of closure and interior are always monotone and idempotent. Here is one corollary to the Craig lemma. Whenever $[\psi \Rightarrow \varphi]$ is the unit of Lindenbaum algebra $\mathcal{A}_W$, then $\overline{[\psi]} \leq [\varphi]$. Let us keep in mind that the closure operation here depends on ψ and φ. Likewise $[\psi] \leq \underline{[\varphi]}$. The interpolant that was computed in our proof of the Craig lemma was the interior of $[\varphi]$ with respect to the algebra generated by common variables of ψ and φ.

Even more appealing (this time to enthusiasts of real analysis) is the separation form of the Craig lemma. Namely, if $a, b \in \mathcal{A}_W$ and $a \wedge b = \bot$ then for some $c \in \mathcal{A}_V$, $a \leq c$ and $c \wedge b = \bot$.

4.4 Exercises

1. Check that the formula

$$(p \vee \neg q) \wedge r \Rightarrow (p \wedge \neg s) \vee (\neg q \vee s)$$

   is a tautology. Using the technique discussed above, find an interpolant.
2. If ϑ_1 and ϑ_2 are two interpolants for $\psi \Rightarrow \psi$, then so is $\vartheta_1 \vee \vartheta_2$.
3. If ϑ_1 and ϑ_2 are two interpolants for $\psi \Rightarrow \psi$, then so is $\vartheta_1 \wedge \vartheta_2$.

4. Conclude that there is a strongest and a weakest interpolant for $\psi \Rightarrow \varphi$ (assuming the lattter formula is a tautology).

5. Look at the (trivial) example of $p \wedge q \Rightarrow p \vee q$. Even though the strongest and weakest interpolants do not bring anything new, there are non-trivial interpolants.

Chapter 5

Complete sets of functors

5.1 Beyond De Morgan functors 74
5.2 Tables 75
5.3 Field structure in *Bool* 78
5.4 Incomplete sets of functors, Post classes 83
5.5 Post criterion for completeness 85
5.6 If-then-else functor 88
5.7 Exercises 90

In this chapter we study the following question: Was the choice of functors of logic ($\neg, \wedge, \vee, \Rightarrow$, and $\equiv$) arbitrary, or was it somehow forced on us? Could we select other functors? If so, how? In fact, what is a functor?

It turns out that our choice of functors is pretty arbitrary since they were introduced in the historical development of logic rather that by some sort of devise. Later on, it will turn out that *specific* algorithms are tailored to specific representations. But we do not know what representations may be optimal for future algorithms, and so we need to have some sort of general techniques for representation. A word of caution: the issues of representation of formulas in various sets of functors have been studied extensively, especially by electrical engineers. There was an entire industry making a living out of research in this area. This included large research groups in the former Soviet Union. It is fair to say that we will only scratch the surface here.

We will call the functors in the set $\{\neg, \wedge, \vee, \Rightarrow, \equiv\}$ *De Morgan* functors.

There is a danger of overformalizing the entire subject. We will try to be as formal as possible, but without huge inductive definitions, and complex proofs by induction. For a moment, let us assume that we are given a set $\mathcal{C} = \{c_1, \ldots, c_k\}$. The elements of $\mathcal{C}$ are called functors. We will assume that we have associated with each $c_j \in \mathcal{C}$ its *arity* i_j (so we know how many arguments c_j has) and also that we have associated with each symbol c_j (of arity i_j) a function $f_j : Bool^{i_j} \to Bool$. If we *really* wanted to be formal, then we would define a functor as a pair $\langle c_j, f_j \rangle$. We hope that such level of formalization is not needed. We do not really do anything special – we just follow the scheme we pursued when we defined the semantics of formulas built of $\neg, \wedge, \vee, \Rightarrow$, and $\equiv$. Again, the formulas are ordered labeled trees, except that now we require i_j children of a node labeled with c_j. These trees have various textual representations (infix, prefix, etc.) We will soon deal with a prefix, not infix representation in an example.

5.1 Beyond De Morgan functors

Let us fix some set $\mathcal{C}$ of functors. As we look at textual (i.e., string) representations, we have the set $Form_{Var}^{\mathcal{C}}$ of formulas based on $\mathcal{C}$. Of course we will drop the subscript Var to save on the effort.

Once we define the set $Form^{\mathcal{C}}$, we can define semantics. The semantics once more will be given by valuations of propositional variables occurring in the formula. The functors c_j will now be interpreted by the Boolean functions f_j whose availability we required above.

To put this in the suitable perspective (but not to do too much) let us consider an example.

Example 5.1

We will have $\mathcal{C}$ consisting of just one binary functor that we will call *NAND*. This functor, intuitively, assigns to the pair (x, y) the value of the formula $\neg(x \wedge y)$. But this is intuition; formally we just have a symbol *NAND*, which is interpreted by the Boolean function *NAND* (we even use the same symbol for both the linguistic entity and for Boolean function). The truth table for *NAND* is presented in Table 5.1.

TABLE 5.1: Truth table for *NAND*

x	y	$NAND(x, y)$
0	0	1
0	1	1
1	0	1
1	1	0

Here is an example of a formula φ (now in prefix notation). This formula is built using only $NAND$.

$$NAND(NAND(x, x), NAND(x, y)).$$

Let us evaluate φ at the valuation v such that $v(x) = 0$ and $v(y) = 1$. The result of the evaluation is 0, i.e., $v(\varphi) = 0$. In other words, the valuation v does *not* satisfy the formula φ. □

Our example 5.1 should convince the reader that we are not doing anything exotic, just lifting the techniques of logic based on De Morgan functors to the more general case.

Now, once we have assigned semantics to the formulas built out of other sets of functors (remember such semantics requires interpretation of functors as Boolean functions), we can define the notions such as satisfiability, tautology, etc. Specifically, a tautology as before is an expression that is evaluated by every valuation as 1. As before, the constants $\bot$ and $\top$ are evaluated as 0 and 1, respectively, regardless of the valuation. We just note in passing that, as usual, the satisfaction depends only on the values assigned to variables actually occurring in a formula, and other nice things that applied earlier to formulas apply in the present context as well.

5.2 Tables, complete sets of functors

We now formalize the notion of a *table*. A table T is just the listing of a function f from $Bool^n \to Bool$. But $Bool^n$ (under reasonable assumptions, such as fixing the order of propositional variables $x_1, \ldots, x_n$) is nothing more than the set of all valuations of propositional variables $\{x_1, \ldots, x_n\}$. Therefore a table (in n variables) has $n+1$ columns, 2^n rows. Each row has $n+1$ entries, and its first n entries is a valuation. Different rows have different prefixes of the first n entries, that is, represent different valuations of variables. The last entry in the row is the value assigned by f to the first n entries.

A table of a formula $\varphi \in Form^{\mathcal{C}}$ is the table that assigns the value $v(\varphi)$ to the valuation v. We denote this table by T_φ.

So now it is quite clear how we are going to formalize the notion of a complete set of functors. Namely, a set $\mathcal{C}$ of functors is *complete* if for *every* table T, there is a $\varphi \in Form^{\mathcal{C}}$ such that $T = T_\varphi$. We then say that the formula φ *represents* the table T. Let us observe that a given table may be represented by more than one formula. Completeness of $\mathcal{C}$ means that every table is represented by some formula φ that includes only functors from $\mathcal{C}$.

Let us come back to the example and see what table is associated with the formula $NAND(NAND(x,x), NAND(x,y))$. There are two propositional variables, so the table will have four rows. We show the table for this formula in Table 5.2.

TABLE 5.2: The table for the formula $NAND(NAND(x,x), NAND(x,y))$

x	y	$NAND(NAND(x,x), NAND(x,y))$
0	0	0
0	1	0
1	0	1
1	1	1

We hope the reader does not trust us, and checks the correctness of this table herself.

Now, it is not clear that our original set of functors $\{\neg, \wedge, \vee, \Rightarrow, \equiv\}$ (what we called De Morgan functors) is complete. It is, indeed, the case, but not surprisingly, it requires a proof. First, we have the following truly obvious fact.

PROPOSITION 5.1
If C, C_1 are two sets of functors, C is complete, and $C \subseteq C_1$ then C_1 is also complete.

So now we shall prove that $\{\neg, \wedge, \vee\}$ is a complete set of functors and will conclude that $\{\neg, \wedge, \vee, \Rightarrow, \equiv\}$ is a complete set of functors.

PROPOSITION 5.2
The set $\{\neg, \wedge, \vee\}$ is a complete set of functors. Thus the set of De Morgan functors is complete.

Proof: We proceed by induction on the number of propositional variables in a formula. If that number is 0, then the formula takes constant value; it is 0, or 1. Then let us select a variable, say x, and observe that the formula $x \wedge \neg x$ always takes the value 0, while $x \vee \neg x$ takes always the value 1. Thus we have a representation for tables with the constant last column. Next, let us see how a table with $n+1$ variables looks. It is, in fact, computed out of two tables T_0 and T_1 each of n propositional variables. To form a table out of these two tables, we prepend the arguments of the first table T_0 with 0, prepend the arguments of the second one (T_1) with 1, and take the union. It should be clear that *every* table of $n+1$ arguments can be formed in this way. By inductive assumption, $T_0 = T_{\psi(x_2,\ldots,x_{n+1})}$ and $T_1 = T_{\vartheta(x_2,\ldots,x_{n+1})}$, for some formulas ψ and ϑ from $Form^{\{\neg,\wedge,\vee\}}$. Now define this formula $\varphi(x_1,\ldots,x_{n+1})$:

$$(\neg x_1 \wedge \psi(x_2,\ldots,x_{n+1})) \vee (x_1 \wedge \vartheta(x_2,\ldots,x_{n+1})).$$

The formula φ has only functors $\neg, \wedge, \vee$ since the same was true about ψ and ϑ. We leave to the reader to check that $T = T_\varphi$. □

It turns out that we can find an even smaller set of functors that is complete. But we first need another fact.

PROPOSITION 5.3
Let C_1 and C_2 be two sets of functors. Let us assume that C_2 is a complete set of functors, and that for every functor $c \in C_2$ there is a formula $\varphi \in Form^{C_1}$ such that the table T_c is identical with the table T_φ. Then C_1 is also a complete set of functors.

Proposition 5.3 really says that if all functors from C_2 are definable in terms of functors of C_1, and C_2 is complete, then so is C_1. We leave the proof, which is a version of the substitution lemma (except that it is more tedious), to the reader.

COROLLARY 5.1

Let C_1 and C_2 be two sets of functors. Let us assume that C_2 is a complete set of functors, and that for every functor $c \in C_2$ there is a formula $\varphi \in Form^{C_1}$ such that the formula $c(x_1, \ldots, x_n) \equiv \varphi(x_1, \ldots, x_n)$ is a tautology. Then C_1 is a complete set of functors.

We use Corollary 5.1 to show that $\{\neg, \wedge\}$ is a complete set of functors. Indeed, all we need to do is to show that the functor $\vee$ is definable in terms of $\neg$ and $\wedge$. That is, we need to find a formula φ that involves $\neg$ and $\wedge$ only, so that

$$(x \vee y) \equiv \varphi$$

is a tautology. One such formula φ is $\neg(\neg x \wedge \neg y)$. We leave to the reader the simple task of checking that, indeed, the tables for $x \vee y$ and for $\neg(\neg x \wedge \neg y)$ have the same last column. □

In this, and analogous ways, we find the following facts.

PROPOSITION 5.4

(a) $\{\neg, \wedge\}$ is a complete set of functors.

(b) $\{\neg, \vee\}$ is a complete set of functors.

(c) $\{\Rightarrow, \bot\}$ is a complete set of functors.

(d) $\{\wedge, +, \top\}$ is a complete set of functors.[1]

The proof of (a) was outlined above, and the proof of (b) is quite similar.
Proof of (c). The following are tautologies (we expect the reader to check them): First, the formula $(\neg x) \equiv (x \Rightarrow \bot)$ is a tautology. Second, $(x \vee y) \equiv ((x \Rightarrow \bot) \Rightarrow y)$ is a tautology. Now Corollary 5.1 and (b) entail completeness of the set $\{\Rightarrow, \bot\}$.
Proof of (d). Again we use Corollary 5.1. We have the following tautologies: $\neg x \equiv (x + \top)$, and $(x \vee y) \equiv ((x \wedge y) + x + y)$. This last one requires checking, which we leave to the reader. □

The careful inspection of the proof of Proposition 5.4 indicates that we really are dealing with two different concepts of completeness (a weaker one, where constants are assumed to be made available, and a stronger one, where this assumption is not made). Let us observe that $\{\neg, \wedge\}$ is complete in the strong sense, while $\{\Rightarrow\}$, as shown in (c), is complete in the weaker sense because $\{\Rightarrow, \bot\}$ is complete.

There are two binary functors that by themselves are complete (in the stronger sense.) One is the functor $NAND$ discussed above. Indeed, let us look at the following tautologies: $(x \vee y) \equiv NAND(NAND(x, x), NAND(y, y))$, and $(\neg x) \equiv NAND(x, x)$. These tautologies show that $NAND$ alone defines both $\neg$ and $\vee$, and so it is complete.

---
[1] The reader will recall that for reasons not clear at the time, we defined the functor $+$ in Chapter 2.

The other binary functor that is complete by itself is a functor called *NOR* with the table shown in Table 5.3. We leave to the reader the task of finding the appropriate witnessing tautologies.

TABLE 5.3: Truth table for *NOR*

x	y	$NOR(x,y)$
0	0	1
0	1	0
1	0	0
1	1	0

It is easy to find all sets of binary functors that are complete. We need to be a bit careful, because the negation functor (which is unary) appears as *two*, formally different, binary functors (one that assigns to (x, y) the value $\neg x$, and the other that assigns $\neg y$). As we observed above, a superset of a complete set of functors is again complete. So if we want to learn all complete sets of functors then we need to list only *inclusion-minimal* complete sets. It so happens that there are 36 of such minimal sets, and none of these has more than 3 functors. We will return to the problem of minimal complete sets of functors in Section 5.7 Exercises.

5.3 Field structure in *Bool*, Boolean polynomials

This section treats the set *Bool* as a field, a kind of structure considered in algebra, and looks at some consequences. We will use slightly different notation here. The reason is that we touch on the subject traditionally treated by algebraists, and we will abide by their conventions here. Therefore we will write the relational structure $\langle Bool, +, \wedge, \bot, \top \rangle$ as $\langle Bool, +, \cdot, 0, 1 \rangle$. The multiplication operation $\cdot$ will be the conjunction $\wedge$. In algebra the multiplication symbol $\cdot$ is often omitted in the product. We will try not to omit it, though.

The idea is that once we check that the structure $\langle Bool, +, \cdot, 0, 1 \rangle$ is a field, we will be able to use algebra in our arguments.

For the sake of completeness let us recall that a structure $\langle F, +, \cdot, 0, 1 \rangle$ is a field if $+$ defines a commutative group structure in F (with 0 as the neutral element), $\cdot$ defines a commutative group structure in $F \setminus \{0\}$ with the neutral element 1, distributivity of multiplication w.r.t. addition holds, and $0 \cdot x = 0$ for all x. If the reader is not familiar with the concept of a group, s/he should consult any basic text on abstract algebra.

PROPOSITION 5.5
The structure $\langle Bool, +, \cdot, 0, 1 \rangle$ is a field.

Proof: Only the distributivity is not completely trivial. That is, we need to check that for all $x, y,$ and z, $x \cdot (y + z) = (x \cdot y + x \cdot z)$. This can be done by looking at the table involving both the left- and right-hand sides. We list both truth tables in a single Table 5.4. We see that the penultimate and last columns in that table are identical.

TABLE 5.4: Distributivity in *Bool*

x	y	z	$y+z$	$x \cdot y$	$x \cdot z$	$x \cdot (y+z)$	$(x \cdot y) + (x \cdot z)$
0	0	0	0	0	0	0	0
0	0	1	1	0	0	0	0
0	1	0	1	0	0	0	0
0	1	1	0	0	0	0	0
1	0	0	0	0	0	0	0
1	0	1	1	0	1	1	1
1	1	0	1	1	0	1	1
1	1	1	0	1	1	0	0

The field $\langle Bool, +, \cdot, 0, 1 \rangle$ is usually denoted by Z_2 as it is the field of remainders of integers modulo 2. The characteristic of the field Z_2 is 2, i.e., for all x, $x + x = 0$. Moreover, for all x in Z_2, $x \cdot x = x$.

We will now show a uniqueness of representation of Boolean formulas over Z_2 as polynomials. This is done under the additional assumptions (but this is common in algebraic considerations), namely we will assume that variables are ordered (for instance in the order $x_1 \preceq x_2 \preceq \ldots \preceq x_n$). The reduction rules ($x \cdot x = x$, $x \cdot 1 = x$) imply that we can assume that monomials (products of variables) have all variables with exponent 1, and that the constants do not occur in polynomials except as a constant term. Moreover, since $x + 0 = x$, $x \cdot 0 = 0$, the only case 0 will be listed in a polynomial is when that polynomial is 0. To make our assumptions explicit: *we will assume that monomials are reduced monomials, that is, the exponents of variables are 1, that there is no repetition of variables in a monomial, and that the variables occur in their order $\preceq$.* To give an example, $x_1 \cdot x_2 \cdot x_4$ is a monomial according to the rules above, but $x_1 \cdot x_3 \cdot x_1 \cdot x_4$ is not a monomial, for we could move second x_1 to the front, and then reduce the exponent.

Next we order monomials. A monomial m is a constant, and then its rank is 0, or it is of the form

$$x_{i_1} \cdot \ldots \cdot x_{i_k},$$

with $k \geq 1$, and then the *rank* of monomial m is k.

Here is how we order monomials:

1. $m_1 \preceq m_2$ if the rank of m_1 is bigger than the rank of m_2.
2. When $m_1 = x_{i_1} \cdot \ldots \cdot x_{i_k}$ and $m_2 = x_{j_1} \cdot \ldots \cdot x_{j_k}$ have same rank then $m_1 \preceq m_2$ if the sequence $\langle i_1. \ldots i_k \rangle$ precedes the sequence $\langle j_1. \ldots i_k \rangle$ lexicographically.

The following statement is a consequence of the fact that lexicographical ordering of strings over a finite ordered alphabet is a linear ordering.

PROPOSITION 5.6
The relation $\preceq$ is a linear order of monomials.

When we have a polynomial $p(x_1, \ldots, x_n)$, we will list its monomials in their ordering $\preceq$. This is a linguistic issue, not a semantic issue: we are after specific syntactic representation of formulas.

So now, when we write a polynomial it is a *list* of monomials (no repetitions of monomials), ordered according to their ordering $\preceq$. Our goal is to show that this specific representation of a formula is unique. Thus we are able to assign to any formula $\varphi(x_1, \ldots, x_n)$ its unique representation as a polynomial (with the convention on listing of monomials). We will call polynomials written using this convention (reduced monomials, listed in the order $\preceq$) *standard polynomials* (also known as *Zhegalkin polynomials*). Now, observe that a standard polynomial is a formula (over the set of functors $\{+, \wedge, 0, 1\}$). There is only one standard polynomial having the occurrence of 0: it is the constant 0 itself. In Proposition 5.4(d) we proved that the set of functors $\{+, \cdot, 1\}$ is complete. The following fact follows from this observation.

PROPOSITION 5.7
For every formula φ there is a polynomial f such that $\varphi \equiv f$ is a tautology.

Proof: In our proof of Proposition 5.2 we represented a formula $\varphi(x_1, \ldots, x_{n+1})$ as

$$(\neg x_1 \wedge \psi(x_2, \ldots, x_{n+1})) \vee (x_1 \wedge \vartheta(x_2, \ldots, x_{n+1})).$$

Assuming that we have polynomials f_1, f_2 that represent ψ, ϑ, respectively, we get a representation of φ as

$$((1 + x_1) \cdot f_1) \vee (x_1 \cdot f_2).$$

Then we use the representation of disjunction $x_1 \vee x_2$ as $x_1 \cdot x_2 + x_1 + x_2$. Now we have an expression equivalent to φ which involves only $+, \cdot,$ and constants. We now transform it (using familiar algebraic rules) to a polynomial. □

So we know that every formula is equivalent to a polynomial. We will now show that when we bring that polynomial into the standard form, the resulting polynomial, which is a standard polynomial, is unique.

To this end, we will need a bit of terminology. A *tail* of standard polynomial f is any postfix of its listing. For instance $x_1 + x_2$ is a tail of a polynomial $x_1 \cdot x_2 + x_1 + x_2$.

When we have two standard polynomials f_1 and f_2 they have a longest common tail. For instance, the polynomials $x_1 \cdot x_2 + x_1 + x_2$ and $x_1 \cdot x_2 + x_2$ have longest common tail x_2.

Our goal now is to find, for any pair of different standard polynomials f_1 and f_2, a specific assignment $(a_1, \ldots, a_n)$ so

$$f_1(a_1, \ldots, a_n) \neq f_2(a_1, \ldots, a_n).$$

Here is what we do. Given standard polynomials f_1 and f_2 such that $f_1 \neq f_2$, it is possible that both are constants. As they are different, one of them is 0, the other 1, and they are different at any point.

If at least one of f_1 and f_2 is not constant, their longest common tail cannot coincide with both of them because then they must coincide. We will select a latest monomial m which occurs in either f_1 or in f_2 but not in their longest common tail. In particular, m is a lower degree monomial that occurs in one but not in the other. Without loss of generality, we can assume that that monomial m occurs in f_1 but not in f_2 (it cannot occur in both, because, being latest, it would have to be in their common tail).

Let t be the longest common tail of f_1 and f_2. Let us form auxiliary polynomials $g_1 = f_1 + t$, and $g_2 = f_2 + t$. Now we have the following obvious fact.

PROPOSITION 5.8
The longest common tail of g_1 and g_2 is 0.

We will now find an assignment $(a_1, \ldots, a_n)$ so that

$$g_1(a_1, \ldots, a_n) \neq g_2(a_1, \ldots, a_n).$$

This, together with the obvious facts that $f_1 = g_1 + t$, and $f_2 = g_2 + t$ implies that $f_1(a_1, \ldots, a_n) \neq f_2(a_1, \ldots, a_n)$.

To this end, we take the least monomial in g_1. The first case is when it is a constant. It cannot be 0, because then $g_1 = 0$, which is not the case, as it has a monomial not in g_2. If $m = 1$, then it is a constant term of g_1, but not of g_2, because the common tail of g_1 and g_2 is 0. Then $g_1(0, \ldots, 0) = 1$, while $g_2(0, \ldots, 0) = 0$ and so we already have a desired assignment.

Now, let us assume that m is not constant. Then m is of the form:

$$x_{i_1} \cdot \ldots \cdot x_{i_k}$$

with $k \geq 1$. We define an assignment $(a_1, \ldots, a_n)$ as follows:

$$a_i = \begin{cases} 1 & \text{if } i \in \{i_1, \ldots, i_k\} \\ 0 & \text{otherwise.} \end{cases}$$

We claim that all monomials in g_1 except m take the value 0 under the assignment $(a_1, \ldots, a_n)$, and that all monomials in g_2 take the value 0 under the assignment

$(a_1, \ldots, a_n)$. Indeed, all monomials m' in g_1, $m' \neq m$, must have an occurrence of a variable that is not in m. This is totally obvious if the rank of m' is bigger than the rank of m (simply, m' has more occurrences of variables) and is also immediate if they have the same rank (for they are different). But then m' is evaluated as 0 under the assignment $(a_1, \ldots, a_n)$. On the other hand, the monomial m is evaluated as 1 by $(a_1, \ldots, a_n)$. Thus $g_1(a_1, \ldots, a_n) = 1$.

Now, let us look at the polynomial g_2. By our choice of m we know that m does not occur in g_2, and that all the monomials in g_2 have the rank greater or equal to that of m. Reasoning as in the previous paragraph, we conclude that all the monomials in g_2 are evaluated as 0 under the assignment $(a_1, \ldots, a_n)$. Thus $g_2(a_1, \ldots, a_n) = 0$. In other words $g_1(a_1, \ldots, a_n) \neq g_2(a_1, \ldots, a_n)$.

But now, as $f_1 = g_1 + t$, and $f_2 = g_2 + t$, $f_1(a_1, \ldots, a_n) \neq f_2(a_1, \ldots, a_n)$. □

Thus we get the following fact.

PROPOSITION 5.9

Let the order of variables be fixed. Then for every formula φ there is a unique standard polynomial f_φ so that $\varphi \equiv f_\varphi$ is a tautology.

Proof: We showed before, in Proposition 5.7, that there is a polynomial f such that $\varphi \equiv f$ is a tautology. But there is only one standard polynomial f' so that $f \equiv f'$, and so that f' is the unique standard polynomial equivalent to φ. □

COROLLARY 5.2

Let f be a standard polynomial in variables $x_1, \ldots, x_n$. If $f \not\equiv 0$, then for some assignment $(a_1, \ldots, a_n)$, $f(a_1, \ldots, a_n) = 1$.

Proof: Either f has the constant term 1, and then $f(0, \ldots, 0) = 1$, or if not, it has constant term 0 and at least one monomial of non-zero rank. But for such a polynomial we constructed an assignment which makes it 1. □

Corollary 5.2 implies the following (just consider polynomial $f = g + 1$).

COROLLARY 5.3

Let g be a standard polynomial in variables $x_1, \ldots, x_n$. If $g \not\equiv 1$, then for some assignment $(a_1, \ldots, a_n)$, $g(a_1, \ldots, a_n) = 0$.

Thus non-constant standard polynomials take both the value 0 and the value 1 at suitably chosen arguments.

Our proof of Proposition 5.9 allows us to count standard polynomials in n variables: there is 2^{2^n} of them. We can find the same result directly (and with this fact we can find an alternative proof for Proposition 5.7). See Section 5.7 Exercises.

5.4 Incomplete sets of functors, Post classes

Our goal (which will be realized in the next section) is to show a criterion for completeness of sets of functors. Before we do so, we need to investigate specific five classes of functors that are incomplete. Those classes are called *Post classes* after the American logician E. Post.

The argument for incompleteness of all five cases will be almost identical. The scheme of the argument will be this: we will exhibit a property of tables $\mathcal{P}$. Then we will show that if all functors from a set of functors $\mathcal{C}$ have the property $\mathcal{P}$ then the table of any formula built of functors in $\mathcal{C}$ has a table defining a functor having the property $\mathcal{P}$. The last element of the proof is that there are tables defining functors that do *not* have property $\mathcal{P}$.

So here are the five Post classes. To introduce the first class, we need a definition. We say that an n-ary functor c with its table T_c is *self-dual* if whenever the row

$$(a_1, \ldots, a_n, a) \in T_c$$

then also

$$(1 + a_1, \ldots, 1 + a_n, 1 + a) \in T_c.$$

In other words, if we negate the inputs in c, then the result will be negated output of c. We will call the class of all self-dual functors **S**.

The second class of functors is the class of *monotone* functors. Recall that $Bool^n$ has its *Post ordering*, which is the product order of the natural order of truth values. Then we say that an n-ary functor c is *monotone* if whenever $(a_1, \ldots, a_n) \leq_p (b_1, \ldots, b_n)$ and $(a_1, \ldots, a_n, a) \in T_c$ and $(b_1, \ldots, b_n, b) \in T_c$, then $a \leq b$. We will call the class of monotone functors **M**.

The third class of functors we consider consists of *linear* functors. Those are functors that are represented by linear standard polynomials, that is polynomials where all monomials are constants, or single variable.[2] We will call the class of all linear functors **L**.

The fourth class of functors we consider is the class of 1-*consistent* functors. Those are functors c with its table T_c containing the row $(1, \ldots, 1, 1)$. We will denote those by $\mathbf{F_1}$.

The fifth class of functors we consider is the class of 0-*consistent* functors. Those are functors c with its table T_c containing the row $(0, \ldots, 0, 0)$. We will denote those by $\mathbf{F_0}$.

None of the classes considered above consists of all functors. For instance, the $\wedge$ functor is not self-dual (as is easily seen looking at the rows $(0, 1, 0)$ and $(1, 0, 0)$). Likewise $\vee$ is not self-dual. The functor $\neg$ is not monotone. The functor $\wedge$ is not

---

[2]The need for looking at linear polynomials, thus polynomials in general, is what forced us to consider the polynomial representation of functors here, and not in the chapter on normal forms: after all, the standard polynomial *is a normal form* of a formula.

linear, as its standard polynomial is $x_1 \cdot x_2$ which is not a linear polynomial. The constant function 1 is not in $\mathbf{F}_0$, and the constant function 0 is not in $\mathbf{F}_1$.
We will say that a formula φ is self-dual (resp. monotone, linear, $\mathbf{F}_1$, $\mathbf{F}_0$,) if its table is self-dual (resp. monotone, linear, $\mathbf{F}_1$, $\mathbf{F}_0$).

PROPOSITION 5.10

1. If all functors in $\mathcal{C}$ are self-dual then the table of any formula of $Form^{\mathcal{C}}$ is self-dual.
2. If all functors in $\mathcal{C}$ are monotone then the table of any formula of $Form^{\mathcal{C}}$ is monotone.
3. If all functors in $\mathcal{C}$ are linear then the table of any formula of $Form^{\mathcal{C}}$ is linear.
4. If all functors in $\mathcal{C}$ are in $\mathbf{F}_1$ then the table of any formula of $Form^{\mathcal{C}}$ is in $\mathbf{F}_1$.
5. If all functors in $\mathcal{C}$ are in $\mathbf{F}_0$ then the table of any formula of $Form^{\mathcal{C}}$ is in $\mathbf{F}_0$.

Proof: In each case the argument is by induction on the height of the tree of the formula φ. Let us see how it works in the case of (1). Let c be the label of the root of the tree of the formula φ. Assuming c is a k-ary functor, our formula is $c(\varphi_1, \ldots, \varphi_k)$. Writing the variables of φ explicitly we have

$$\varphi(1+x_1, \ldots, 1+x_n) = c(\varphi_1(1+x_1, \ldots, 1+x_n), \ldots \varphi_k(1+x_1, \ldots, 1+x_n)) =$$
$$c(1 + \varphi_1(x_1, \ldots, x_n), \ldots, 1 + \varphi_k(x_1, \ldots, x_n)) =$$
$$1 + c(\varphi_1(x_1, \ldots, x_n), \ldots, \varphi_k(x_1, \ldots, x_n)) = 1 + \varphi(x_1, \ldots, x_n)$$

First equality was just writing explicitly what φ was. The second equality used inductive assumption for each of formulas $\varphi_1, \ldots, \varphi_k$, and the last one used the fact that the functor c is self-dual.
All the remaining arguments are equally simple, and follow the same line of argument (with respect to the corresponding classes of functors). □
Now, we have the following corollary.

COROLLARY 5.4
Let $\mathcal{C}$ be a class of functors.

1. If $\mathcal{C} \subseteq \mathbf{S}$ then $\mathcal{C}$ is incomplete.
2. If $\mathcal{C} \subseteq \mathbf{M}$ then $\mathcal{C}$ is incomplete.
3. If $\mathcal{C} \subseteq \mathbf{L}$ then $\mathcal{C}$ is incomplete.
4. If $\mathcal{C} \subseteq \mathbf{F}_1$ then $\mathcal{C}$ is incomplete.

5. If $C \subseteq \mathbf{F}_0$ then C is incomplete.

Proof: (1) We found that $\vee$ is not self-dual, and so, *a fortiori* cannot be defined out of C.
(2) We found that $\neg$ is not monotone, and so cannot be defined out of C.
(3) We found that $\wedge$ is not linear, and so cannot be defined out of C.
(4) We found that the constant $\bot$ is not in F_1, and so, cannot be defined out of C.
(5) Similarly, we found that the constant $\top$ is not in F_0, and so, *a fortiori* cannot be defined out of C. □

It follows that the set of functors $\{\wedge, \vee\}$ is not complete (as it consists of monotone functors), and that the set $\{\Rightarrow\}$ is not complete, as $\Rightarrow$ is in $\mathbf{F}_1$.

5.5 Post criterion for completeness

We will now prove a criterion characterizing complete sets of functors.

THEOREM 5.1 (Post theorem)
Let C be a set of functors. Then C is complete if and only if C is not included in $\mathbf{S}$, C is not included in $\mathbf{M}$, C is not included in $\mathbf{L}$, C is not included in $\mathbf{F}_1$, and C is not included in $\mathbf{F}_0$.

Proof: The necessity of our condition follows from Corollary 5.4. Indeed, if C is included in any of these five classes of functors then C is not complete.
So let us assume that C is not included in either of the five classes. This means that there are five functors in C (not necessarily different) f_s, f_m, f_l, f_0, and f_1 such that: $f_s \notin \mathbf{S}$, $f_m \notin \mathbf{M}$, $f_l \notin \mathbf{L}$, $f_0 \notin \mathbf{F}_0$, and $f_1 \notin \mathbf{F}_1$.
All we need to do is to define $\neg$ and $\wedge$ in terms of C. Our first goal is to define $\neg$ and constant functors, 0 and 1. Here is what we do. Given a functor c let us define the functor $\hat{c}(x)$ as a unary functor resulting from substituting for all variables of c the same variable, x. What happens here is that we look at the two rows of the table for c: the one determined by the assignment $(0, \ldots, 0)$ and the one determined by the assignment $(1, \ldots, 1)$. The resulting table has just two rows: the first row is $(0, f(0, \ldots, 0))$, and the other row $(1, f(1, \ldots, 1))$.
Let us see what are the functors $\hat{f}_0(x)$ and $\hat{f}_1(x)$. Let us look first at $\hat{f}_0(x)$. We know that $f_0(0, \ldots, 0) = 1$ (because $f_0 \notin \mathbf{F}_0$). On the other hand either $f_0(1, \ldots, 1) = 0$, or $f_0(1, \ldots, 1) = 1$. In the first case the functor $\hat{f}_0$ is $\neg$, in the second $\hat{f}_0$ is the constant 1. Reasoning in the same way but this time about $\hat{f}_1$, we see that either $\hat{f}_1$ is $\neg$, or it is the constant 0.
Putting these two facts together we find that two cases are possible.

(a) One of f_0, f_1 defines negation.

(b) $\{f_0, f_1\}$ defines both constants.

In case (a) we already have the negation functor. We will use the function f_s (and negation) to define constants. Let us recall that f_s is not self-dual. This means that for some assignment $(a_1, \ldots, a_n)$

$$f_s(a_1, \ldots, a_n) = f_s(1 + a_1, \ldots, 1 + a_n).$$

Let us fix this sequence $(a_1, \ldots, a_n)$. Some of a_i's are 0, and others are 1. Here is how we define a new functor h, of one variable. To facilitate notation we will write: $x + 0$ for x, and $x + 1$ for $\neg x$. We then define new functor h as follows:

$$h(x) = f_s(x + a_1, \ldots, x + a_n).$$

What is $h(0)$? It is $f_s(0 + a_1, \ldots, 0 + a_n)$. It is easy to see that it is precisely $f_s(a_1, \ldots, a_n)$. On the other hand $h(1)$ is $f_s(1 + a_1, \ldots, 1 + a_n)$. By the choice of the function f_s and the assignment $(a_1, \ldots, a_n)$ we find that $f_s(1+a_1, \ldots, 1+a_n) = f_s(a_1, \ldots, a_n) = h(0)$. Thus $h(0) = h(1)$. That is, the functor h is constant. Now, regardless whether h is 0, or 1 we can use negation to define the other constant. Thus in case (a) we now have both negation and constants.

In case (b) we have constants. We need to define negation. We have at our disposal the function f_m which is not monotone. That means that we have two sequences of values $(a_1, \ldots, a_n)$ and $(b_1, \ldots, b_n)$ such that

$$(a_1, \ldots, a_n) \leq_p (b_1, \ldots, b_n)$$

but $f_m(a_1, \ldots, a_n) = 1$ and $f_m(b_1, \ldots, b_n) = 0$ (the only possible case contradicting monotonicity). What happens here is that whenever $a_i = 1$, b_i is also 1. We claim that we can impose a stronger constraint on the sequences $(a_1, \ldots, a_n)$ and $(b_1, \ldots, b_n)$, namely that $(a_1, \ldots, a_n) \leq_p (b_1, \ldots, b_n)$, $(a_1, \ldots, a_n)$ and $(b_1, \ldots, b_n)$ differ in *exactly* one place and also that we have the properties: $f_m(a_1, \ldots, a_n) = 1$ and $f_m(b_1, \ldots, b_n) = 0$. For otherwise, whenever sequences $(a_1, \ldots, a_n)$ and $(b_1, \ldots, b_n)$ differ in only one place (and $(a_1, \ldots, a_n) \leq_p (b_1, \ldots, b_n)$)

$$f_m(a_1, \ldots, a_n) \leq_p f_m(b_1, \ldots, b_n).$$

But n is an integer. Therefore we can gradually increase values of $a_1, \ldots, a_n$ increasing one by one those values where a_i is 0, but b_i is 1. In a finite number (bounded by n) of steps we will reach $(b_1, \ldots, b_n)$. But then $f_m(b_1, \ldots, b_n) = 1$, a contradiction. So now we have two sequences $(a_1, \ldots, a_n)$ and $(b_1, \ldots, b_n)$. They differ on one position only (say i). For that i, $a_i = 0$ and $b_i = 1$. Finally, $f_m(a_1, \ldots, a_n) = 1$ and $f_m(b_1, \ldots, b_n) = 0$. Now, let us recall that we are in case (b), that is, we have constants. We define new functor $h(x)$ as follows. For every variable x_j, $j \neq i$, we substitute the constant a_j (that is b_j - it is the same thing). For the variable x_i we substitute x. Let us compute $h(0)$. It is $f_m(a_1, \ldots, a_{i-1}, 0, a_{i+1}, \ldots, a_n)$, that is $f_m(a_1, \ldots, a_n)$, i.e., 1. On the other hand, $h(1)$ is $f_m(a_1, \ldots, a_{i-1}, 1, a_{i+1}, \ldots, a_n)$, that is $f_m(b_1, \ldots, b_{i-1}, 1, b_{i+1}, \ldots, b_n)$, that is $f_m(b_1, \ldots, b_n)$, that is 0. In other word, h is the negation functor.

Complete sets of functors

Thus, after considering our alternative (a) and (b) we now know that we can define both constants, and also negation. All we need to define is conjunction. We are going to use for this purpose the nonlinear functor f_l (and the constants and the negation functor, already available).

So let f_l be a nonlinear functor in our set of functors C. That functor must have at least two variables; all functions of 0 or 1 variables are linear. Let us look at the standard polynomial g representing f_l. It must have at least one monomial of rank at least 2. Since we can always substitute variables for variables in formulas, we can assume that g has a monomial of the form $x_1 \cdot x_2 \cdot \ldots$. Now we use algebra (as taught in high schools) to get the following representation of the standard polynomial g:

$$x_1 \cdot x_2 \cdot h_1(x_3, \ldots, x_n) + x_1 \cdot h_2(x_3, \ldots, x_n) + x_2 \cdot h_3(x_3, \ldots, x_n) + h_4(x_3, \ldots, x_n).$$

The polynomial $h_1(x_3, \ldots, x_n)$ is a non-zero standard polynomial (in variables x_3, $\ldots, x_n$). This requires checking; we leave this fact to the reader. Therefore we can use Corollary 5.2 and find an assignment $(a_3, \ldots, a_n)$ so that $h_1(a_3, \ldots, a_n) = 1$. When we substitute the assignment $(a_3, \ldots, a_n)$ to polynomials h_2, h_3, and h_4, we find that the functor f_l with $a_3, \ldots, a_n$ substituted for $x_3, \ldots, x_n$ is one of eight functors: $x_1 \cdot x_2$, or $x_1 \cdot x_2 + 1$, or $x_1 \cdot x_2 + x_2$, or $x_1 \cdot x_2 + x_2 + 1$, or $x_1 \cdot x_2 + x_1$, or $x_1 \cdot x_2 + x_1 + 1$, or $x_1 \cdot x_2 + x_1 + x_2$, or $x_1 \cdot x_2 + x_1 + x_2 + 1$. Instead of dealing with these eight cases separately (which certainly can be done), let us write this standard polynomial (which we call $k(x_1, x_2)$) in its general form:

$$k(x_1, x_2) = x_1 \cdot x_2 + a \cdot x_1 + b \cdot x_2 + c$$

with $a, b, c \in Bool$. We make a substitution to the polynomial k as follows. We substitute $x_1 + b$ for x_1, $x_2 + a$ for x_2. Moreover, we add to $k(x_1 + b, x_2 + a)$ a constant term $a \cdot b + c$. This is certainly justified, because when a constant b is 0 we really do nothing with x_1, and when $b = 1$ we negate x_1, but we already found that the negation functor is definable from C. We reason similarly about adding a to x_2, and adding $a \cdot b + c$ to the polynomial $k(x_1 + b, x_2 + a)$. Now let us do a bit of algebra.

$$k(x_1 + b, x_2 + a) + a \cdot b + c =$$
$$(x_1 + b) \cdot (x_2 + a) + a \cdot (x_1 + b) + b \cdot (x_2 + a) + c + a \cdot b + c =$$
$$x_1 \cdot x_2 + b \cdot x_2 + a \cdot x_1 + a \cdot b + a \cdot x_1 + a \cdot b + b \cdot x_2 + a \cdot b + c + a \cdot b + c.$$

It is quite clear that each term in our polynomial, except $x_1 \cdot x_2$, occurs an even number of times, so we find that

$$k(x_1 + b, x_2 + a) + a \cdot b + c = x_1 \cdot x_2.$$

Thus we found a definition of $x_1 \cdot x_2$ (that is $x_1 \wedge x_2$) in C, and since $\neg$ is also definable, the set of functors C is complete. $\square$

Theorem 5.1 gives rise to an algorithm for testing completeness of a set of functors C. All we need to do is to test each of functors in C for self-duality, monotonicity, linearity, 1-consistency, and 0-consistency. If each of these tests fails on at least one of functors in $\mathcal{C}$, then $\mathcal{C}$ is complete.

We will now use Post Theorem 5.1 to describe Boolean functions that define all functions. We already encountered two such functions, both binary, namely *NAND* and *NOR*. Historically, the first such function, *NOR*, was discovered by Sheffer, and so such functions are called *Sheffer-like functions*.

PROPOSITION 5.11
An n-ary Boolean function f is Sheffer-like if and only if $f \notin \mathbf{F}_0$, $f \notin \mathbf{F}_1$, and $f \notin \mathbf{S}$.

Proof: Taking $\mathcal{C} = \{f\}$, we see that if f is Sheffer-like, i.e., $\{f\}$ is complete, then $f \notin \mathbf{F}_0$, $f \notin \mathbf{F}_1$, and $f \notin \mathbf{S}$.

To show the converse implication, in view of Theorem 5.1, all we need to do is to show that f is not monotone, and that f is not linear.

To see the non-monotonicity, let us observe that $f(0, \ldots, 0) = 1$, and $f(1, \ldots, 1) = 0$. This immediately implies non-monotonicity since $(0, \ldots, 0) \leq_p (1, \ldots, 1)$, but $f(1, \ldots, 1) < f(0, \ldots, 0)$.

To see that f is not a linear function, let us assume that f is linear. Then f is of the form:
$$f(x_1, \ldots, x_n) = x_{i_1} + \ldots + x_{i_k} + \varepsilon.$$

Now, $f(0, \ldots, 0) = 1$. Therefore $\varepsilon = 1$. Hence, since $f(1, \ldots, 1) = 0$, the integer k (the number of variables in standard polynomial of f) must be odd. But f is not self-dual. Therefore, for some choice of $(a_1, \ldots, a_n) \in Bool^n$,
$$f(a_1, \ldots, a_n) = f(\bar{a}_1, \ldots, \bar{a}_n).$$

This means that
$$a_{i_1} + \ldots + a_{i_k} + 1 = \bar{a}_{i_1} + \ldots + \bar{a}_{i_k} + 1.$$

But the last equality means that
$$a_{i_1} + \ldots + a_{i_k} = \bar{a}_{i_1} + \ldots + \bar{a}_{i_k}.$$

This, however, implies that k is an even integer. Thus we found a contradiction, $f \notin \mathbf{L}$, and so $\{f\}$ is complete, i.e., f is Sheffer-like, as desired. □

5.6 If-then-else functor

So far we have looked mostly at binary functors. Here is one important ternary one. This functor, complete in a weak sense, is the "if-then-else" functor, often denoted

TABLE 5.5: *ite* functor

x	y	z	$ite(x,y,z)$
0	0	0	0
0	0	1	1
0	1	0	0
0	1	1	1
1	0	0	0
1	0	1	0
1	1	0	1
1	1	1	1

as *ite*, or *ITE*. See Table 5.5. We observe that, since *ite* is 1-consistent, *ite* is not complete by itself. We already saw, briefly, the *ite* functor in Section 3.7. Intuitively $ite(x,y,z)$ means: "if x is $\top$, then the value is y, otherwise (i.e., when $x = \bot$), the value is z.

PROPOSITION 5.12
The set of functors $\{ite, \bot, \top\}$ is complete.

Proof: Here are two tautologies: $(\neg x) \equiv ite(x, \bot, \top)$ and $(x \wedge y) \equiv ite(x, y, \bot)$. Now, the assertion follows from Corollary 5.1. □

The ternary functor *ite* has various appealing properties. The most important of those is a kind of distribution property. Recall that in Section 2.4 we discussed substitutions of formulas into other formulas. Here is an interesting property of the operation *ite*.

PROPOSITION 5.13 (Shannon theorem)
Let φ be a propositional formula and let x be a variable. Then for all variables y and for all formulas ψ and ϑ the equivalence

$$\varphi\left(\begin{matrix} x \\ ite(y,\psi,\vartheta) \end{matrix}\right) \equiv ite(y, \varphi\left(\begin{matrix} x \\ \psi \end{matrix}\right), \varphi\left(\begin{matrix} x \\ \vartheta \end{matrix}\right))$$

is a tautology.

Proof: By the substitution lemma (Proposition 2.16) under any valuation v, when $v(y) = 1$ then the left-hand side evaluates to $v(\varphi(\begin{smallmatrix}x\\\psi\end{smallmatrix}))$, and when $v(y) = 0$ then the left-hand side evaluates to $v(\varphi(\begin{smallmatrix}x\\\vartheta\end{smallmatrix}))$. This is the value to which the right-hand-side evaluates. □

The set consisting of the functor *ite* is not complete as the the functor *ite* is 1-consistent. For that same reason the set $\{ite, \top\}$ is also not complete, but the set $\{ite, \top, \bot\}$ is complete. This last fact can be checked directly (see Section 5.7 Exercises), or by explicit definition of $\neg$ and $\vee$ out of $\{ite, \top, \bot\}$.

5.7 Exercises

1. In the proof of Proposition 5.2 we assigned to two tables of $n-1$ variables one table of n variables. Use this to show that the number t_n of tables in n Boolean variables satisfies the recursion:
$$t_{n+1} = t_n^2$$
with the initial condition $t_1 = 4$. Use this recursion to establish another proof of the fact that $t_n = 2^{2^n}$.

2. Given a functor h of n Boolean variables, the functor h_c has a table that to every valuation $(a_1, \ldots, a_n)$ assigns the value $1 + h(a_1, \ldots, a_n)$. That is, we flip the last column of T. We call that functor h_c a *complementary* functor for h. Prove that the assignment $h \mapsto h_c$ is a one-to-one and "onto" mapping of Boolean functors of n variables.

3. Given a functor h of n Boolean variables, the functor h_d has a table that to every valuation $(a_1, \ldots, a_n)$ assigns the value $1 + h(1 + a_1, \ldots, 1 + a_n)$. We call that functor h_c a *dual* functor for h. Prove that the assignment $h \mapsto h_d$ is a one-to-one and "onto" mapping of Boolean functors of n variables.

4. With the terminology of Problem (3), a self-dual functor is a functor h such that $h = h_d$. Prove that a self-dual functor is uniquely determined by its "lower-half," that is, the part of its table where $x_1 = 0$. It is also determined by its "upper-half," that is, the part of its table where $x_1 = 1$. Find the connection between that "lower-half" and the "upper-half."

5. Let s_n be the number of *self-dual* functors of n Boolean variables. Prove that for all n, $s_{n+1} = t_n$.

6. In this problem we look at specific functions and their duals. What is the dual to $x_1 \vee x_2$? What is the dual to $x_1 \wedge x_2$? What is the dual to $x_1 \Rightarrow x_2$?

7. The functor *maj*, which appears in a number of places in this book, is defined by
$$maj(x_1, x_2, x_3) = (x_1 \wedge x_2) \vee (x_1 \wedge x_3) \vee (x_2 \wedge x_3).$$
Compute the dual to *maj*, both in symbolic form, and also its table.

8. What is dual of the ternary functor *ite*?

9. We encountered the monotone functors of n variables. The "closed form" for the number of such monotone functors of n Boolean variables is not known at the time of writing of this book. Nevertheless, this number (also called the *Dedekind number*, d_n) can be studied. Show that d_n is the number of antichains in the set $\mathcal{P}(\{1, \ldots, n\})$.

10. Is the function *maj* discussed above monotone?
In problems below we fix the order of variables $x_1, \ldots, x_n$.

11. Find the standard polynomial of the function *maj*.
12. Compute the number of reduced monomials of rank k whose variables are among $x_1, \ldots, x_n$.
13. Compute the number of all non-zero reduced monomials whose variables are among $x_1, \ldots, x_n$.
14. Compute the number of standard polynomials over variables $x_1, \ldots, x_n$.
15. Use the result of problem (14) to show that every formula is equivalent to exactly one standard polynomial.
16. Show that there is precisely 2^{n+1} standard linear polynomials in variables $x_1, \ldots, x_n$.
17. How many Boolean functions of n variables are both in $\mathbf{F}_0$ and in $\mathbf{F}_1$?
18. How many Boolean functions of n variables are both in $\mathbf{F}_0$ and in $\mathbf{S}$?
19. How many Boolean functions of n variables are both in $\mathbf{F}_1$ and in $\mathbf{S}$?
20. How many Boolean functions of n variables are in $\mathbf{F}_0$, $\mathbf{F}_1$, and in $\mathbf{S}$?
21. Use your solution of the problem 5, and of the problems 17–20, to compute the number of Sheffer-like functions of n variables. Before you do this, refresh your memory of the so-called *inclusion-exclusion* formula (any reasonable introductory text on combinatorics should do).
22. Given a function $f : Bool^n \to Bool$, its *derivative* with respect to x_i, $\frac{\partial f}{\partial x_i}$, is defined as:

$$\frac{\partial f}{\partial x_i} = f(x_1, \ldots, x_{i-1}, 1, x_{i+1}, \ldots x_n) + f(x_1, \ldots, x_{i-1}, 0, x_{i+1}, \ldots x_n)$$

While it may look at first glance strange, it is quite reasonable, when you look at the polynomial representation of f. For instance, the derivative of

$$f(x, y, z) = x \cdot y \cdot z + x + z$$

with respect to x is, of course, $y \cdot z + 1$ (as we were taught in calculus). Check that, indeed, $y \cdot z + 1 = f(0, y, z) + f(1, y, z)$.

23. Explain why the symbolic computation (as the derivative of a polynomial) and the computation of $\frac{\partial f}{\partial x_i}$ coincide.
24. Find the derivative of the majority function $maj(x, y, z)$ with respect to variable z.
25. Find the derivative of $ITE(x, y, z)$ with respect to variable z.
26. When g is a linear function, what can we find about derivatives of g?
27. Prove that
$$\frac{\partial (f + g)}{\partial x_i} = \frac{\partial f}{\partial x_i} + \frac{\partial g}{\partial x_i}.$$

28. In calculus they taught us other rules for differentiation. Which of these apply in Boolean context?
29. Prove Reed-Muller identities for Boolean functions:
    (a) $f(x_1,\ldots,x_n) = x_i \cdot \frac{\partial f}{\partial x_i} + f|_{x_i=0}$.
    (b) $f(x_1,\ldots,x_n) = \bar{x}_i \cdot \frac{\partial f}{\partial x_i} + f|_{x_i=1}$.
30. Use the first Reed-Muller identity to get a different (and simpler) proof of uniqueness of representation of Boolean functions by standard polynomials.
31. Since the use of Reed-Muller identity for the uniqueness of representation of Boolean functions by standard polynomials is so simpler, what was the benefit of the argument we gave above?

Chapter 6

Compactness theorem

6.1 König lemma 93
6.2 Compactness, denumerable case 95
6.3 Continuity of the operator Cn 99
6.4 Exercises 100

This chapter of our book contains the proof of the compactness theorem for propositional logic. The special case we study is based on clausal logic (a minor limitation; compactness of clausal logic entails compactness of full propositional logic, see below). What happens here is that we restrict ourselves to the situation where the underlying language has only a denumerable number of formulas (clauses in our case). This limits the number of variables to finite or infinite denumerable. In such circumstances we can enumerate with integers all formulas (clauses in our case) and so we can enumerate all clauses over a given set of variables. This limitation allows for use of a very powerful tool, known as *König lemma*. In the next section we will prove König lemma. Then, with this tool we will show how the compactness property for *denumerable* CNFs is a relatively simple corollary to that result.

It should be stated that the proof of the compactness theorem we present here hides rather than elucidates the fundamental principles related to compactness of propositional logic. In fact, one can give a general argument for compactness of propositional logic that does not depend on the cardinality of the set of clauses. Such an argument uses, however, additional tools and goes beyond the scope of our book.

6.1 König lemma

König lemma asserts that an infinite, finitely splitting tree must possess an infinite branch. Here are a number of definitions to make this precise.

We will assume that the reader is familiar with basic concepts such as string over a given alphabet, concatenation, initial segment, etc. We will use the symbol ε to denote the empty string. We will use the symbol $\frown$ for the concatenation operation. The string consisting of one symbol a will be denoted by $\langle a \rangle$.

Let A be an alphabet. A string s is an *initial segment* of a string t if for some string u, $t = s \frown u$. $lh(s)$ is the length of string s. A *tree* is a set T of finite strings over A

such that whenever $t \in T$ and s is an initial segment of t then $s \in T$. The elements of T will be referred to as *nodes*.

Given a tree T, the node $s \in T$ is an *immediate successor* of a node $t \in T$ if for some $a \in A$, $s = t^\frown \langle a \rangle$. The set of immediate successors of t in T is denoted by $I_T(t)$. The subscript T is dropped if it is clear from the context.

A tree T is *finitely splitting* if every string $s \in T$ has only finitely many immediate successors in T. We do not exclude the case when a string s has no successors at all. Thus, T is finitely splitting if for all $t \in T$, $I(t)$ is finite.

An infinite branch through T is an *infinite* string $S \in A^\omega$ such that for all n, the initial segment $S \mid_n$ belongs to T.

We are now ready to formulate and prove König lemma (see also [KM84]).

LEMMA 6.1 (König lemma)
If T is an infinite tree in which every node has finitely many immediate successors, then T possesses an infinite branch.

Proof: A bit of notation is still needed. Given the node $s \in T$, the set of successors of s, $T(s)$, is defined as the least set Z containing all immediate successors of s and such that whenever $t \in Z$, then all immediate successors of t belong to Z. It is easy to see that for a given node s the set $T(s)$ is well-defined (namely as the intersection of all sets H which contain immediate successors of s and are closed under immediate successors.) Of course we can also use the fixpoint theorem to show that $T(s)$ exists.

Now, let us call a node t *potentially infinite* if the set $T(t)$ is infinite. Clearly, the root of our tree (which is the empty string, ε), is potentially infinite, because T is infinite. Now, the course of our argument is clear. What we show is that in a finitely splitting tree, for a given potentially infinite node s there is always at least one immediate successor t of s such that t is also potentially infinite.

Let us see what is the dependence between the set $T(s)$ and the sets $T(t)$ for t immediate successors of s. We have:

$$T(s) = I(s) \cup \bigcup_{t \in I(s)} T(t).$$

What this identity says is that u is a (proper) successor of the node s in the tree T if it is either an immediate successor of s in T or a (proper) successor in T of an immediate successor of s. We will not prove this fact, leaving it to the reader.

Since in a finitely splitting tree the set $I(s)$ is finite for each s, it follows that if s is potentially infinite then at least one of its immediate successors also has to be potentially infinite.

Now, the construction is inductive. We start with the root; it is potentially infinite. Assume t_n is already defined and potentially infinite. Select as t_{n+1} an immediate successor of t_n which is also potentially infinite. It is clear that the infinite string S such that $S \mid_n = t_n$ is well-defined and is an infinite branch through T. □

It will be convenient to use König lemma in a slightly different form. We say that a node s is of rank n if $lh(s) = n$. We now have:

COROLLARY 6.1
If a tree T is finitely splitting and has nodes of arbitrarily large rank, then T possesses an infinite branch.

It is customary to give an example showing that the assumption of finite splitting is needed for existence of infinite branches.

Example 6.1
Let A be N, the set of natural numbers, and let T consist of the empty string, ε, together with strings $\langle n \rangle$ for all $n \in N$. Then T is an infinite tree, but it has no branch of any length > 1. □

6.2 Compactness of propositional logic, denumerable case

Let us recall that a clause is a formula of the form

$$C : p_1 \vee \ldots \vee p_k \vee \neg q_1 \vee \ldots \vee \neg q_l,$$

where p_is and q_js are propositional variables, $k, l \geq 0$. A clausal theory is a set of clauses. The conjunctive normal form theorem (Proposition 3.7) implies that every theory is equivalent to a clausal theory. We will prove the compactness theorem for clausal theories. The proof has the advantage of stripping immaterial details (that occur when we talk about arbitrary formulas). We will then use the compactness for clausal logic to prove compactness of full propositional logic.

First of all, we need to define what we mean by compactness. Here is what we do. Let F be a collection of clauses. We say that F is *finitely satisfiable* if for every *finite* subset F_0 of F, F_0 is satisfiable.

PROPOSITION 6.1 (Compactness of clausal logic, denumerable case)
Let F be an infinite denumerable collection of clauses. Then F is finitely satisfiable if and only if F is satisfiable.

Proof: Let us observe that the implication $\Leftarrow$ is entirely obvious. Thus all we need to show is the implication $\Rightarrow$. To this end we suppose that F is denumerable, infinite, and finitely satisfiable. Let us observe that F cannot contain the empty clause, as this clause (and thus all sets containing it) is unsatisfiable.

First of all we fix the ordering of the collection F into type ω,

$$F = \langle C_n \rangle_{n \in N}.$$

Since F is denumerable, the set of variables Var occurring in the clauses of F is finite or denumerable. In either case the set $Form_{Var}$ is infinite denumerable. To build a tree T we need an alphabet. That alphabet consists of two parts. First, we have clauses of F. Second, we have literals occurring in F.

The tree T will be finitely splitting. During the construction, in nodes of even rank we will be putting strings that end with a literal. In nodes of odd rank we will be putting strings that end with clauses from F.

We start, of course, with the empty string. Now, assume that we already know the nodes of rank $k-1$. Two cases need to be considered.

Case 1: $k = 2n+1$. That is, we defined all the nodes on the *even* level $k-1 = 2n$. We assume that these preceding nodes, *if defined*, were strings ending with C_n. Here is what we do at level k. Assume we extend the node s. Let C_n (which is the last symbol in s) be

$$l_1^n \vee \ldots \vee l_{h_n}^n,$$

then s gets h_n immediate successors. Those successors are $s^\frown \langle l_j^n \rangle$ for all j, $1 \le j \le h_n$. Clearly, in this stage of construction, every node is extended by a finite number of immediate successors (as each C_n is a clause, thus finite).

Case 2: $k = 2n+2$. We proceed here differently, and more cautiously. Let s be a node on the level $k-1$. This node is a string, ending in a literal. Take the set F_s consisting of all objects occurring in s. Those are either clauses from F (actually, $C_0, \ldots, C_n$) or literals (which occur in the even places of s). Thus we can treat set F_s as a set of clauses (after all, literals are just unit clauses.) This set, F_s, may be satisfiable or not. If this set F_s is unsatisfiable, the node s is *not* extended at all. If the set F_s is satisfiable, we extend s by just one immediate successor, namely $s^\frown \langle C_{n+1} \rangle$. Our construction ensures that the nodes visited at this stage have zero or one immediate successor.

We conclude that the tree T whose construction we just outlined is finitely splitting. Indeed, nodes at the odd level had at most one successor, while the nodes on the even level had a finite number of successors but at least one.

Our first task is to see that the tree T has an infinite branch. Of course, we want to use Corollary 6.1, but to this end we have to show that there are elements of arbitrary big rank in T.

So let us consider some $n \in N$. Consider $F_0 = \{C_0, \ldots, C_n\}$. The set F_0 is a finite subset of F. Hence it is satisfiable. Let v be a valuation satisfying F_0. Recall that C_j, $0 \le j \le n$ looks like this:

$$l_1^j \vee \ldots \vee l_{p_j}^j.$$

Therefore, for each j, $0 \le j \le n$, there is h_j, $1 \le h_j \le p_j$, so that $v(l_{h_j}^j) = 1$. But then the string

$$C_0 \, l_{h_0}^0 \, C_1 \, l_{h_1}^1 \, \ldots \, C_{n-1} \, l_{h_{n-1}}^{n-1} \, C_n$$

is a node in the tree T, as can be easily checked by induction on the initial segments of s, with the valuation v serving as a witness of satisfiability. Since n can be arbitrarily big, we just proved the assumptions of Corollary 6.1.

So now we know that the tree T possesses an infinite branch. Let S be that infinite branch. The set of the symbols of the alphabet on that branch (clauses or literals) contains all clauses in F. It also contains plenty of literals. We will now show how those literals can be used to construct a valuation w satisfying F.

The first thing we consider is the question of the complementary literals. Specifically, we ask the following question: "Is it possible that a pair of complementary literals, say l and $\bar{l}$, occur as last terms of nodes in the branch S?" If the answer is positive, then for some m and n with $m < n$ and for some literal l there are two initial segments of S, s_1 and s_2 as follows:

$$s_1: C_0\, l_{h_0}^0\, C_1\, l_{h_1}^1\, \ldots\, C_{m-1}\, l_{h_{m-1}}^{m-1}$$

and

$$s_2: C_0\, l_{h_0}^0\, C_1\, l_{h_1}^1\, \ldots\, C_{n-1}\, l_{h_{n-1}}^{n-1}.$$

Both these strings are segments of the branch S, but $l_{h_{m-1}}^{m-1} = l$ and $l_{h_{n-1}}^{n-1} = \bar{l}$. But this last string s_2 extended s_1. This means that there was a valuation v making all literals of the string s_2 true. But this is a contradiction, since $v(l) = 1$ and also $v(\bar{l}) = 1$. So, we just established that we cannot have a pair of complementary literals occur as symbols in the branch S.

Now, it is simple to define a valuation w on all variables $a \in \mathit{Var}$ as follows:

$$w(a) = \begin{cases} 1 & \text{if } a \text{ is a last element of a string of even length in branch } S \\ 0 & \text{if } \neg a \text{ is a last element of a string of even length in branch } S \\ 0 & \text{if } a \text{ does not appear in last element of a string of even length in } S. \end{cases}$$

The last case was needed because, although a variable a may appear in a clause in F, neither a nor $\neg a$ has to appear as the literal which we selected at some even stage to extend a string.

It should be clear that w is a valuation; by our discussion above there is no clash between the first and second cases. The third case was the "default," thus we conclude that the assignment w is well-defined.

So now, all we need to do is to check that w satisfies F. To this end let $C_n \in F$. Then there is a string

$$s_1: C_0\, l_{h_0}^0\, C_1\, l_{h_1}^1\, \ldots\, C_{n-1}\, l_{h_{n-1}}^{n-1}\, C_n\, l_{h_n}^n$$

in S. By construction of the tree T, $l_{h_n}^n \in C_n$. By the definition of w, $w(l_{h_n}^n) = 1$. Thus $w \models C_n$. Since n was arbitrary, we proved that $w \models F$, as desired. □

There are other arguments for compactness, some of which are based on the corollary we state below (of course the argument cannot use the proof we give, in such case).

COROLLARY 6.2 (Lindenbaum theorem)

Let Var be a denumerable set of propositional variables. If $F \subseteq Form_{Var}$ is a finitely satisfiable set of clauses then there is a maximal set of clauses G such that $F \subseteq G$ and G is satisfiable (thus finitely satisfiable).

Proof: Let F be finitely satisfiable. Let $v \models F$ (such v exists by Proposition 6.1). Treating v as a consistent set of literals, thus unit clauses, we define $G' = Th(v)$. Let G be the set of clauses in G'. We claim that G is the desired maximal satisfiable set of clauses extending F. Surely G contains F as $v \models F$. Moreover G is satisfiable as $v \models G$. Now, if $C \notin G$, then $v \not\models C$. Therefore $v \models \neg C$ and $\neg C \in G'$. But there is no valuation satisfying both C and $\neg C$, and so $G \cup \{C\}$ is unsatisfiable. Thus if $G \subset H$, then H is unsatisfiable. □

Let us call a set of clauses G *complete* if for every clause $C = l_1 \vee \ldots \vee l_k$ either $C \in G$ or for all j, $1 \leq j \leq k$, $\bar{l}_j \in G$. We then get the following corollary, often also called the Lindenbaum theorem.

COROLLARY 6.3

Let Var be a denumerable set of propositional variables. If $F \subseteq Form_{Var}$ is a finitely satisfiable set of clauses then there is a complete set of clauses H such that $F \subseteq H$.

Proof: Let G be a maximal satisfiable set of clauses extending F. If v_1, v_2 are two valuations satisfying G then it must be the case that $v_1 = v_2$. Otherwise, for some literal l, $v_1(l) = 1$ whereas $v_2(l) = 0$, i.e., $v_2(\bar{l}) = 1$. but then $G \cup \{l\}, G \cup \{\bar{l}\}$ are both satisfiable, and since G is satisfiable, at least one of these sets cannot be G. This is a desired contradiction and so there is exactly one valuation v such that $v \models G$. Now, let
$$H = \{C : C \text{ is a clause and } v \models C\}.$$

So now we claim H is complete. Let $C = l_1 \vee \ldots \vee l_k$, $k > 0$. If $C \notin H$, i.e., $v \notin C$ then $v \models \bar{l}_1 \wedge \ldots \wedge \bar{l}_k$, therefore $v \models \bar{l}_j$, $1 \leq j \leq k$, i.e., $\bar{l}_j \in H$, $1 \leq j \leq k$, as desired. □

Finally, we derive the compactness of full propositional logic (denumerable case only) from the compactness of *clausal* propositional logic. To this end, let T be a denumerable infinite set of propositional formulas over the set Var of variables. Let us assume that T is finitely satisfiable, that is, every finite subset $T_0 \subseteq T$ is satisfiable. For every $\varphi \in T$, let F_φ be a (finite) set of clauses such that F_φ is equivalent to φ, that is, for every valuation v, $v \models \varphi$ if and only if $v \models F_\varphi$. Let us form $F_T = \bigcup_{\varphi \in T} F_\varphi$. Then, since T is denumerable, and each F_φ is finite, F_T is denumerable infinite. We claim that F_T is finitely satisfiable. Indeed, let $F_0 \subseteq F_T$ be finite. Then there is a finite set $T_0 \subseteq T$ such that $F_0 \subseteq \bigcup_{\varphi \in T_0} F_\varphi$. Since T_0 is satisfiable, $\bigcup_{\varphi \in T_0} F_\varphi$ is satisfiable. Therefore F_0 is satisfiable. But now we can use Proposition 6.1. Since we just proved that the set of clauses F_T is finitely satisfiable, we conclude that F_T is satisfiable. Let v be a valuation satisfying F_T. Then for each

$\varphi \in T$, $v \models F_\varphi$. Therefore, for every $\varphi \in T$, $v \models \varphi$. Hence $v \models T$ and so T is satisfiable. The argument given above proves the following.

COROLLARY 6.4 Compactness theorem
Let Var be a set of propositional variables, and let T be a denumerable infinite set of formulas of $\mathcal{L}_{Var}$. Then T is finitely satisfiable if and only if T is satisfiable.

6.3 Continuity of the operator Cn

We are now in a position to prove the continuity of the consequence operator Cn. Let us recall that in Chapter 2 we proved the basic properties of that operator, including its monotonicity and idempotence (Proposition 2.26).

First we will prove a version of the compactness theorem that involves the operator Cn.

PROPOSITION 6.2
Let T be a set of propositional formulas and φ a propositional formula. Then $\varphi \in Cn(T)$ if and only if for some finite $T_0 \subseteq T$, $\varphi \in Cn(T_0)$.

Proof. Since the operator Cn is monotone (Proposition 2.26(2)), the implication $\Leftarrow$ holds.

So now let us assume that $\varphi \in Cn(T)$. We can assume that T is infinite, otherwise we could take T_0 equal to T. So let us assume that $\varphi \in Cn(T)$, but for every finite $T_0 \subseteq T$, $\varphi \notin Cn(T_0)$. This means that for every finite $T_0 \subseteq T$, $T_0 \cup \{\neg\varphi\}$ is satisfiable. We show that $T \cup \{\neg\varphi\}$ is satisfiable. To this end, all we need to do is to show that for every finite subset $T' \subseteq T \cup \{\neg\varphi\}$ is satisfiable. But

$$T' \subseteq T'' \cup \{\neg\varphi\} \subseteq T \cup \{\neg\varphi\}$$

for some finite $T'' \subseteq T$. But $T'' \cup \{\neg\varphi\}$ is satisfiable, and so T' is also satisfiable. So $T \cup \{\neg\varphi\}$ is finitely satisfiable. Now, by Corollary 6.4, $T \cup \{\neg\varphi\}$ is satisfiable. Let v be a satisfying valuation for $T \cup \{\neg\varphi\}$. Then $v \models \neg\varphi$. But $v \models T$, and so $v \models \varphi$, since $\varphi \in Cn(T)$. This is, of course, a contradiction. □

So now we can prove the desired continuity of the Cn operator.

PROPOSITION 6.3 (Continuity theorem)
The operator Cn is continuous. That is, for every increasing sequence of sets of formulas $\langle T_m \rangle_{m<\omega}$, $Cn(\bigcup_{m \in N} T_m) = \bigcup_{m \in N} Cn(T_m)$.

Proof: By the monotonicity of operator Cn, the inclusion $\supseteq$ follows. So now let us assume that a formula φ belongs to $Cn(\bigcup_{m\in N} T_m)$. By Proposition 6.2 there is a finite set $T \subseteq \bigcup_{m\in N} T_m$ such that $\varphi \in T$. But since T is finite and the sequence $\langle T_m \rangle_{m<\omega}$ is increasing, for some n, $T \subseteq T_n$. Now, by monotonicity of Cn, $\varphi \in Cn(T_n)$. Thus, φ belongs to the right-hand side. Since φ was arbitrary, we are done. □

6.4 Exercises

1. Let T be a finitely splitting tree that possesses only finitely many infinite branches. Show that there must exist an integer n such that all nodes of level at least n have at most one immediate successor.

2. Construct a finitely splitting tree T possessing denumerably many infinite branches.

3. Construct a finitely splitting tree T possessing continuum many infinite branches.

4. Construct an infinite set of clauses F such that every two-element subset of F is satisfiable, but F is not satisfiable.

5. Let n be a positive integer. Generalizing problem (4) construct a collection F such for every $F' \subseteq F$, $|F'| \leq n$, F' is satisfiable, but F is not satisfiable.

6. If $\langle F_n \rangle_{n\in N}$ is any denumerable family of sets of clauses, construct an increasing family of sets of clauses $\langle G_n \rangle_{n\in N}$ so that

$$Cn\left(\bigcup_{n\in N} F_n\right) = \bigcup_{n\in N} G_n.$$

Chapter 7

Clausal logic and resolution

7.1	Clausal logic	102
7.2	Resolution rule	107
7.3	Completeness results	110
7.4	Query-answering with resolution	113
7.5	Davis-Putnam lemma	117
7.6	Semantic resolution	119
7.7	Autark and lean sets	124
7.8	Exercises	132

We have several goals in this chapter. First, we shall investigate clausal logic and its satisfiability problem. As a consequence of the permutation lemma, we will see how special clauses, called *constraints*, relate to satisfiability. We also investigate unsatisfiability and its strong version, minimal unsatisfiability. The next goal of this chapter is to introduce the so-called resolution rule of proof, and then discuss the issue of the completeness of the resolution. The issue that interests us is what happens when we use the resolution rule. We will establish two basic completeness results for reasoning with clauses. Those are Theorems 7.2 (completeness theorem for resolution) and 7.8 (completeness of resolution refutation). Then we will discuss the scheme where a given CNF formula F is a *knowledge base* and clauses are queries. Here the idea is that the knowledge base F answers "*yes*" to query C if and only if $F \models C$. Of course, this is an obvious generalization of the propositional representation of relational databases.

We will then discuss a fundamental result, called the Davis-Putnam lemma, that allows for reduction of the number of variables in a set of clauses tested for satisfiability. This fact (in a stronger form) will be the foundation of the so-called DPLL algorithm. The reason why we prove the Davis-Putnam lemma here is that we will apply it in a proof of the completeness of a very restricted form of resolution, called semantic resolution with the ordering of variables. Finally, we discuss the properties of autarkies and related concepts of autark sets and lean sets.

7.1 Clausal logic, satisfiability problem and its basic properties

In Section 3.3, Proposition 3.6, we saw that every propositional formula is equivalent to a formula in conjunctive normal form. In effect, every formula is equivalent to a set of clauses. Recall that a *clause* is a formula of the form

$$p_1 \vee \ldots \vee p_m \vee \neg q_1 \vee \ldots \vee \neg q_n.$$

It follows that for every formula φ there is a finite set of clauses G_φ such that for every valuation v, $v \models \varphi$ if and only if for all clauses $C \in G_\varphi$, $v \models C$.

Thus, clearly, for every set of propositional formulas F there exists a set of clauses G, such that for every valuation v, $v \models F$ if and only if $v \models G$. Namely, $G = \bigcup_{\varphi \in F} G_\varphi$. The importance of this obvious fact is that every propositional theory has precisely the same models as a theory consisting of clauses. Moreover, if the theory F is finite, then so is its clausal representative, G. This means that if we want to represent some problem by means of a propositional theory F so that solutions to the problem are in one-to-one correspondence to the satisfying valuations of F, then we can, among many possible representations, choose F to be clausal, i.e., consisting of clauses. For that reason and the fact that rules for manipulating clauses are particularly simple, we will be interested in clausal logic, that is, logic where formulas are collections of clauses. From the point of view of knowledge representation it is, as we have just seen, not a limitation.

We will be dealing with collections of clauses (mostly finite, but not always). Usually we will use symbols such as F, G, also S, possibly with indices, to denote sets of clauses. As before we will be using the term CNF or CNF formula; it is the same thing, namely a finite collection of clauses

A clause is called a *constraint* if it is of the form:

$$\neg q_1 \vee \ldots \vee \neg q_n.$$

Later on, when we discuss Horn logic, we will encounter constraints again. For the moment, we just use them for characterizing satisfiability. We will use Permutation lemma 2.23 to give a necessary and sufficient condition for satisfiability of a set of clauses.

It will be convenient to say that a literal l *belongs* to the clause C if for some clause D, C is $D \vee l$. We write it as $l \in C$. Let us recall that we treat the clauses as sets of literals; the order of literals in a clause is not important unless we specifically order clauses. A small warning: we thought about valuations as sets of literals, too. This may lead to some confusion. It is very easy when the reader is "combinatorially minded," because a valuation satisfies a set of clauses if it is a *hit set* for that set of clauses represented as a family of sets. So maybe confusion can be minimized. Once we think about clauses as sets of literals, the inclusion of clauses is very natural; it is just subsumption. We will write $C_1 \subseteq C_2$ to mean that every literal l occurring in C_1

occurs in C_2. The empty clause is the clausal representation of $\bot$ and is, of course, unsatisfiable.

PROPOSITION 7.1
Let F be a collection of nonempty clauses. Then F is satisfiable if and only if there exists a consistent permutation of literals π such that $\pi(F)$ contains no constraints.

Proof: First, let us observe that a clausal theory F, where all clauses are nonempty and no clause is a constraint, is satisfiable. What is the model? The entire set *Var*, or in terms of valuations, the valuation that assigns 1 to every atom. Thus, if $\pi(F)$ contains no constraints then $\pi(F)$ is satisfiable. But then Proposition 2.23 tells us that F is satisfiable.

Conversely, let us assume that v is a valuation, $v \models F$. Given a literal l, recall that $|l|$ is a propositional variable underlying l. Let us define a mapping of $\pi_v : Lit \to Lit$ as follows:

$$\pi_v(l) = \begin{cases} l & \text{if } v(|l|) = 1 \\ \bar{l} & \text{if } v(|l|) = 0. \end{cases}$$

Then, if p is a propositional variable and $v(p) = 1$ then $\pi_v(p) = p$ and $\pi_v(\neg p) = \neg p$. On the other hand, if $v(p) = 0$ then $\pi_v(p) = \neg p$ and $\pi_v(\neg p) = p$. Clearly, p is a consistent permutation of literals. Now, we claim that $\pi_v(F)$ contains no constraints. Indeed, let $C \in F$. Since $v \models C$, there is a literal $l \in C$ such that $v(l) = 1$. Two cases are possible.

Case 1: $l = p$ for some variable p. Then, as $v(p) = 1$, $\pi_v(p) = p$ and since $\pi_v(p) \in \pi_v(C)$, $\pi_v(C)$ is not a constraint.

Case 2: $l = \neg p$ for some variable p. Then, as $v(p) = 0$, $\pi_v(p) = \neg p$, thus $\pi_v(\neg p) = p$. But since $\pi_v(p) \in \pi_v(C)$, $\pi_v(C)$ is again not a constraint.

Thus $\pi_v(F)$ contains no constraints. □

A set F of clauses is *unsatisfiable* if there is no valuation v such that $v \models F$. We have an immediate corollary to Proposition 7.1.

COROLLARY 7.1
Let F be a set of clauses. Then F is unsatisfiable if and only if for every consistent permutation of literals π, $\pi(F)$ contains at least one constraint.

Among unsatisfiable sets of clauses, a special role is played by *minimally unsatisfiable* sets of clauses. A set of clauses F is minimally unsatisfiable if F is unsatisfiable, but all proper subsets G of F are satisfiable. We will see later that minimally unsatisfiable sets of clauses F have a resolution refutation (derivation of the empty clause) that uses all clauses of F as premises (for the moment we do not know what derivations are, but they will appear later in this chapter).

The compactness theorem implies that every minimally unsatisfiable set of clauses (or any formulas, for that matter) is finite. We will now find an easy property characterizing minimally unsatisfiable sets of clauses.

PROPOSITION 7.2
Let F be an unsatisfiable set of clauses. Then the following are equivalent:

1. F is minimally unsatisfiable.
2. For every $C \in F$ there is a consistent permutation of literals π such that $\pi(C)$ is a unique constraint in $\pi(F)$.

Proof: Let us assume that F is minimally unsatisfiable and $C \in F$. Then $F \setminus \{C\}$ is satisfiable. Then there is a permutation π such that $\pi(F \setminus \{C\})$ contains no constraint. But $\pi(F)$ must contain a constraint since F is unsatisfiable. Thus, that unique constraint must be $\pi(C)$.
Conversely, let us assume that F is unsatisfiable, $C \in F$, and let π be a permutation such that $\pi(C)$ is the unique constraint in $\pi(F)$. Then $\pi(F \setminus \{C\})$ is satisfiable (for it does not contain a constraint). Thus $F \setminus \{C\}$ is satisfiable, and since we assumed F is unsatisfiable, and C is an arbitrary clause in F, F must be minimally unsatisfiable.
$\square$

Our next goal is to show that minimally unsatisfiable sets of clauses have a surprising property: they must have contain more clauses than variables. In the literature this fact is called the "Tarsi lemma." We will now show this fact. It will require some information from graph theory. It is not our goal here to study graph theory; we need to assume *some* knowledge. So here is what we will assume without proof, and refer the reader to any reasonable handbook on graph theory for its proof (if s/he so desires). First, a couple of definitions. A bipartite graph is a triple $G = \langle X, Y, E \rangle$ where X, Y are two disjoint sets of vertices, and E is the set of edges. Each edge e in E is incident with X and with Y. That is such edge starts in X and ends in Y. Next, we say that a set of edges $E' \subseteq E$ is called a *matching* if no edges in E' are incident, i.e., do not share a vertex. One way of thinking about matchings is that those are partial one-to-one functions from (subsets of) X to Y, except that values must be chosen consistently with E. This way of treating matchings will be used below. We will often write M, possibly with indices to denote matchings. A *cover* for a bipartite graph G is a set $H \subseteq X \cup Y$ such that every edge of E is incident with some vertex of H. Now, among matchings in G there are those of maximum size (for we are dealing with finite graphs). Likewise, there are covers for G of minimum size. Both those facts are obvious. Let $m(G)$ be the size of maximum size matching, and $c(G)$ be the size of the minimum size cover. Here is an old (but still amazing) theorem due to König.

THEOREM 7.1 (König theorem)
If G is a bipartite graph, then $c(G) = m(G)$.

To confuse the matter even more, let us observe that there was another König lemma we used in the proof of compactness (and that other one we proved).

We will not prove Theorem 7.1. But we will prove the rest of combinatorial facts needed to prove a desired fact on minimally unsatisfiable CNFs. Before we do, the reader should ask why we suddenly started to discuss issues not seemingly related to logic. However, it is quite clear that there is one bipartite graph that can be naturally associated with a CNF F. We take as X the set F (of clauses), and as Y the set Var of variables. We create an edge between the clause C and variable x if x occurs (positively or negatively) in C. The resulting graph G_F is a bipartite graph and, eventually, we will use König theorem to prove some interesting properties of that graph, especially if F is minimally unsatisfiable. Now, convinced that the bipartite graphs have something to do with CNFs, we return to combinatorics.

Let $G = \langle X, Y, E \rangle$ be a bipartite graph. A subset $A \subseteq X$ is *matchable into* subset $B \subseteq Y$ if there exists a matching M in G such that A is included in the domain of M and the range of M is included in B. Likewise, we talk about subsets B of Y matchable into $A \subseteq X$. Here, we want the range of M equal to B with the domain of M contained in A. Since we think about matchings as partial one-to-one functions (one-to-one is the key issue here), when M is a matching in $\langle X, Y, E \rangle$, then M^{-1} is also a matching, except (maybe we overformalize it a bit) it is a matching in $\langle Y, X, E \rangle$.

Here is a combinatorial fact we will use.

LEMMA 7.1
If $G = \langle X, Y, E \rangle$ is a bipartite graph, M is a matching in G such that $|M| = m(G)$, C is a cover of G such that $|C| = c(G)$ and $A = C \cap X$, $B = C \cap Y$, then M maps A into $Y \setminus B$ and M^{-1} maps B into $X \setminus A$.

Proof: We define two subsets of the matching M:

$$M_1 = \{e \in M : e \text{ is incident with } A\}$$

and

$$M_2 = \{e \in M : e \text{ is incident with } B\}.$$

Then, because C is a cover, $M = M_1 \cup M_2$. Next, since M is a matching, $|M_1| = |A|$, and $|M_2| = |B|$. We have the following equalities and inequalities:

$$m(G) = |M| \leq |M_1| + |M_2| = |A| + |B| = |C| = c(G) = m(G).$$

The first equality follows from our choice of C. Then we have an inequality which holds because $M = M_1 \cup M_2$. Next equality follows from our above estimates of $|M_1|$ and $|M_2|$. The next equality follows because X and Y (thus any of their respective subsets) are disjoint. Then, the last equality is König theorem (Theorem 7.1). Ultimately, it follows that $|M| = |M_1| + |M_2|$.

But for any two subsets M_1 and M_2 of M,

$$|M| + |M_1 \cap M_2| = |M_1| + |M_2|.$$

Thus, it must be the case that $M_1 \cap M_2 = \emptyset$. But now, since $M_1 \cap M_2$ is empty, no edge in the matching M which originates in A ends in B, and no edge of M ending in B originates in A! Thus M matches A into $Y \setminus B$, and M^{-1} matches B into $X \setminus A$, as desired. □

We get the following corollary which will be used to get the result on the minimally unsatisfiable sets of clauses.

COROLLARY 7.2
If $G = \langle X, Y, E \rangle$ is a finite bipartite graph, then there is a cover C for G such that $C \cap X$ is matchable into $Y \setminus (C \cap Y)$ and $C \cap Y$ is matchable into $X \setminus (C \cap X)$.

This corollary is true for infinite bipartite graphs as well, but the infinite case is of no interest for our considerations in this book.

Recall now that for a set of clauses F we constructed a bipartite graph G_F with two parts. One was F itself, the other was $Var(F)$. Here is the crucial property of G_F.

PROPOSITION 7.3

1. *If G_F possesses a matching M with the domain of M being entire F, then F is satisfiable.*

2. *If F is minimally unsatisfiable, then in G_F there is a matching M with the range of M being the entire $Var(F)$.*

Proof: First we prove (1). Since M is a matching with the domain of M being entire F, for each clause C of F we have a variable $M(C)$ so that (a) $M(C)$ occurs in C, (b) if $C_1 \neq C_2$ then $M(C_1) \neq M(C_2)$. But now we define the following valuation v by setting:

$$v(x) = \begin{cases} 1 & \text{if } x = M(C) \text{ and } x \text{ occurs in } C \text{ positively} \\ 0 & \text{otherwise.} \end{cases}$$

It is now clear that $v \models F$. For let $C \in F$. Then, by construction, $v(M(C))$ is 1. Thus (1) is proved.

(2) Let us assume that F is minimally unsatisfiable. In order to prove our assertion, we will use Lemma 7.1. Since G_F is a bipartite graph there is a cover D for the graph G_F, $D \subseteq F \cup Var(F)$, D splitting into $A = D \cap F$, $B = D \cap Var(F)$, A matchable into $Var(F) \setminus B$, B matchable into $F \setminus A$. We will prove that B is the entire $Var(F)$. So let us assume $B \neq Var(F)$. This means that there are variables that are in $Var(F)$ but not in B. But $Var(F)$ is the set of variables of F. This means that there are clauses that contain variables which do not belong to B. In the context of the graph G_F this means that there are edges that end in points which are not in B. But D was a cover. Therefore such edges must start in a clause belonging to A. In

particular this means that $A \neq \emptyset$. So now, let us look at the set $F \setminus A$. Since $A \neq \emptyset$, $F \setminus A \subset F$ and since F is minimally unsatisfiable, $F \setminus A$ is satisfiable.

Now, let us ask this question: is it possible that a clause in $F \setminus A$ contains a variable in $Var(F) \setminus B$? The answer to this question is "no." The reason is that D is a cover; every edge either starts in a node of A or ends in a node of B. Therefore all clauses of $F \setminus A$ have variables from B only! So, on the one hand we know that $F \setminus A$ is satisfiable, on the other we know that all variables of $F \setminus A$ are in B. Therefore, we have a partial valuation v_1 with the domain B so that v_1 satisfies $F \setminus A$.

But now we use Corollary 7.2; A is matchable into $Var(F) \setminus B$. This means that we have a partial valuation v_2, with $Dom(v_2) = Var(F) \setminus B$, so that v_2 satisfies A. Now, v_1 and v_2 have disjoint domains. Thus $v_1 \cup v_2$ is a valuation. But then $v_1 \preceq_k v_1 \cup v_2$ and $v_2 \preceq_k v_1 \cup v_2$. Hence $v_1 \cup v_2$ satisfies $F \setminus A$ and $v_1 \cup v_2$ satisfies A. Thus $v_1 \cup v_2$ satisfies F, a contradiction. □

COROLLARY 7.3 (Aharoni and Linial)
If F is minimally unsatisfiable then $|Var(F)| \leq |F|$.

Proof: By Proposition 7.3, $Var(F)$ is matchable into F, thus $|Var(F)| \leq |F|$. □
Thus, indeed, we proved that minimally unsatisfiable sets of clauses have more clauses than variables.

7.2 Resolution rule, closure under resolution

First, the *resolution rule* itself is an execution of the following partial binary operation on clauses:

$$\frac{l \vee C_1 \qquad \bar{l} \vee C_2}{C_1 \vee C_2}.$$

In other words, if clause $D_1 = l \vee C_1$ contains a literal l and $D_2 = \bar{l} \vee C_2$ contains its dual $\bar{l}$ then $Res(D_1, D_2)$ is defined, and results in eliminating $l, \bar{l}$ from D_1, D_2, resp., and conjoining the rest of the literals into a single clause. If there is more than one such pair of dual literals then the resulting clause is a tautology and does not give us any constraint on the putative valuation. For that reason, we will assume that all clauses under consideration are non-tautological, i.e., do not contain a complementary pair of literals. Also, we will assume that when resolution is executed, the duplicate literals are eliminated. For instance resolving $p \vee q \vee s$ with $\bar{p} \vee q \vee \bar{t}$ results in $q \vee s \vee t$.

Even though the operation $Res(\cdot, \cdot)$ is partial, for every set of clauses F we can define the closure of F under resolution. Indeed, for a CNF F, there exists a set G of clauses that contains F and is closed under resolution. Specifically, it is easy to see that there exists a *least* set of clauses G satisfying the following conditions:

1. $F \subseteq G$.

2. Whenever D_1 and D_2 are two clauses in G and the operation $Res(D_1, D_2)$ is executable and results in a non-tautological clause, then $Res(D_1, D_2)$ belongs to G.

We can prove the above using the Knaster-Tarski fixpoint theorem (Proposition 1.2). Indeed, let F be fixed, and let us define an operator $res_F(\cdot)$ in the complete lattice of subsets of the set of clauses as follows:

$$res_F(G) = F \cup \{Res(C_1, C_2) : C_1, C_2 \in F \cup G$$
$$\text{and } Res(C_1, C_2) \text{ is defined and non-tautological}\}.$$

It is easy to see that the operator $res_F(\cdot)$ is monotone; $G \subseteq H$ implies $res_F(G) \subseteq res_F(H)$. At this stage we know that the least fixpoint of res_F exists. We will denote this fixpoint by $Res(F)$ and call it *closure of F under resolution*.

Let us look more closely at the fixpoint $Res(F)$. We will find an alternative characterization of $Res(F)$ by means of *derivations*. A derivation $\mathcal{D}$ of a clause C from set of clauses F is a labeled binary tree (which we will now invert, leaves will be at the top, not at the bottom). The leaves are labeled with clauses of F. The internal nodes are labeled with resolvents (i.e., results of resolution) of the labels of the parents. The root is labeled with C. Denote the set of all clauses that have a derivation from F by Der_F. We then have the following result.

PROPOSITION 7.4
The sets $Res(F)$ and Der_F coincide.

Proof: It is quite clear that Der_F is closed under the application of resolution rule and contains F. In other word Der_F is a fixpoint of the operation $Res(\cdot)$, and so $Res(F) \subseteq Der_F$. Conversely, by induction on the height of the binary tree serving as derivation we prove that all clauses in Der_F are in $Res(F)$. Thus, the equality follows. □

The alternative characterization of the least fixpoint via derivations can be used to show that the operator $res_F(\cdot)$ is continuous, that is, for every $\subseteq$-increasing family of sets of clauses $\langle X_n \rangle_{n \in N}$,

$$res_F(\bigcup_{n \in N} X_n) = \bigcup_{n \in N} res_F(X_n).$$

Thus the fixpoint $Res(F)$ is reached by the iteration of the operator $res_F(\cdot)$ in at most ω steps. Moreover, the variables in $Res(C_1, C_2)$ occur either in C_1 or in C_2 (or in both). Since we do not allow for repetition of literals in clauses (after all we think about clauses as sets, not bags, of literals) the closure of a finite set of clauses under resolution is finite.

PROPOSITION 7.5

If a valuation v satisfies clauses C_1 and C_2 and the operation $Res(C_1, C_2)$ is executable, then v satisfies $Res(C_1, C_2)$.

Proof: If $Res(C_1, C_2)$ can be executed then there are two clauses D_1 and D_2 and a literal l so that C_1 is $D_1 \vee l$, C_2 is $D_2 \vee \bar{l}$. If $v(l) = 1$, then $v(\bar{l}) = 0$, thus $v \models D_2$ and so $v \models D_1 \vee D_2$, i.e., $v \models Res(C_1, C_2)$. If $v(l) = 0$ then $v \models D_1$ and so again $v \models D_1 \vee D_2$, i.e., $v \models Res(C_1, C_2)$. □

Thus we proved that the resolution rule is *sound*. That is, if a valuation v satisfies all clauses of F, then it must also satisfy all clauses in Der_F (an easy induction on the height of the tree of the derivation is needed). We formulate this corollary to Proposition 7.5 explicitly.

COROLLARY 7.4

If v is a valuation satisfying a set of clauses F then v satisfies $Res(F)$. Thus, if F is a satisfiable set of clauses, then so is $Res(F)$.

Now, the question arises if Res is *complete*. In other words, if F semantically entails C, is it the case that $C \in Res(F)$? The answer to this question is an obvious "no". Let us build a most trivial counterexample. Let F consist of just one clause consisting of a single atom, a and let D be $a \vee b$. Then, obviously $F \models D$, but the $Res(F) = F$ and so $D \notin Res(F)$.

Next, we make the following observation. Given a non-tautological clause D, there is a unique valuation v which is defined on variables occurring in D and such that $v \models \neg D$ (we set all variables occurring in D to the opposite polarity to that in D). For the purposes of this book, we call it the *canonical valuation* making D false.

Now, it turns out that we are not completely out of our luck with the completeness of resolution. First, let us observe the following useful fact (closely related to Proposition 3.9).

LEMMA 7.2

Let us assume that C, D are non-tautological clauses.

1. If $D \subseteq C$ then $D \models C$.
2. Conversely, if $D \models C$ then $D \subseteq C$.

Proof: (1) is obvious.
(2) Assume $D \not\subseteq C$. Then there is a literal l occurring in D but not in C. Consequently, the literal $\bar{l}$ does not occur in D because D is non-tautological. We consider the canonical valuation v making C false. Then, since $l \notin C$ either v is not defined on the atom $|l|$ at all, or $v(l) = 1$ (this last case happens when $\bar{l}$ occurs in C). In the first case we extend the canonical valuation making C false by setting $v(l) = 1$ and arbitrary values on other atoms occurring in D but not in C. In the second case just extend v arbitrarily to all variables occurring in D. We call the constructed valuation

v'. Then, because v' extends v, it is the case that $v(C) = 0$. But in either case of our construction, $v'(l) = 1$ so $v'(D) = 1$. But then $D \not\models C$, a contradiction. □

Let F be a CNF, i.e., a set of clauses (Generally, we use CNF, set of clauses, and clause set interchangeably.) A clause C is a *resolution consequence* of F if $C \in Res(F)$. A clause C is a *minimal resolution consequence* of F if C is an inclusion-minimal resolution consequence of F. That is, C is a resolution consequence of F but no proper subset of C is a resolution consequence of F.

7.3 Completeness results

In this section we will investigate the issues related to the completeness of resolution and related topics.

THEOREM 7.2 (Quine theorem)
A non-tautological clause C is a consequence of a set of clauses F if and only if there is a clause D such that D is a resolution consequence of F and $D \subseteq C$.

Before proving Theorem 7.2 let us see its consequences. First of all, we observe the corollary following from the fact that clauses are finite collections of literals.

COROLLARY 7.5
A non-tautological clause C is a consequence of a CNF F if and only if for some minimal resolution consequence D of F, $D \subseteq C$.

Yet another consequence is:

COROLLARY 7.6 (Completeness theorem for resolution)
Let F be a CNF. Then F is satisfiable if and only if $\emptyset \notin Res(F)$.

Proof: The implication $\Rightarrow$ follows from the soundness of the resolution rule.
To see $\Leftarrow$, assume $\emptyset \notin Res(F)$. Select a new propositional variable $p \notin Var(F)$ and make this query: "Does $F \models p$"? If it is the case, then some element of $Res(F)$ must subsume the unit clause p. But $Var_{Res(F)} = Var_F$. So the only possible clause from $Var_{Res(F)}$ subsuming p is empty. This is a contradiction. Therefore, $F \not\models p$ and so F is satisfiable. □

By *subsumption rule* we mean the following rule of proof:

$$\frac{C}{C \vee D}$$

This rule is sometimes called *weakening* or *absorption*. The following is obvious.

PROPOSITION 7.6
The subsumption rule is sound. That is, for every valuation v, if $v \models C$, then $v \models C \vee D$.

Proposition 7.6 says that the proof system consisting of resolution and subsumption is sound. But now we have to ask the same question we asked above: "Is the proof system consisting of resolution and subsumption complete?" The answer to this question is positive. The meaning of Theorem 7.2 is precisely this. Formally, we have the following.

PROPOSITION 7.7
The proof system consisting of resolution and subsumption rules is complete for clausal logic. That is, given a set of clauses F and a clause C, $F \models C$ if and only if there exists a derivation (using resolution and subsumption rules) that proves C. We can limit such proof to admit a single application of subsumption rule.

Now, we are finally ready for the *proof of Theorem 7.2*. First, let us observe that it is enough to prove Theorem 7.2 for finite sets F. Indeed, by the compactness theorem (Theorem 6.1), if $F \models C$ then for some finite $F' \subseteq F$, $F' \models C$. But then, having proved Theorem 7.2 for finite CNFs, we know that for some finite $F' \subseteq F$, for some resolution consequence of F', D, $D \subseteq C$. But the operator $Res(\cdot)$ is monotone: more clauses, more consequences! Thus a resolution consequence D of F' that is included in C is a resolution consequence of F.

Clearly, if C extends some clause D which is a resolution consequence of F then every valuation v satisfying F also satisfies D (soundness of the resolution rule, Proposition 7.5) and so also satisfies C (soundness of the subsumption rule, Proposition 7.6). Thus we proved the implication $\Rightarrow$.

The implication $\Leftarrow$ is more subtle. Following our discussion above, it is enough to prove that a non-tautological consequence of a *finite* F must contain as a subset a resolution consequence of F.

So let us suppose that F is finite, a clause C is its consequence, but our implication $\Leftarrow$ is false. Thus, formally,

(a) $C \in Cn(F)$.

(b) C does not contain a subset that is in $Res(F)$.

Now comes a delicate moment. Let us consider the set of propositional variables occurring in F and in C. Let us call it V. Then, our assumption is that there is a clause C' having only variables in V such that:

(c) $C' \in Cn(F)$.

(d) C' does not contain a subset that is in $Res(F)$.

Indeed, C is such clause C'.

Since there are only finitely many clauses where all variables are in V (recall that no repetition of literals is allowed in a clause), we can select a *longest* (i.e., weakest) clause C' making (c) and (d) true. Without loss of generality, we can assume that C is that C' (we could select C to satisfy the conditions (a) and (b) and the additional maximality condition upfront).

With possible renaming we can assume that the set V is $\{x_1, \ldots, x_n\}$. The first question we ask is this. Is it possible that all variables of V occur (positively or negatively) in C? If this is the case, then (again without loss of generality) C looks like this:
$$x_1 \vee \ldots \vee x_k \vee \neg x_{k+1} \vee \ldots \vee \neg x_n.$$

Let us look at the canonical valuation making C false. This valuation v evaluates atoms $x_1, \ldots, x_k$ as false, and atoms $x_{k+1}, \ldots, x_n$ as true. Since $v \not\models C$, and C is a consequence of F it must be the case that for some $D \in F$, $v \not\models D$, in other words, $v \models \neg D$. Let us look at the clause D. Here it is:
$$x_{i_1} \vee \ldots \vee x_{i_p} \vee \neg x_{j_1} \vee \ldots \vee \neg x_{j_r}.$$

Since all variables in V occurred in C, it must be the case that, already, v is defined on all variables occurring in D. But $v \models \neg D$ so we have:

(I) $v(x_{i_1}) = \ldots = v(x_{i_p}) = 0$.

(II) $v(x_{j_1}) = \ldots = v(x_{j_r}) = 1$.

This, however, means that
$$\{x_{i_1}, \ldots, x_{i_p}\} \subseteq \{x_1, \ldots, x_k\}$$
and
$$\{x_{j_1}, \ldots, x_{j_r}\} \subseteq \{x_{k+1}, \ldots, x_n\},$$
that is, $D \subseteq C$. Consequently some clause from F (and thus from $Res(F)$) subsumes C. Since our assumption was that there is no such D, we just showed that C does *not* involve all atoms in V.

Let x_s be the first atom *not* occurring in C (either positively or negatively!). Since C entails weaker clauses $x_s \vee C$ and $\neg x_s \vee C$, it is the case that both $x_j \vee C$ and $\neg x_j \vee C$ belong to $Cn(F)$. But by our construction, C was the longest clause E involving only variables in V and such that E belongs to $Cn(F)$ and E does not contain a clause in $Res(F)$. And certainly both clauses $x_s \vee C$ and $\neg x_s \vee C$ are longer than C. Therefore, it must be the case that there are clauses D_1 and D_2 such that D_1, D_2 belong to $Res(F)$ and:
$$D_1 \subseteq x_s \vee C, \qquad D_2 \subseteq \neg x_s \vee C.$$

Let us look at D_1. Is it possible that D_1 does not contain x_s? No, because in such case $D_1 \subseteq C$ and this was not the case since C did not contain a subset which is in

$Res(F)$. Likewise, it must be the case that $\neg x_s \in D_2$. But with these facts we find that D_1 and D_2 are resolvable. In other words, since x_s has no occurrence (positively or otherwise) in C, $D_1 = x_s \vee E_1$, and $E_1 \subseteq C$. Likewise, $D_2 = \neg x_1 \vee E_2$ and $E_2 \subseteq C$. But then $Res(D_1, D_2) = E_1 \vee E_2 \subseteq C$. But $Res(D_1, D_2) \in Res(F)$ so we found that C contains a subset D in $Res(F)$, a contradiction. □

We will use Theorem 7.2 to get a completeness result about the resolution rule alone. First, let us observe that if C is a clause then $\neg C$ is semantically equivalent to a collection of literals (i.e., unit clauses). When we write $\neg C$ we mean this collection of unit clauses. Now, looking at $\neg C$, we observe that $C = \bigvee_{l \in \neg C} \bar{l}$. Indeed, $l \in \neg C$ if and only if for some literal m of C, $l = \bar{m}$, that is for some m of C, $m = \bar{l}$. But then $C = \bigvee_{l \in \neg C} \bar{l}$.

Given a CNF F and a clause C we say that C follows from F by *resolution refutation* if the closure under resolution of $F \cup \neg C$ contains an empty clause (i.e., is inconsistent).

PROPOSITION 7.8
Let F be a CNF, and let C be a clause. Then $F \models C$ if and only if C follows from F by resolution refutation.

Proof: First, assume that $F \models C$. Then, by Theorem 7.2 there is a clause D in $Res(F)$ such that $D \subseteq C$. So now, since the operator $Res(\cdot)$ is monotone, $Res(F) \subseteq Res(F \cup \neg C)$[1]. In particular $D \in Res(F \cup \neg C)$. But since $D \subseteq C$, negation of *every* literal occurring in D belongs to $Res(F \cup \neg C)$. Thus, by repeated application of the resolution rule we get the empty clause $\emptyset$ out of D and the set of literals $\neg C$. Conversely, assume that $Res(F \cup \neg C)$ contains the empty clause, $\emptyset$. Let v be an arbitrary valuation satisfying F. Then it is not the case that $v \models \bigwedge_{l \in \neg C} l$. In other words, $v \models \neg \bigwedge_{l \in \neg C} l$, that is $v \models \bigvee_{l \in \neg C} \bar{l}$. But then $v \models C$, as desired. □

7.4 Query-answering, computing the basis of the closure under resolution

Assume that we want to use a resolution/subsumption proof system as a tool for answering queries. In this scheme the formula F is a database that is stored on the system. The queries are clauses. The query C is answered *"yes"* if $F \models C$, and *"no"* otherwise.

In this scheme of things, the completeness of resolution refutation as a proof system means that when we get a query C, we can transform C into a set of literals $\neg C$ (recall that the negation of a clause reduces to a set of unit clauses), and then run

---
[1] Let us recall that $\neg C$ is a set of (unit) clauses.

the resolution engine. If we ever compute the empty clause, $\emptyset$, we answer "*yes.*" otherwise (when we saturate F with resolution and do not get $\emptyset$) we answer "*no.*"

This is fine (ignoring the space requirements related to the size of the closure under resolution), but it means that we do, repeatedly, the same thing over and over again. The issue now is if we could do something better. The idea is similar to tabling or memoizing.

Here is the proposed alternative: Let us compute the $Res(F)$ once and store it. Then, once we are asked a query C, we check if any of the precomputed clauses (i.e., an element of $Res(F)$) is included in C. If so, the answer is "*yes.*" Otherwise, by Theorem 7.2, the answer is "*no.*"

Now, the issue is if we have to compute the entire $Res(F)$ to use this scheme. Of course, the answer is *no*; otherwise there would be no point in discussing it.

What we will do is to construct a much smaller subset (which we will call Min) of $Res(F)$, with the same completeness property: $C \in Cn(F)$ if and only if some $D \in Min$ is included in C. So the query processing algorithm goes like this: First, precompute the "basis" Min. Then, given a query C check if some element of the basis Min is included in C. If it is the case, answer "*yes*," otherwise, answer "*no.*"

To justify the proposed approach we first need to prove that such a simple set exists, then we need an algorithm to construct it.

We will now define a *basis* of $Res(F)$. The basis of F will be a set G of clauses with the following properties:

1. $G \subseteq Res(F)$.
2. $Cn(F) = Cn(G)$.
3. G forms an antichain, i.e., for $C_1, C_2 \in G$, if $C_1 \subseteq C_2$ then $C_1 = C_2$.
4. Every element of $Res(F)$ is subsumed by some element of G.

Here is the meaning of this definition. First, we want our basis to be a subset of $Res(F)$. Second, we want a subset that generates the entire $Cn(F)$. Third, we want an antichain, i.e., the set of inclusion-incompatible elements (i.e., with no subsumption between the clauses). Finally, we want the property that all elements of $Res(F)$ are subsumed by the elements of the basis.

PROPOSITION 7.9
Let F be a set of clauses. Then there exists a unique basis G for F.

Proof: Given F, construct the resolution closure $Res(F)$. Now, let Min be the collection of all inclusion-minimal clauses in $Res(F)$. We claim that Min satisfies conditions (1)–(4) and is the unique set of clauses satisfying these conditions.

(1) is obvious by construction.

(2) Since $Min \subseteq Res(F) \subseteq Cn(F)$, $Cn(Min) \subseteq Cn(F)$. Conversely, if $C \in Cn(F)$ then, by Theorem 7.2 some clause $D \in Res(S)$ is included in C, and so an inclusion-minimal clause $D \in F$ is included in C. But $D \in Min$, thus, by soundness of subsumption rule, $C \in Cn(Min)$.

(3) Minimal elements in any family of sets form an antichain.
(4) Assume D is an element of $Res(F)$. Then, since D is finite, there is an inclusion-minimal element of $Res(F)$, say C, such that $C \subseteq D$. But then, by definition, $C \in Min$.

Now, we need to show that Min is the unique basis for F. To this end, let G be a basis for F. Assume $G \neq Min$.
Case 1: $G \setminus Min \neq \emptyset$. Choose $C \in G \setminus Min$. Then, there is $D \in Min$ such that $D \subset C$. But G is a basis, and $D \in Cn(F)$. Hence there is $E \in G$, $E \subseteq D$. Consequently, $E \subset C$, a contradiction with the fact that G is an antichain.
Case 2: $Min \setminus G \neq \emptyset$. Select $C \in Min \setminus G$. Then, there is $D \in G$ so that D subsumes C. $D \neq C$ because $C \notin Min$. But $D \in Cn(F)$, so there is $E \in Min$ such that $E \subseteq D$. But then $E \subset C$, and both $C, E \in Min$, again contradiction. □

Given a set of clauses F we can, in the polynomial time in size of F, eliminate subsumption in F while preserving consequence. Here is how we do this efficiently. We sort the clauses of F according to their sizes, and within the fixed length ordering the clauses lexicographically (with x_i preceding $\neg x_i$). Then, for each clause we need to eliminate those clauses in this ordering that it subsumes. All these clauses must appear later in the ordering. It is clear that this procedure runs in time $O(s^2)$ where s is the size of F. The output of this procedure will be an antichain. We can assume that F is an antichain, that is, subsumption-free.

We could construct the unique basis Min as follows: first construct $Res(F)$ and then construct its basis. But we will do this differently.

(1) Start with a CNF F. Due to the discussion above, we can assume that F is subsumption-free.
(2) Non-deterministically, select from F a pair of resolvable clauses C_1 and C_2 which has not been previously resolved.
(3) If $Res(C_1, C_2)$ is subsumed by some other clause in F or is a tautology, do nothing and select the next pair.
(4) If $D : Res(C_1, C_2)$ is *not* subsumed by some other clause in F, do two things:
(a) Compute $R := \{E \in F : D \subseteq E\}$. Eliminate from F all clauses subsumed by D, that is $F := F \setminus R$.
(b) Set $F := F \cup \{D\}$.
(5) Do this until F does not change.

We first observe that the construction described by the above pseudocode has two invariants. The first one is that F is an antichain (Step (1), and then we leave to the reader checking that after each iteration of the loop (4) F is still an antichain). Here is the second one: Let F_0 be the input. Then, after each iteration, $Cn(F) = Cn(F_0)$. To see this, assume that F_1 is the content of the variable F before the execution of the loop, and F_2 is the content of F after the execution of the loop.

If $F_2 = F_1$, then the invariant is preserved. So now, assume that $F_2 = (F_1 \setminus R) \cup \{C\}$ and $Cn(F_1) = Cn(F)$. We will show that $Cn(F_2) = Cn(F)$. Indeed, since $F_1 \subseteq Cn(F)$, $F_1 \setminus R \subseteq Cn(F)$. Moreover, $C \in Cn(F)$ and so $F_2 \subseteq Cn(F)$. Then, $Cn(F_2) \subseteq Cn(F)$. Conversely, $R \subseteq Cn(\{C\})$ (because all clauses in R are subsumed by C). Therefore we have:

$$Cn(F_1) = Cn((F_1 \setminus R) \cup R) \subseteq Cn((F_1 \setminus R) \cup Cn(\{C\}))$$
$$\subseteq Cn((F_1 \setminus R) \cup \{C\}) = Cn(F_2).$$

Next, we need to see that the construction terminates. Here is why this happens. Indeed, all we need to show is that once a clause has been removed from F at some iteration, it will not be put in again. To this end, we show another invariant: at each iteration the closure under subsumption can only increase (thus once a clause has been subsumed after some loop it will always be subsumed). Indeed, assume that a clause D is subsumed by the set of clauses F_1 (the value of the variable F before the execution of the loop). Let F_2 be the value of the variable F after the iteration of the loop. If $F_2 = F_1$ there is nothing to prove. If $F_2 = (F_1 \setminus R) \cup \{C\}$), then either D is subsumed by a clause in $F_1 \setminus R$, and so is still subsumed by a clause in F_2, or D is subsumed by a clause in R. But C subsumes every clause in R, thus if $E \in R$, $E \subseteq D$, then $C \subseteq E \subseteq D$. Thus we proved invariance, the set of subsumed clauses grows, and the construction halts, eventually.

But did we compute a basis? We did. The reason is that, after we terminate, the resulting set of clauses held in the variable F satisfies the property that $Cn(F) = Cn(F_0)$ where F_0 is the input, and that for every clause $C \in Res(F)$ there is $D \in F$ such that D subsumes C. Thus for each $E \in Cn(F_0)$ there is $D \in F$ such that D subsumes E and we are done.

Having established the virtues of preprocessing the knowledge base (finding the basis, and reducing the entailment to subsumption, i.e., inclusion), let us now play the other side and see the virtues of *not* preprocessing.

To this end, we will see that there is a powerful fact that tells us that it is beneficial to process on the case-by-case basis.

Specifically, given a CNF F, and a clause C, define

$$F/C = \{D : D \in F \text{ and } D \text{ can not be resolved with } C\}.$$

So, what is F/C? It consists of those clauses in F where variables occurring in C do not occur at all, or they occur, but with the same polarity as in C.

Now, we have the following pruning lemma:

LEMMA 7.3
Let F be a CNF, and C a clause. Then, $F \models C$ if and only if $F/C \models C$.

Proof: The implication $\Leftarrow$ is obvious.
For the other implication, assume $F/C \not\models C$. Then, for some valuation v, $v(F/C) = 1$ but $v(C) = 0$. It must be the case that the valuation v evaluates all atoms occurring in C to the sign opposite their sign in C (i.e., v must contain the canonical valuation falsifying C). But $F \models C$. Thus there must be a clause $D \in F$ so that $v(D) = 0$. Since $v(F/C) = 1$ it must be the case that $D \notin F/C$. Since $D \in F \setminus (F/C)$ it must be the case that D can be resolved with C. That means that for some variable x_i, x_i

occurs both in C and in D, but with opposite signs. Now, we have two cases.
Case 1: x_i occurs in C positively. Then, $\neg x_i \in D$. Moreover, recall $v(x_i) = 0$, Thus $v(\neg x_i) = 1$, $v(D) = 1$ and we get a contradiction.
Case 2: x_i occurs in C negatively. Then $x_i \in D$. Then, $v(x_i) = 1$, $v(D) = 1$ and we get a contradiction. □

We get the following corollary.

COROLLARY 7.7
Assume a CNF F is finite and C is a non-tautological clause containing all variables occurring in F. Then, $F \models C$ if and only if some clause in F subsumes C.

Proof: The implication $\Leftarrow$ is obvious.
To see $\Rightarrow$, assume $F \models C$. Then, by Lemma 7.3, $F/C \models C$. Can F/C be empty? No, because the empty set of clauses does not entail any non-tautological clause. Thus let $D \in F/C$. Then, for every variable x_i occurring in D, the polarity of x_i in C cannot be opposite. But C contains all variables occurring in D so the polarity of those variables in D and in C is the same. Thus $D \subseteq C$, as desired. □

Lemma 7.3 suggests another strategy for query answering. Namely, once we have F and the query C is preprocessed, eliminate from F all clauses containing the literal $\bar{l}$ with $l \in C$. This clearly speeds up the matters.

Except that when we look at the resolution refutation we see that since we add all literals $\bar{l}$ for $l \in C$ the clauses containing such literals $\bar{l}$ will be immediately eliminated in the DPLL search which we will encounter in Chapter 8.

7.5 Reduct by a literal, the Davis-Putnam lemma

We now discuss a technique for reducing the number of variables in a set of clauses. The key lemma below will form the basis for one of fundamental algorithms for testing satisfiability. It will be convenient (as we did in Section 7.1) to think about a clause $C := l_1 \vee \ldots \vee l_k$ as a set of literals $\{l_1, \ldots, l_k\}$. In this way some operations on clauses will reduce to set operations.

By a *reduct of a set F of clauses by a literal l* we mean the following set of clauses G:

$$\{C \setminus \{\bar{l}\} : C \in F \wedge l \notin C\}.$$

So what is the reduct G? It is just the effect of assuming l to be true. Indeed, assume that $C \in F$. If $l \in C$ then once l is assumed to be true, C is satisfied. As concerns the remaining clauses, these either do not mention the underlying variable at all, or if they do, the resolution with a unit clause $\{l\}$ can be performed, resulting in $C \setminus \{\bar{l}\}$.

Let us observe that a literal l determines, in effect, two reducts: one with respect to l and another with respect to $\bar{l}$. Given F, we will denote these reducts by F_l and $F_{\bar{l}}$, respectively. Here is the key property of these reducts.

LEMMA 7.4 (Davis–Putnam lemma)
Let F be a set of non-tautological clauses, and let l be a literal. Then, F is unsatisfiable if and only if both F_l and $F_{\bar{l}}$ are unsatisfiable.

Proof: First, assume that at least one of $F_l, F_{\bar{l}}$ is satisfiable. Without loss of generality assume F_l is satisfiable. Since F consists of non-tautological clauses, F_l has no occurrence of $l, \bar{l}$. Since F_l is satisfiable, let $v \models F_l$. Define now a valuation w as follows[2]

$$w(m) = \begin{cases} v(m) & m \notin \{l, \bar{l}\} \\ 1 & m = l \end{cases}$$

Then, as $l, \bar{l}$ do not appear in G at all, for $C \in F$, as long as $l \notin C$, $w \models C$. But if $l \in C$, then $w \models C$ by construction. This completes the proof of $\Rightarrow$.

Now, assume that both F_l and $F_{\bar{l}}$ are unsatisfiable, but v is a valuation satisfying F. Without loss of generality we can assume $v(l) = 1$. Let us look at F_l. If $C \in F_l$ two cases are possible. First, it is possible that $C \in F$ and $l, \bar{l} \notin C$. Then, clearly, $v \models C$. The second case is that $C = D \setminus \{\bar{l}\}$ with $\bar{l} \in D$, $D \in F$. Then, $v \models D$. But $v(\bar{l}) = 0$ because $v(l) = 1$. Thus one of remaining literals in D, i.e., one in C, is evaluated by v as 1. But then $v \models C$. Thus $v \models F_l$, a contradiction. This completes the proof. □

COROLLARY 7.8
Let F be a set of non-tautological clauses, and let l be a literal. Then, F is satisfiable if and only if at least one of $F_l, F_{\bar{l}}$ is satisfiable.

We get a little bit more of our consideration of reducts. Recall that a literal l is *pure* in a set of clauses F if no clause of F contains $\bar{l}$. When l is pure in F, the reduct F_l is included in F. This implies the following fact.

PROPOSITION 7.10
Let F be a set of clauses. If l is a pure literal in F then F is satisfiable if and only if F_l is satisfiable. Moreover, if v is a valuation satisfying F_l, then a valuation w defined by

$$w(m) = \begin{cases} v(m) & m \notin \{l, \bar{l}\} \\ 1 & m = l \end{cases}$$

---
[2] We define here the valuation v on the entire set of literals at once. Technically, valuations are functions mapping Var into $Bool$. But since the valuations extend uniquely to the set of all literals, we can define valuations on the entire set of literals Lit as long as we maintain the consistency condition: $v(\bar{l}) = \overline{v(l)}$.

satisfies F.

Let us observe that Proposition 7.10 tells us what to do when we encounter a pure literal in a set of clauses F that we test for satisfiability: turn l to *true*, and eliminate clauses containing l. On a related theme, we recall that in Proposition 2.8 we proved that whenever l is pure literal in F then $\{l\}$ is an autarky for F.

In Chapter 8 we will return to the reduct and use it in algorithms for satisfiability testing.

7.6 Semantic resolution

Can we do something to limit the tremendous growth of the closure under resolution, $Res(F)$, without compromising the completeness property (Corollary 7.6)?

A lot of effort has been expended to do just this. In the early period of the development of studies of satisfiability an immense amount of work was spent on this topic, but there are current papers dealing with this problem as well. The scheme is always similar: take some version of resolution (say r) to limit the number of clauses that can be derived, and show the analogue of Corollary 7.6 holds for r. In other words make sure that there is (hopefully) much fewer clauses that can be derived, but if the empty clause $\emptyset$ can be derived using resolution, then such restrictive closure, using r still derives $\emptyset$. In this section we will do this for a scheme that uses two elements to limit available resolutions. The first element will be a valuation v that allows us to split the input formula into two parts: those clauses that are true under v and those clauses that are false under v. We will require that the first (left) input to an application of resolution be *always* true under v. Likewise, we will require that the second input to the resolution rule be *always* false under the input valuation v. The second element limiting the acceptable resolutions will be a linear ordering $\prec$ of variables. The idea is to limit the resolutions by insisting that the literal used for resolution (the one that is being eliminated) be largest in the ordering $\prec$ in the second (right) input. By insisting that the first input be true under v and the other false, we brake the symmetry of the resolution - we put a limitation on the inputs. This is a small limitation. The second one, namely the requirement that a predefined ordering $\prec$ must be consulted every time we want to resolve, is a strong limitation. This approach to limitation of resolution is due to Slagle [Sl67].

We start with a simple observation on satisfiability to see that the first part of our strategy makes sense. Since our goal is to have one of the inputs of the resolution rule false under the input valuation v, let us ask how big a limitation is it. Given a set of clauses F, split F into two classes F_1 and F_0 according to validity under F: F_1 consists of clauses in F true under v and F_0 of those false. Here is an observation.

PROPOSITION 7.11
If F is unsatisfiable, then both F_0 and F_1 are nonempty.

Proof: Clearly, if F is unsatisfiable F_0 must be nonempty (for otherwise $v \models F$.) If F_1 is empty, then consider valuation $\tilde{v}$ defined by $\tilde{v}(p) = 1 - v(p)$ for all p. Here v is the input valuation that is used to define F_0 and F_1. It is easy to see that $\tilde{v}$ must satisfy F. □

We will always eliminate tautological clauses (that is, ones that contain pairs of complementary literals). This clearly does not change satisfiability. Now, let us look at the orderings of variables. If $\prec$ is a linear ordering of variables, then, in each non-tautological clause C, $\prec$ orders the literals. How? Namely, if we assign to a literal l its underlying variable $|l|$, then the induced ordering of literals in the clause C is $l \prec_C m$ if $|l| \prec |m|$. Since C is non-tautological $\prec_C$ is well-defined.

Now, we will define the restricted resolution rule $Res_{v,\prec}(\cdot,\cdot)$. As indicated before this rule will not act on any pair of clauses; there will be limitations. We require that several things happen for an ordered pair of clauses $\langle D, E \rangle$ to apply the limited resolution rule:

(a) $v(D) = 1$.

(b) $v(E) = 0$.

(c) D, E are resolved on the literal largest in the ordering $\prec_E$.

If the result of this limited form of resolution is defined, we denote it by $res_{v,\prec}(D, E)$. This definition clearly breaks the symmetry of resolution and limits the results.

Before we move further, let us look at the left input of the restricted resolution $res_{v,\prec}$. Since the right input E is supposed to be false under v, all the literals in E are falsified by v. Thus their duals are true under v. But what about the result of the application of resolution, $Res_{v,\prec}(D, E)$? Here both things may happen; it may be either true or false under v. When the clause that is false under v is computed, we will not be able to use it as the left input of any application of $Res_{v,\prec}$ (but of course we will be able to use it as a right input).

Although we limited greatly the applicability of resolution, we still resolve, just that we accept fewer results. Let v and $\prec$, and a set of clauses F be fixed. There is a monotone operator $res_{F,v,\prec}$ such that the least fixpoint of $res_{F,v,\prec}$ is the least set of clauses containing F and closed under the limited resolution rule $Res_{v,\prec}$. Just for the completeness of our presentation here is the definition of $res_{F,v,\prec}$.

$$res_{F,v,\prec}(G) = \{C : C \in F \vee \exists_{D,E}(D \in F \cup G \wedge \\ E \in F \cup G \wedge v(D) = 1 \wedge v(E) = 0 \wedge C = Res_{v,\prec}(D,E))\}.$$

PROPOSITION 7.12
The operator $res_{F,v,\prec}$ is monotone. Thus it possesses a least fixpoint.

Clausal logic and resolution

The elements of the least fixpoint of $res_{F,v,\prec}$ can be characterized in a proof-theoretic way as possessing proofs (derivations) using resolution rule $Res_{v,\prec}$. In Figure 7.1 we see what this derivation looks like. The input R_0 belongs to F, and must be true under v. The inputs $C_0, \ldots, C_{j-1}$ are derived using the limited resolution rule $Res_{v,\prec}$. They are all false in v. The outputs $R_0, R_1, \ldots, R_{j-1}$ are true under v. The clauses R_i are $Res_{v,\prec}(R_{i-1}, C_{i-1})$, for $1 \leq i \leq j$. We will call $Res_{v,\prec}(F)$ the least fixpoint of the operator $res_{F,v,\prec}$.

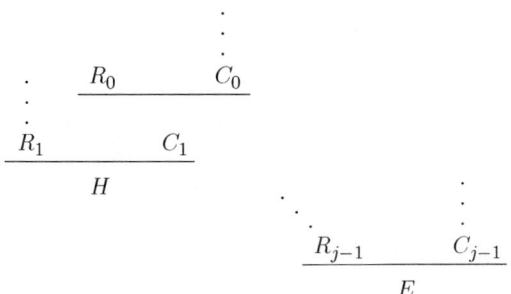

FIGURE 7.1: The derivation of a clause E using a limited resolution rule $res_{v,\prec}$

We are now ready to prove the main result of this section.

PROPOSITION 7.13
Let v be a valuation, $\prec$ a linear ordering of variables, and let F be a set of non-tautological clauses. Then, F is unsatisfiable if and only if $Res_{v,\prec}(F)$ contains the empty clause $\emptyset$.

Proof: Clearly, since $Res_{v,\prec}(F) \subseteq Res(F)$, if $Res_{v,\prec}(F)$ contains $\emptyset$ then so does $Res(F)$ and so F is unsatisfiable. By completeness theorem 7.6, if F is unsatisfiable then $Res(F)$ contains the empty clause. Thus all we need to show is that if $Res(F)$ contains the empty clause, then so does $Res_{v,\prec}(F)$. In other words, if there is a resolution proof of $\emptyset$ out of F, we want to find one that uses the restricted form of the resolution, $Res_{v,\prec}$, only. First, let us observe that the compactness theorem allows us to limit to the finite F. Indeed, if the set of clauses F is unsatisfiable, then there is a finite $G \subseteq F$ which is unsatisfiable and so, if we have the result proved for finite sets of clauses, $Res(G)$ contains $\emptyset$, but then $Res_{v,\prec}(F)$ is even bigger and so it also contains $\emptyset$. So we assume that F is finite. One advantage of this is that we can use induction on the size of Var_F.

The base case is really simple. If there is just one variable, say p, we can select $l \in \{p, \neg p\}$ which is false under v. Say it is p. Then, since F is inconsistent and there is just one variable, both p and $\neg p$ are in F and we just resolve with the unit clause p as the second input. This is a valid resolution derivation of $\emptyset$ using the limited resolution because p is largest in $\prec_{\{p\}}$.

So now assume $|Var_F| = n + 1$, and that our proposition is valid for all sets of clauses with the variable set of size at most n. Our proof will require manipulation of derivations such as the one presented in Figure 7.1.

There will be two cases in our proof.

Case 1: Let us assume that F contains a unit clause l such that $v(l) = 0$. Let G be the reduct of F by l, that is, the following set of clauses.

$$G := \{C \setminus \{\bar{l}\} : C \in F, l \notin C\}.$$

Can G be satisfiable? The set F is not satisfiable. Thus G is unsatisfiable because $G = F_l$ and by Lemma 7.4 F_l is unsatisfiable. Moreover, the variable $|l|$ does not occur in G. Thus inductive assumption applies, and there is a derivation of $\emptyset$ using just clauses of G, and only the restricted form of the resolution $Res_{v,\prec}$. It looks like the derivation in Figure 7.1, except that E is $\emptyset$.

We will now show that under the assumptions of Case 1, $Res_{v,\prec}(F_l) \subseteq Res_{v,\prec}(F)$. This fact is proved by induction on the height of derivation of a clause $C \in Res_{v,\prec}(F_l)$. Recall that each clause $C \in Res_{v,\prec}(F_l)$ has a derivation, that is, there is a binary tree whose leaves are labeled by clauses of F_l and internal nodes are valid applications of $res_{v,\prec}$ to labels of parents.

Basis: Let $C \in F_l$. If $C \in F$, then we are done. Otherwise $C \cup \{\bar{l}\}$ belongs to F. Since $v(l) = 0, v(\bar{l}) = 1, v \models C \cup \{\bar{l}\}$. Thus $C \cup \{\bar{l}\}$ can serve as a left input. Also, since $v(l) = 0$ and l is unique in $\{l\}$, $\{l\}$ can serve as right input. But then C has a derivation.

Inductive step. Assume $E = res_{v,\prec}(R,C)$, R, C belong to $Res_{v,\prec}(F_l)$. By inductive assumption both R and C belong to $Res_{v,\prec}(F)$. Let $\mathcal{D}_1, \mathcal{D}_2$ be derivations of R, C, respectively, using F. We now combine the derivations $\mathcal{D}_1, \mathcal{D}_2$ and $res_{v,\prec}(R,C)$ into a derivation $\mathcal{D}$ of E from F. Clearly this derivation uses only valid applications of $res_{v,\prec}$.

Once we proved the inclusion result, we apply it to our inductive assumption. Since $\emptyset \in Res_{v,\prec}(F_l), \emptyset \in Res_{v,\prec}(F)$. This completes the proof of Case 1.

Case 2: Now, we will deal with the second case, where there is no such unit false under F. This second case uses a "dirty trick." We essentially reduce to Case 1. There will be more proof transformations. This reduction to Case 1 implies that we cannot "short-cut" and do the Case 2 alone (which at the first glance looks more general than Case 1).

Let us select the $\prec$-least variable, say p. Then we select $l \in \{p, \neg p\}$ so that $v(l) = 0$. Once again, let us reduce F by a literal, in this case $\bar{l}$. That is, we have the set G of clauses as follows.

$$G := \{C \setminus \{l\} : C \in F, \bar{l} \notin C\}.$$

As before, because F is unsatisfiable, G is unsatisfiable (Lemma 7.4). Since G has fewer variables, there is a derivation of $\emptyset$ out of G using only $Res_{v,\prec}$. Now, let us restore each input from F to what it was before the reduction. We have to be careful in two ways. First, we will have to restore l not only to inputs, but also, as we resolve, to resolvents. What is more subtle here, we need to convince ourselves that the applications of the resolution were valid (for we allow *only* applications of $Res_{v,\prec}$). As we restore the presence of l in clauses, let us observe that now the arguments to resolution will have, possibly, l in them. But if such input was the second one (that is where we care) it will look like this: it will be either C_i or $C_i \vee l$. In either case this clause is false under v. But it is not the end of our troubles. For in the application of the rule $res_{v,\prec}$ we allowed the application of the resolution under $\prec$-largest of the literals. Fortunately, we selected l so it is in the bottom of $\prec$ – it will *never* interfere with the application of the $Res_{v,\prec}$ – what was largest is still largest when we brought l back. So we restored l, we resolved as we were resolving in the original derivation, and what do we get? Two cases are, actually, possible:

Subcase 2.1: Nothing was restored as we tried to put back l. In this case we already have a valid derivation of $\emptyset$ from F, and we are done.

Subcase 2.2: l has been restored in at least one place. Since only l was restored (and the clauses containing $\bar{l}$ are not present in the derivation, the resulting derivation derives l, not $\emptyset$. Thus we have a derivation (using $Res_{v,\prec}$) of l from F. Now, let us shift gears and look at a new set of clauses, $H := F \cup \{l\}$. This set H has at most $n+1$ variables. This may appear to be bad news. But H has a unit clause false under v! Moreover H is unsatisfiable because it is bigger than F and F is unsatisfiable, This throws us back to Case 1. We know that there is a derivation of $\emptyset$ from H, let us call it $\mathcal{D}_0$. But there is an unpleasant hitch. This derivation $\mathcal{D}_0$ allows us to use as inputs not only clauses from F, but also l. This is bad news, because we want a derivation using inputs from F alone. But l itself has a derivation from F alone. We will call that derivation of $\{l\}$ from F, $\mathcal{D}_1$. So now our goal is to combine $\mathcal{D}_0$ and $\mathcal{D}_1$ into a single derivation $\mathcal{D}$ with inputs from F alone (and using only the rule $res_{v,\prec}$). First, let us look at clauses C_i, $1 \leq i \leq j-1$, in the derivation $\mathcal{D}_0$. We can now modify their derivations in $F \cup \{l\}$ so that they involve only F. We modify the derivation $\mathcal{D}_0$ by substituting these derivations from F alone. So now we have handled all "left-side" inputs. But we are not finished, yet. Specifically, we need to handle the clause R_0. Either R_0 belongs to F and we do not have to do anything (because we now have a derivation of $\emptyset$ using inputs of F alone), or R_0 is l. In this final case we reuse the derivation $\mathcal{D}_1$, replacing l in $\mathcal{D}_0$ by this derivation $\mathcal{D}_1$, which is only from F (and uses only the rule $Res_{v,\prec}$). Thus we got our new derivation $\mathcal{D}$ of $\emptyset$ from F, using only $Res_{v,\prec}$. This completes the proof. □

Let us observe that if we wanted to limit resolution so that the first input is true under v, and the second false, then the completeness result still holds. The reason is that such form resolution computes more than $Res_{v,\prec}$, but less than Res. So if F is inconsistent, then such semantic resolution without the ordering still derives $\emptyset$. Similar argument shows that limiting the resolution to resolving on the largest literal in a specific ordering is also complete.

7.7 Autark sets and lean sets

We will now discuss an elegant result, due to O. Kullmann. This result provides a surprising connection between resolution and autarkies. The connection comes via the set of clauses touched by an autarky. Specifically, given a nonempty partial valuation v, and a set of clauses F, we associate with v and F the set of clauses in F which are touched by v. Formally we define

$$\alpha_F(v) = \{C \in F : Var_C \cap Var_v \neq \emptyset\}.$$

In other words, $\alpha_v(F)$ is the set of clauses in F that are touched by v. We then say that $F' \subseteq F$ is an *autark set with a witness* v if v is an autarky, and $F' = \alpha_F(v)$. An autark subset of F is one that has a witness. It follows from the fundamental properties of autarkies that autark subsets have a desirable property: if F' is an autark subset of F, then F is satisfiable if and only if $F \setminus F'$ is satisfiable. Next, we observe that the witness of an autark subset is not, in general, unique. One reason is that when F is satisfiable then F itself is its own autark subset and every satisfying valuation for F is a witness.

Let us recall the operation $\oplus$ on partial valuations. $v \oplus v' = v \cup \{l \in v'; \bar{l} \notin v\}$. We established that the set of autarkies is closed under the operation $\oplus$. Moreover, for all v, v', $v \preceq_k v \oplus v'$. An important property of the operation $\oplus$ is the following.

LEMMA 7.5
For all partial valuations v_1, v_2 and for all sets of clauses F,

$$\alpha_F(v_1 \oplus v_2) = \alpha_F(v_1) \cup \alpha_F(v_2).$$

Proof: We show inclusion $\subseteq$. First, let C be a clause, $C \in \alpha_F(v_1 \oplus v_2)$. Then, C is touched by $v_1 \oplus v_2$. If v_1 touches C then C belongs to the right-hand side. So let us assume that v_1 does not touch C. Then, no variable of v_1 occurs in C. But C is touched by $v_1 \oplus v_2$ so one of the variables of $\{l \in v_2 : \bar{l} \notin v_1\}$ touches C. In particular v_2 touches C and so C belongs to the right-hand side.

Conversely, let C belong to $\alpha_F(v_1) \cup \alpha_F(v_2)$. If $C \in \alpha_F(v_1)$ then C is touched by $v_1 \oplus v_2$. If $C \in \alpha_F(v_2) \setminus \alpha_F(v_1)$ then for some l, $|l| \in Var_{v_2} \setminus Var_{v_1}, l \in C$. But then $C \in \alpha_F(v_1 \oplus v_2)$, as desired. □

We then get the following fact.

COROLLARY 7.9
The collection of autark subsets of a CNF F is closed under finite unions.

Next, we have the following fact concerning autarkies.

LEMMA 7.6

Let F be a CNF. Let $\langle v_\alpha \rangle_{\alpha < \beta}$ be a chain of autarkies for F so that $\alpha_1 < \alpha_2$ implies $v_{\alpha_1} \preceq_k v_{\alpha_2}$. Then, the least upper bound of the chain $\langle v_\alpha \rangle_{\alpha < \beta}$ exists, and is an autarky for F.

Proof: The least upper bound of a $\preceq_k$-chain of partial valuations always exists. It is just the set theoretic union of that sequence of partial valuations (when we think about those as consistent sets of literals it is plainly obvious). So, let $v = \bigcup_{\alpha < \beta} v_\alpha$ with α limit (otherwise the conclusion is trivial). We claim that v is an autarky for F. Indeed, let v touch $C \in F$. Then, because C is finite, there is $\alpha < \beta$ such that v_α touches C. Then $(v_\alpha)_3(C) = 1$, and thus $v_3(C) = 1$, as $v_\alpha \preceq_k v$. □

Now, the Zorn lemma immediately entails the following corollary (which is obvious in finite case).

COROLLARY 7.10

If F is a CNF, then F possesses a maximal autarky.

Let us observe, however, that this autarky does not need to be non-empty. There are sets of clauses (we will see one below) with no nonempty autarkies.

PROPOSITION 7.14 (Kullmann)

Let F be a set of clauses. Then, F possesses a largest autark subset.

Proof: Let v be a maximal autarky for F and let $F' = \alpha_F(v)$. We claim that F' is largest autark. To this end, let F'' be any autark subset of F, and let v'' be its witness autarky, that is $F'' = \alpha_F(v'')$. Let us define

$$v' = v \oplus v''.$$

Now, since both v, v'' are autarkies and the collection of autarkies is closed under the operation $\oplus$, v' is an autarky. But v is a maximal autarky and $v \preceq_k v'$. Thus $v = v'$. That is, $v = v \oplus v''$. But then $\alpha_F(v) = \alpha_F(v) \cup \alpha_F(v'')$ (Lemma 7.5), that is $F' = F' \cup F''$, i.e., $F'' \subseteq F'$, as desired. □

We will denote the largest autark subset of F by A_F.

We recall that a set of clauses is *minimally unsatisfiable* if F is unsatisfiable but every proper subset $F' \subset F$ is satisfiable. Clearly, compactness theorem entails that minimally unsatisfiable sets of formulas are finite. Better yet, we have the following fact.

PROPOSITION 7.15

If a set of clauses G is minimally unsatisfiable, then every resolution derivation of $\emptyset$ out of G must involve all clauses of G.

Proof: If there is a resolution derivation $\mathcal{D}$ of $\emptyset$ such that some clauses of G do not occur as premises of $\mathcal{D}$, then let G' be the set of premises of $\mathcal{D}$. Then, G' is unsatisfiable, and as G' is a proper subset of G, G is not minimally unsatisfiable. $\square$

We now define P_F as the union of all minimally unsatisfiable subsets of F. By Proposition 7.15, P_F consists precisely of those clauses in F which are premises in some derivation of $\emptyset$ (different clauses may occur in different derivations). We will call a clause C belonging to F *plain* (in F) if C is a premise of some derivation of $\emptyset$ from F. We will call a subset of $F' \subseteq F$ *lean* if it consists of clauses plain in F. It is quite obvious that P_F is the largest lean subset of F.

We will now prove several lemmas and a corollary that will bring us closer to the fundamental result on the connection of autark subsets and resolution. First we have a fact of separate interest.

LEMMA 7.7 (Van Gelder)
If v is an autarky for F, then v is an autarky for $Res(F)$.

Proof: By induction on the height of derivation tree of a clause. If that height is 0, then the proof is obvious. So let us assume that D is the result of application of resolution to D_1, D_2, both $D_1, D_2 \in Res(F)$ and that we resolved on variable x. Moreover let us assume that v touches D.
We can assume that $D_1 = C_1 \vee x, x \notin C_1$, and that $D_2 = C_2 \vee \neg x, \neg x \notin C_2$.
Case 1: $x \in Dom(v)$. We assume $v(x) = 1$ (the case of $v(x) = 0$ is similar). Because of our case, v touches D_2. But $v(\neg x) = 0$ so for some literal $l \in C_2$, $v(l) = 1$. But then $l \in D$ and so $v_3(D) = 1$.
Case 2: $x \notin Dom(v)$. Since v touches D for some literal $l \in C_1 \vee C_2$, v touches l. Without the loss of generality, $l \in C_1$. Thus $l \in D_1$. Thus v touches D_1 and since D_1 has a shorter derivation, v evaluates D_1 as 1. Thus for some literal m in D_1, $v(m) = 1$. But $m \neq x, m \neq \neg x$ because x is not in the domain of v. Thus $m \in D$, and so $v_3(D) = 1$. $\square$

Next we have the following fact.

LEMMA 7.8
Let F be a collection of clauses and let v be an autarky for F. Let T be a resolution tree whose conclusion is a clause D. If v evaluates any of the premises of the tree T as 1, then v evaluates D as 1. That is $v_3(D) = 1$.

Proof: We proceed by induction on the height of the tree T, $ht(T)$. If $ht(T) = 0$, that is, T is a single node labeled with D, then the assertions are obviously true.
Hence let us assume that T is the effect of composing two trees (which are determined by parents of the root of the tree T), T_1 and T_2. Let us assume that the conclusion of T_1 is D_1 and the conclusion of T_2 is D_2. Let the variable on which we made the resolution of D_1 and D_2 be x. Without the loss of generality we can assume that for some clause C which is a premise of T_1, $v_3(C) = 1$. By inductive assumption we assume that $v_3(D_1) = 1$.

Again without the loss of generality we can assume that $x \in D_1$ (the case when $\neg x \in D_1$ is very similar). We need to consider two cases.
Case 1: $x \in v$. Then, v touches D_2 because $\neg x \in D_2$. Since v is an autarky for F, by Lemma 7.7, $v_3(D_2) = 1$. But then

$$v_3(D_2 \setminus \{\neg x\}) = 1.$$

But $D_2 \setminus \{\neg x\} \subseteq D$, so $v_3(D) = 1$.
Case 2: $x \notin v$. Then, since $v_3(D_1) = 1$, $v_3(D_1 \setminus \{x\}) = 1$, and as before $v_3(D) = 1$.
□

We then have the following corollary.

COROLLARY 7.11
Let A_F be the largest autark subset of F. Then $P_F \cap A_F = \emptyset$. In particular $A_F \subseteq F \setminus P_F$.

Proof: Let $C \in A_F$. Then, for a maximal autarky v of F, $v_3(C) = 1$. But then for every resolution derivation T where C serves as a premise, v evaluates the conclusion of T as 1, so that conclusion cannot be $\emptyset$.
□

Our next fact corresponds to the familiar property that contradictory theory proves everything. Here, in the clausal logic and using resolution, we have a similar property.

LEMMA 7.9
Let F be a set of clauses that form a premise set of a resolution derivation T of $\emptyset$. Let $V = Var_F$ and $L = Lit_V$. Then for every literal $l \in L$, there is a resolution derivation tree T_l with the following properties:

1. *All premises of T_l occur among the premises of T.*

2. *The conclusion of T_l is the unit clause $\{l\}$.*

Proof: First, let us observe that T cannot possess a pure literal m. For in such case $\{m\}$ is an autarky for F, and then, since m occurs in some premise of T, m must evaluate the conclusion of T as 1, contradicting Lemma 7.8. Now, T is a tree and we can associate with each node of T the distance from the root. Then, given a variable x we associate with x the number $d(x)$ which is the minimum of distances of nodes in which we resolve on x. Every variable that occurs in any premise of T must be eventually eliminated by an application of resolution. Thus, for every $x \in V$, $d(x)$ is defined. By induction on $d(x)$ we show that both units $\{x\}$ and $\{\neg x\}$ possess resolution derivations out of F. This is entirely obvious if $d(x) = 0$ because this means that x is the last variable on which we resolved, and the clauses in the parents of the root had to be unit clauses $\{x\}$ and $\{\neg x\}$. So now assume that the assertion is valid for variables y with $d(y) < d(x)$. There is a node n with the distance of n equal to $d(x)$ where x is a variable on which we resolve. The

parents of node n are labeled with $C_1 = l_1 \vee \ldots \vee l_k \vee x$, $C_2 = m_1 \vee \ldots \vee m_r \vee \neg x$. All literals $l_1, \ldots, l_k, m_1, \ldots, m_r$ are eventually resolved afterwards. Therefore $d(l_1) < d(x), \ldots, d(l_k) < d(x), d(m_1) < d(x), \ldots, d(m_r) < d(x)$. By inductive assumption there are derivations $T_{1,1} \ldots, T_{1,k}$ of the units $\{\bar{l}_i\} \ldots, \{\bar{l}_k\}$. Likewise there are derivations $T_{2,1} \ldots, T_{2,r}$ of the units $\{\bar{m}_i\} \ldots, \{\bar{m}_r\}$. Those derivations use only premises from F. Combining derivations $T_{1,1} \ldots, T_{1,k}$ with the derivation of $l_1 \vee \ldots \vee l_k \vee x$ we get a derivation of the unit $\{x\}$. Analogously we derive $\{\neg x\}$.
□

COROLLARY 7.12
Let $x \in Var_{P_F}$. Then, both $\{x\}, \{\neg x\}$ belong to $Res(P_F)$.

Proof: Let $x \in Var_{P_F}$. Then, there is a resolution derivation tree T such that the conclusion of T is $\emptyset$ and for some clause C among premises of T, x occurs positively or negatively in C. But all premises of T belong to P_F, and now the assertion follows from Lemma 7.9. □

We will now describe the technique (due to Kullmann) of *crossing out* variables from a clause. Given a clause C and a set of variables X, crossing out X from C results in a clause that does not mention variables from X or literals from $\bar{X}$. Formally, $C \star X = C \setminus (X \cup \bar{X})$. Likewise, $F \star X$ is $\{C \star X : C \in F\}$.

Here is a fundamental relationship of autark subsets and crossing out operation.

PROPOSITION 7.16
A subset $F' \subseteq F$ is autark if and only if for $X = Var_{F \setminus F'}$, $F' \star X$ is satisfiable.

Proof: First, let us assume that $F' \subseteq F$ is an autark subset of F and let v be a witnessing autarky. Thus F' consists of those clauses C in F which v touches (and thus satisfies). We define
$$F'' = F' \star X,$$
where $X = Var_{F \setminus F'}$. What kind of clauses are in F''? We crossed out, in all clauses of F', variables that occur outside of F'. But F' was an autark subset of F with a witness v. Therefore those variables in X cannot occur in the domain of v (for if they were, the corresponding clauses must be in F' as v is a witness for F'). Thus crossing out of those variables does not affect the evaluation by v. Consequently any extension of v to a complete valuation of variables of $F' \star X$ satisfies $F' \star X$.

Conversely, let us assume that for $X = Var_{F \setminus F'}$, $F' \star X$ is satisfiable. Let v be a satisfying valuation for $F' \star X$. The assignment v is defined on the set of variables $Var_{F' \star X}$. We claim that v is a witnessing autarky to the fact that F' is an autark subset of F. Thus two items need to be proved.

(a) v is an autarky for F.

(b) F' is the set of clauses touched by v.

Proof of (a). Let us assume that v touches $C \in F$. That is $Var_v \cap Var_C \neq \emptyset$. By definition, v is defined only on variables occurring in clauses of F', not outside. This means that the clause C contains some variables that do not occur outside of F'. When we cross out variables of X, C results in a clause $C' \in F' \star X$. Then, v satisfies C'. But then $v_3(C) = 1$ because $C' \subseteq C$.

Proof of (b). We need to show that F' is precisely the set of clauses touched by v. Two inclusions need to be shown.

(i) If C is touched by v then C must contain a variable which occurs only in F'. Thus C cannot be outside of F'.

(ii) Conversely, if $C \in F'$ then v satisfies the cross-out $C \star X$. Thus v touches $C \star X$, and so v touches C as well. □

Now, we have the following property of P_F.

LEMMA 7.10
Let $X = Var_{P_F}$. Then, for every clause C in F there is a resolution derivation T of $C \star X$ such that all premises of T are in $P_F \cup \{C\}$.

Proof: Let $C \in F$. Then, C can be written as $C_1 \vee C_2$ where C_1 consists of literals in Lit_{P_F}, and C_2 is $C \star X$. But for every literal l in C_1, there is a resolution derivation with premises in P_F that derives the unit $\{\bar{l}\}$ (Lemma 7.12). We can now construct a derivation of C_2 using C and clauses from P_F in the same way as we constructed a resolution derivation in the proof of Lemma 7.9. □

We are ready to prove the crucial fact that settles the form of the largest autark subset of F.

PROPOSITION 7.17
The set $F \setminus P_F$ is an autark subset of F.

Proof: Instead of looking for a witness, we use the characterization of autark subset given in Proposition 7.16. Let $F' = F \setminus P_F$. All we need to prove is that for $X = Var_{P_F}$, $F' \star X$ is satisfiable. Let us, then, assume that $F' \star X$ is unsatisfiable. Then, there exists a resolution derivation of $\emptyset$ that uses as premises clauses in $F' \star X$. This derivation T has in its premises clauses $D_1, \ldots, D_m$ from $F' \star X$, and $m \geq 1$. The clauses $D_1, \ldots D_m$ are, respectively, $C_1 \star X, \ldots, C_m \star X$ for some $C_1, \ldots, C_m$ in F'.

But now we recall that $X = Var_{P_F}$. Then, by Lemma 7.10, $P_F \cup \{C_i\}$ derives D_i, $1 \leq i \leq m$. Let T_i be a corresponding derivation of D_i. Now, combining derivations T_i, $1 \leq i \leq m$, and T we derive $\emptyset$ using clauses C_i, $1 \leq i \leq m$, and P_F. But then C_1 belongs to P_F and to F'. But $F' = F \setminus P_F$, a contradiction. □

Now, we have the following elegant corollary due to O. Kullman.

COROLLARY 7.13
Let F be a set of clauses. The largest autark subset A_F of F is $F \setminus P_F$.

Proof: We just proved (Proposition 7.17) that $F \setminus P_F$ is autark. Since all autark subsets of F are disjoint with P_F, $F \setminus P_F$ is the largest autark subset of F. □

The next corollary characterizes lean sets of clauses.

COROLLARY 7.14

Let F be a set of clauses. Then, F is lean if and only if F has no non-empty autarkies.

Finally we get the original result of O. Kullman in its two forms.

COROLLARY 7.15

Let F be a set of clauses. Then, there is a unique decomposition $F = F_1 \cup F_2$ where F_1 is autark, and F_2 is lean. In other words there is a unique decomposition $F = F_1 \cup F_2$ where F_2 is lean, and $F_1 \star Var_{F_2}$ is satisfiable.

While Corollary 7.15 is interesting it gives us no direct technique for finding autarkies. We will see in Chapter 8 that the problem of existence of non-empty autarkies is as difficult as the satisfiability problem itself.

We will now see that the largest autark set has an interesting relationship with maximal satisfiable subsets of sets of clauses.

Corollary 7.15 tells us that we can decompose an arbitrary set of clauses into A_F, the largest autark subset, and P_F, the largest lean subset of F. But we do not claim that P_F has no satisfiable subsets. There may be subsets G of P_F that do not contain any set of premises of a proof of an empty clause even though each of the elements of G belongs to one such set. We will see an example of such a situation below. With this in mind, we say that a set $G \subseteq F$ is *maximal satisfiable* if G is satisfiable, but for every H, $G \subset H \subseteq F$, H is unsatisfiable. We write MAXSAT_F to denote the family of maximal satisfiable subsets of F. Clearly, when F itself is satisfiable then the family MAXSAT_F consists of F only. But in general it may have many elements. We have the following fact showing that the largest autark subset of F approximates all maximal satisfiable subsets of F.

PROPOSITION 7.18

If F is a set of clauses, A_F its largest autark subset, and G any maximal satisfiable subset of F, then $A_F \subseteq G$.

Proof: We have $F = A_F \cup P_F$, $A_F \cap P_F = \emptyset$. Let G be a maximal satisfiable subset of F. Let $C \in F \setminus G$. Then $G \cup \{C\}$ is unsatisfiable. Then, there is a resolution tree T with premises from $G \cup \{C\}$ such that the conclusion of T is $\emptyset$. C must be among premises of T for otherwise G alone derives $\emptyset$ which contradicts satisfiability of G. Thus $C \in P_F$. Since C is arbitrary, $F \setminus G \subseteq P_F$. Thus $F \setminus P_F \subseteq G$, i.e., $A_F \subseteq G$, as desired. □

COROLLARY 7.16
Let F be a set of clauses and A_F its largest autark subset. Then, $A_F \subseteq \bigcap \text{MAXSAT}_F$.

In general $A_F \neq \bigcap \text{MAXSAT}_F$, as shown by the following example.

Example 7.1
Let F be the set composed of four clauses (two of those are units): $p, \neg p, p \vee r, \neg p \vee \neg r$. Clearly, F is unsatisfiable. Moreover, F has no nonempty autarkies. Indeed, let v be an autarky for F. Clearly, v cannot contain either p or $\neg p$. Now, if r belongs to v then because v touches $\neg p \vee \neg r$ it must be the case that $\neg p \in v$, a contradiction. A similar argument establishes that $\neg r \notin v$. Thus v cannot be nonempty. But since F has no nontrivial autarkies, its largest autark set is empty. On the other hand, it is easy to see that F has two maximal satisfiable subsets, $G_1 = \{p, p \vee r, \neg p \vee \neg r\}$, and $G_2 = \{\neg p, p \vee r, \neg p \vee \neg r\}$. Thus $\bigcap \text{MAXSAT}_F = \{p \vee r, \neg p \vee \neg r\}$ which is a proper extension of the largest autark subset of F. □

Our goal now is to show that from the point of view of computation of maximal satisfiable subsets of F, the set A_F is, in a sense, superfluous. That is, maximal satisfiable subsets of F are in one-to-one correspondence with maximal satisfiable subsets of $F \setminus A_F$, i.e., of P_F. Specifically we have the following fact.

PROPOSITION 7.19
Let F be a set of clauses and A_F, P_F its largest autark and lean subsets.

1. If G is a maximal satisfiable subset of F then $G \setminus A_F$ is a maximal satisfiable subset of P_F
2. If G is a maximal satisfiable subset of P_F then $A_F \cup G$ is a maximal satisfiable subset of F.

Proof: (1) Let G be a maximal satisfiable subset of F. Then, $G \setminus A_F$ is satisfiable. Clearly $G \setminus A_F \subseteq P_F$. We show that $G \setminus A_F$ is a maximal satisfiable subset of P_F. Let $C \in P_F \setminus (G \setminus A_F)$. All we need to show is that $(G \setminus A_F) \cup \{C\}$ is unsatisfiable. We do know that $G \cup \{C\}$ is unsatisfiable because $C \notin G$ and G is maximal satisfiable subset of F. Thus there is a resolution tree T with conclusion $\emptyset$ so that the premises of T are in $G \cup \{C\}$. Such a resolution tree must involve C as one of its premises because G is satisfiable. Moreover, by Lemma 7.8, none of its premises may belong to A_F. This is because A_F is an autark set and if any of the premises of T belongs to A_F, then for the witnessing autarky v, v evaluates the conclusion of T as 1, a contradiction. Thus all premises of T are in $(G \setminus A_F) \cup \{C\}$ and so $(G \setminus A_F) \cup \{C\}$ is unsatisfiable. As C was an arbitrary clause in $G \setminus A_F$ we are done.

(2) If G is a maximal satisfiable subset of P_F we claim that $G \cup A_F$ is a maximal satisfiable subset of F. First we show that $G \cup A_F$ is satisfiable. Otherwise there is

a resolution tree T with conclusion $\emptyset$ with premises in $A_F \cup G$. Since the premises of T must all belong to P_F, none of premises of T belongs to A_F. Thus all of them belong to G, contradicting the fact that G is satisfiable. Now, we show maximality. Let $C \notin G \cup A_F$. Then, $C \in P_F$. Since $C \notin G$, $G \cup \{C\}$ is unsatisfiable (because G was maximal). Thus $A_F \cup G \cup \{C\}$ is unsatisfiable. □

7.8 Exercises

1. (For enthusiasts of combinatorics) Prove Corollary 7.2 for infinite graphs.
2. Complete the argument of Proposition 7.4.
3. Prove the continuity of the operator Res, that is, for an increasing sequence $\langle F_n \rangle_{n \in N}$,
$$Res(\bigcup_{n \in N} F_n) = \bigcup_{n \in N} Res(F_n).$$
4. Show that the equality in (3) does not have to hold for sequences $\langle F_n \rangle_{n \in N}$ which are not increasing.
5. Prove that the resolution is sound.
6. Clear up in your mind the issue of the sets of literals as conjunctions and sets of literals as clauses (disjunctions). For that purpose think about *hit sets*. Let X be a set, and $\mathcal{X}$ be a collection of subsets of X. A set $Y \subseteq X$ is a *hit set* for $\mathcal{X}$ if for all $Z \in \mathcal{X}$, $X \cap Z \neq \emptyset$. Then, it is possible to characterize satisfying valuations as hit sets. How?
7. If a family $\mathcal{X}$ contains the empty set, then $\mathcal{X}$ has no hit set. What does this statement tell us?
8. Given a valuation v, we define a valuation $\bar{v}$ by setting $\bar{v}(p) = 1 + v(p)$, where $+$ is the Boolean addition, for all variables p. We saw $\bar{v}$ under various names in this book already. Let C be a nonempty clause, and v an arbitrary valuation. Prove that $v \models C$, or $\bar{v} \models C$.
9. Prove that if F is an unsatisfiable set of nonempty clauses then for every valuation v there are two clauses C_1 and C_2, both in F such that $v \models C_1$, and $v \not\models C_2$.
10. Let F be an unsatisfiable set of nonempty clauses. Prove the following stronger version of the previous problem. If v is any valuation then there are clauses C_1 and C_2 in F such that $v \not\models C_2$ and v satisfies *all* literals in C_1. In other words, v satisfies *no* literal in C_2, but all literals in C_1.
11. Conclude that sets of clauses consisting of exactly one nonempty clause are satisfiable (which is obvious, anyway).

Chapter 8

Testing satisfiability, finding satisfying assignment

8.1   Table method 133
8.2   Hintikka sets 135
8.3   Tableaux 137
8.4   Davis-Putnam algorithm 144
8.5   Boolean constraint propagation 154
8.6   The DPLL algorithm 158
8.7   Improvements to DPLL? 161
8.8   Reduction of the search SAT to decision SAT 162
8.9   Exercises 163

In this chapter we investigate a number of algorithms for testing satisfiability. We will study four: (a) table method, (b) tableaux, (c) Davis-Putnam algorithm, and (d) Davis-Putnam-Logemann-Loveland algorithm. The first two relate to arbitrary formulas. The other two deal with satisfiability of CNFs. We will also discuss possible improvements to DPLL algorithm and, finally, reduction of search-version of satisfiability problem to repeated application of decision-version of satisfiability. We observe that there are other technologies for testing satisfiability, for instance decision trees and their variation: binary decision diagrams, but we will not study them in this book

8.1   Table method

The table method is based on recursive evaluation of formulas given a valuation. The term "table" relates to customary representation of the problem, as a table. To start, recall that a valuation of the set Var satisfies a formula φ if and only if the partial valuation $v\mid_{Var_\varphi}$ satisfies φ. This means that if we want to test satisfiability of φ then all we need to do is to traverse all valuations of $Var(\varphi)$ in some order, evaluate φ under all these valuations and see if we generate at least once 1 as the value. If it is the case, φ is satisfiable. One implementation of this technique is the table where rows consist of valuations v appended by value of φ at v, $v(\varphi)$. Thus the table would have 2^n rows, where $n = \mid Var(\varphi)\mid$ and $n+1$ columns – first n for variables, the last one for value of φ.

When this method is used, we use the structure of the formula to simplify the process of computation. Recall that formula is nothing but an infix representation of a tree. The leaves of that tree are labeled with variables; the internal nodes are labeled with functors. There may be leaves labeled with the same variables. Fixing the order of children of each node in the tree, and eliminating duplicate nodes, there is a unique order of subformulas of a given formula where all subformulas precede the formula itself. Then the values in a column are computed by looking at truth tables of the functors and consulting columns to the left of the current column. Here is an example.

Example 8.1
Let us consider the formula

$$\varphi = \neg(((p \wedge \neg r) \vee (\neg q \wedge r)).$$

The tree of the formula φ is shown in Figure 8.1. Next, let us look at the

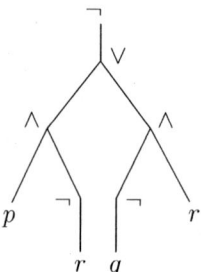

FIGURE 8.1: The tree of formula φ

table of the formula φ presented in Figure 8.1. Clearly, φ is satisfiable and we found several valuations satisfying φ, since there are several entries 1 in the last column of Table 8.1. Table 8.1 allows us also to establish that φ is not a tautology, because there are valuations that do not satisfy φ. □

This is a general observation. At a significant cost; the table size is $2^n \cdot (n+1)$ (not counting the columns corresponding to non-root internal nodes of the tree) we can test satisfiability of φ, unsatisfiability of φ, satisfiability of $\neg \varphi$ and whether φ is a tautology, all at once. Indeed, if the last column contains at least one entry 1, φ is satisfiable. Similarly, if the last column has at least one occurrence of 0 then $\neg \varphi$ is satisfiable. If the entries in the last column are all zeros then φ is not satisfiable, and if all entries in the last column are ones, then φ is a tautology.

The issue that needs to be mentioned is the order in which the valuations of Var_φ are listed. This is done by using some systematic way of listing valuations. Such

TABLE 8.1: The table for formula φ

p	r	q	$\neg r$	$\neg q$	$p \wedge \neg r$	$\neg q \wedge r$	$(p \wedge \neg r) \vee (\neg q \wedge r)$	φ
0	0	0	1	1	0	0	0	1
0	0	1	1	0	0	0	0	1
0	1	0	0	1	0	1	1	0
0	1	1	0	0	0	0	0	1
1	0	0	1	1	1	0	1	0
1	0	1	1	0	1	0	1	0
1	1	0	0	1	0	1	1	0
1	1	1	0	0	0	0	0	1

an arrangement is called Gray code, and there are numerous such listings. We will discuss in Section 8.9 Exercises construction of one particular Gray code.

Let us observe that to test if a formula is satisfiable using table method does not require exponential space (in fact the size of the table is superexponential). Exponential time is what is needed, but only linear space. Indeed we can generate valuations one by one (as a Gray code, say, or as full binary representations of integers in the range $(0..2^n - 1)$) and for each of these strings v test if $v \models \varphi$. As soon as we find such v, we report it, and quit. Similarly, we can test if φ is a tautology, except that here we leave with failure once we find v such that $v \not\models \varphi$.

8.2 Hintikka sets

This section prepares us for the technique in automated theorem proving called *tableaux*. When in the next section we define and construct tableaux, we will need to know that open branches of those tableaux represent satisfiable theories. For that purpose we will see that those open branches have certain desirable properties. These properties are abstracted into the Hintikka sets that we will discuss now.

A set $\mathcal{H}$ of formulas is called *Hintikka set* (after the Finnish-American logician J. Hintikka), if

(H1) $\bot \notin \mathcal{H}$

(H2) For each variable $q \in Var$, $q \notin \mathcal{H}$ or $\neg q \notin \mathcal{H}$

(H3) Whenever $\neg\neg\varphi \in \mathcal{H}$, $\varphi \in \mathcal{H}$

(H4) Whenever $\varphi_1 \vee \varphi_2 \in \mathcal{H}$, $\varphi_1 \in \mathcal{H}$ or $\varphi_2 \in \mathcal{H}$

(H5) Whenever $\neg(\varphi_1 \vee \varphi_2) \in \mathcal{H}$, $\neg\varphi_1 \in \mathcal{H}$ and $\neg\varphi_2 \in \mathcal{H}$

(H6) Whenever $\varphi_1 \wedge \varphi_2 \in \mathcal{H}$, $\varphi_1 \in \mathcal{H}$ and $\varphi_2 \in \mathcal{H}$

(H7) Whenever $\neg(\varphi_1 \wedge \varphi_2) \in \mathcal{H}$, $\neg\varphi_1 \in \mathcal{H}$ or $\neg\varphi_2 \in \mathcal{H}$

The conditions (H1) and (H2) say that there is no *obvious* contradiction in $\mathcal{H}$. (H3)–(H7) are closure properties, forcing simpler formulas in $\mathcal{H}$ whenever more complex formulas are there.

PROPOSITION 8.1 (Hintikka Theorem)
Every Hintikka set $\mathcal{H}$ is satisfiable.

Proof: Given a Hintikka set $\mathcal{H}$, we define $M_\mathcal{H} = \{p \in Var : p \in \mathcal{H}\}$. Now we define a valuation $v_\mathcal{H}$ as follows:

$$v_\mathcal{H}(p) = \begin{cases} 1 & \text{if } p \in M_\mathcal{H} \\ 0 & \text{otherwise} \end{cases}$$

In effect, we inspect the set $\mathcal{H}$ for the variables present in $\mathcal{H}$ and then we perform what "deductive database" researchers call *closed world assumption* on that set $M_\mathcal{H}$, declaring everything else in *Var* false. We will study closed world assumption in Chapter 13. Before we go any further, let us observe that the condition (H2) implies that v we defined above is, indeed, a valuation.

Now, we show the assertion by induction on the rank of formula φ.

Inductive Basis. Here $rk(\varphi) = 1$. There are three subcases. First, when $\varphi = \bot$, we have $v \not\models \bot$, but $\bot \notin \mathcal{H}$, thus the desired property holds. The second case is when $\varphi = \top$. But $v \models \top$ so the desired property holds, regardless of whether $\top$ belongs to $\mathcal{H}$ or not. Finally, we consider the case when $\varphi = p$, for some variable $p \in Var$. Let us look at the definition of v above. Whenever $p \in M_\mathcal{H}$, $v(p) = 1$. But $p \in M_\mathcal{H}$ precisely when $p \in \mathcal{H}$. Thus if $p \in \mathcal{H}$ then $v(p) = 1$, i.e., $v \models p$, as desired.

Inductive step: We will have to consider three cases, corresponding to the top connective of φ.

Case 1: $\varphi = \neg \psi$. This case requires five subcases.

Subcase 1.1: ψ is constant. We leave this case to the reader.

Subcase 1.2: ψ is p for some variable p. That is $\varphi = \neg p$, for $p \in Var$. Since for no $q \in Var$, $q \in \mathcal{H}$ and $\neg q \in \mathcal{H}$, we have $p \notin \mathcal{H}$ since $\neg p \in \mathcal{H}$. Thus $v(p) = 0$, $v \not\models p$, i.e., $v \models \neg p$, as desired.

Subcase 1.3: ψ is $\psi_1 \vee \psi_2$, that is $\varphi = \neg(\psi_1 \vee \psi_2)$. Since $\mathcal{H}$ is a Hintikka set, by condition (H5), $\neg \psi_1 \in \mathcal{H}$, and $\neg \psi_2 \in \mathcal{H}$. But then, each of formulas $\neg \psi_1$, and $\neg \psi_1$ is of smaller rank than $\neg(\psi_1 \vee \psi_2)$ (this actually needs checking, we hope the reader does it!). Now, by inductive assumption, $v \models \neg \psi_1$ and $v \models \neg \psi_2$, so $v \not\models \psi_1$ and $v \not\models \psi_2$, so $v \not\models \psi_1 \vee \psi_2$, so $v \models \neg(\psi_1 \vee \psi_2)$, that is $v \models \varphi$.

Subcase 1.4: ψ is $\psi_1 \wedge \psi_2$, that is $\varphi = \neg(\psi_1 \wedge \psi_2)$. Since $\mathcal{H}$ is a Hintikka set, by condition (H7), $\neg \psi_1 \in \mathcal{H}$, or $\neg \psi_2 \in \mathcal{H}$. Without loss of generality we can assume $\neg \psi_1 \in \mathcal{H}$. Now, by inductive assumption, $v \models \neg \psi_1$, so $v \not\models \psi_1$, so $v \not\models \psi_1 \wedge \psi_2$, so $v \models \neg(\psi_1 \wedge \phi_2)$.

Subcase 1.5: ψ is $\neg \vartheta$. Since $\varphi \in \mathcal{H}$, and $\varphi = \neg\neg\vartheta$, $\vartheta \in \mathcal{H}$. But ϑ is of smaller rank. thus $v \models \vartheta$, $v \not\models \neg\vartheta$, $v \not\models \psi$, thus $v \models \varphi$, as desired.

Case 2: $\varphi = \psi \vee \vartheta$. Then, as $\varphi \in \mathcal{H}$, by condition (H4), $\psi \in \mathcal{H}$ or $\vartheta \in \mathcal{H}$. Without loss of generality we can assume $\psi \in \mathcal{H}$. Then, as the rank of ψ is smaller than the

rank of φ, by inductive assumption, $v \models \psi$. But then $v \models \psi \vee \vartheta$, i.e., $v \models \varphi$.
Case 3: $\varphi = \psi \wedge \vartheta$. Then, as $\varphi \in \mathcal{H}$, by condition (H6), $\psi \in \mathcal{H}$ and $\vartheta \in \mathcal{H}$. But now by inductive assumption $v \models \psi$ and $v \models \vartheta$. But then $v \models \varphi$, as desired.
This completes the argument. □

8.3 Tableaux

The technique of tableaux is based on the analysis of the tree of the formula. Tableaux may be viewed as tests for both satisfiability and provability. Here we focus on satisfiability. It will turn out that certain tableaux (called *open finished tableaux*) witness to satisfiability. But then, if we wanted to see if ψ entails φ then all we need is to test if there is a finished, open tableau for $\psi \wedge \neg\varphi$. Indeed, because if such tableau exists, then ψ does not entail φ. Otherwise the tableau is closed, and ψ entails φ.

The idea of tableaux is that as we analyze the formula, we build a tree, where the nodes are labeled with *signed* formulas. Those signed formulas are of the form $T(\varphi)$ with the intended meaning "φ is true" and of the form $F(\varphi)$ with the intended meaning "φ is false." The point here is that once we label a node of a tableau with some formula, it forces us to *expand* tableau and label the subsequent nodes with other signed formulas. Here is an obvious example. If we label some node with a label $F(\neg\varphi)$ then a child of that node should be labeled with $T(\varphi)$ because if we assert that $\neg\varphi$ is false then we must accept that φ is true. There are 6 tableaux expansion rules. Two of those require prolongation of tableau with a single child; those correspond to negation. Then there are two rules that require prolongation of tableaux with two nodes as a child and a grandchild of a node (those correspond to the true conjunction and false disjunction cases), and finally there are two expansion rules that require splitting. This happens when we develop a node consisting of false conjunction or of true disjunction.

Formally, a tableau is a labeled binary tree that obeys tableaux expansion rules. These rules, stated in Figures 8.2, 8.3, and 8.4, tell us how the labels of descendants of a node depend on the label of a node. Specifically, the expansion rule for negation requires that whenever a node n has a label $T(\neg\varphi)$ then every branch passing through n must have a node n' with label $F(\varphi)$. A similar requirement is made on label $F(\neg\varphi)$. Similar requirements are made for true-conjunction and false-disjunction (see Figure 8.3). A slightly different requirement is made with rules for false-conjunction and true-disjunction (Figure 8.4, where the tree splits). These requirements can be turned into a formal inductive definition of the concept of the tableau. We spare the reader such rigorous definition.

The point of building a tableau is that when a branch $\mathcal{B}$ through a tableau $\mathcal{T}$ is finished (i.e., every formula on it has been expanded) and we did not establish a contradiction,

then the set of formulas

$$\{\varphi : T(\varphi) \text{ is a label of a node on } \mathcal{B}\} \cup \{\neg\varphi : F(\varphi) \text{ is a label of a node on } \mathcal{B}\}$$

is a Hintikka set, thus it is satisfiable.

The rules operate *not* on the tableau in its entirety, but rather on *branches* of the tableau. This means that when dealing with tableaux we must do extensive bookkeeping. We must maintain the set of open branches of the tableau, and the rules are applied to *branches*. Thus when a formula labels a node n in the tableau $\mathcal{T}$, the applicable tableaux expansion rule must be applied *on all open branches* of $\mathcal{T}$ that pass through n. One thing missing in this picture is when do we close a branch. Here the idea is very natural: if we see that a branch has two nodes, one labeled with $T(\vartheta)$ and another labeled with $F(\vartheta)$, then there cannot be a valuation that makes both ϑ and $\neg\vartheta$ true, and so we need to close that branch. It will never extend to a Hintikka set.

All the remarks above tell us that in order to test satisfiability of a formula φ, we have to start a tableau, and put $T(\varphi)$ as the label of the root. Whenever we expand the nodes, we always get nodes labeled by simpler formulas. Therefore the process must, eventually, halt. That is, as we expand branches we add nodes labeled with simpler formulas and so, eventually, a branch either becomes closed (because we have two nodes labeled with contradictory labels), or it becomes finished, because expansion rules have been applied to all formulas on that branch. Moreover, it turns out that when a formula is on a branch, then either that formula is a propositional variable or there is exactly one expansion rule that applies.

We will look at three groups of expansion rules. First, we have rules for negation; one for signed true, another for signed false.

FIGURE 8.2: Expansion rules for negation

Next, we show expansion rules for false-disjunction and true-conjunction. Those will be presented in Figure 8.3.

Finally, we show expansion rules for true-disjunction and false-conjunction. Those are depicted in Figure 8.4.

Thus, for instance, if we have an unfinished tableau $\mathcal{T}$ and we have an unexpanded node n on it marked $T(\neg\varphi)$ then we do two things. First, we extend *every* open (i.e., non-closed) branch passing through n with a new node which we label $F(\varphi)$. Second, we mark the node n as expanded. Likewise, if we have an unexpanded

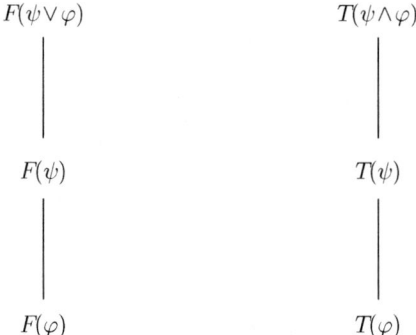

FIGURE 8.3: Expansion rules of true-conjunction and false-disjunction

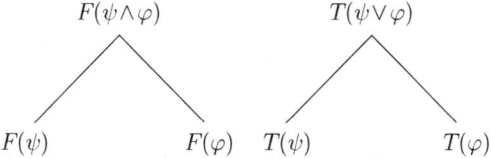

FIGURE 8.4: Expansion rules of false-conjunction and true-disjunction

node n' labeled with $T(\psi \vee \varphi)$, then *every* non-closed branch passing through n' is extended by two, incompatible, nodes. One is labeled with $T(\psi)$ and the other with $T(\varphi)$. Then the node n' is marked as expanded. We hope the reader sees what happens with the remaining cases. As mentioned above, each time we expand the tableau, the labels become simpler. This means that, eventually, all nodes will be expanded, and the tableau will become *finished*. Formally we get the following.

PROPOSITION 8.2
If a tableau T is started by putting in the root a signed formula $l(\varphi)$ (where l is T or F) and applying the tableau expansion rules on all branches, then the tableau T will eventually be finished, and will be of depth at most $2k$ where k is the depth of the tree of the formula φ.

Now, we will build a tableau where we start with a label $T(\varphi)$ in the root of the tableau. Here φ is the formula of our Example 8.1. This tableau is presented in Figure 8.5. Let us observe that the right-most branch is closed.

Before we continue, let us observe that different branches of a completed tableau may share satisfying valuations.

The following should now be clear. Given an open (i.e., non-closed) branch $\mathcal{B}$ through tableau T, we define $S_\mathcal{B}$ as the following set of formulas:

$$S_\mathcal{B} = \{\vartheta : T(\vartheta) \text{ is a label of a node on } \mathcal{B}\} \cup \{\neg\vartheta : F(\vartheta) \text{ is a label of a node on } \mathcal{B}\}.$$

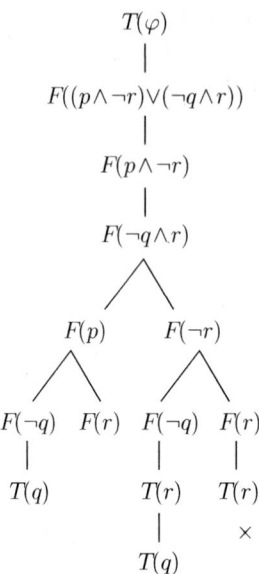

FIGURE 8.5:   The finished tableau for $T(\varphi)$ showing satisfiability of φ

PROPOSITION 8.3
If $\mathcal{B}$ is an open branch in a finished tableau $\mathcal{T}$, then $S_\mathcal{B}$ is a Hintikka set.

Proof: Looking at the conditions for Hintikka sets, we see that since branch $\mathcal{B}$ is open, for no variable p, p and $\neg p$ can belong to $S_\mathcal{B}$. Next, let us assume that the set $S_\mathcal{B}$ contains the formula $\neg\neg\varphi$. Two cases are possible. When $T(\neg\neg\varphi)$ is a label of a node on $\mathcal{B}$, then $F(\neg\varphi)$ is a label of another node on $\mathcal{B}$ (by the first of negation expansion rules), and then $T(\varphi)$ is a label of yet another node on $\mathcal{B}$ (by the second of negation expansion rules). In the other case, when $F(\neg\varphi)$ is the reason why $\neg\neg\varphi$ is in $S_\mathcal{B}$, we just use the second negation expansion rule.
Next, if $\psi \wedge \varphi$ belongs to $S_\mathcal{B}$, then it must be the case that $T(\psi \wedge \varphi)$ is a label of the node on $\mathcal{B}$. Since $\mathcal{B}$ is not closed and $\mathcal{T}$ finished, it must be the case that both $T(\psi)$ and $T(\varphi)$ are labels of nodes on $\mathcal{B}$ and so $\psi \in S_\mathcal{B}$ and $\varphi \in S_\mathcal{B}$. Finally, whenever $\psi \vee \varphi$ belongs to $S_\mathcal{B}$ then by application of one of true-disjunction, or of false-conjunction rule, $\psi \in S_\mathcal{B}$ or $\varphi \in S_\mathcal{B}$, whenever the branch $\mathcal{B}$ is not closed and $\mathcal{T}$ is finished. □

Now, let us look at the way we are going to apply Proposition 8.3. Assume we are given a set of formulas S and we want to establish that S is satisfiable. We start a tableau $\mathcal{T}_S$ by initially putting in the root $T(\bigwedge\{\varphi : \varphi \in S\})$. Then, after repeated application of the true-conjunction rule, all signed formulas $T(\varphi)$, $\varphi \in S$ will be on every open branch when $\mathcal{T}_S$ is finished. If we find such an open branch $\mathcal{B}$ when $\mathcal{T}_S$ is finished, $S \subseteq S_\mathcal{B}$, so, since $S_\mathcal{B}$ is a Hintikka set, S is satisfiable. Let us call $\mathcal{T}_S$ a *canonical tableau* for S. We then have the following.

COROLLARY 8.1
If T_S has an open branch, then S is satisfiable.

But is the tableaux method complete? In fact it is. Here is the argument. Assume that v is a satisfying valuation for S. We say that a node n of T *agrees* with v if the label of n is $T(\varphi)$ and $v \models \varphi$ or if the label of n is $F(\varphi)$, and $v \models \neg\varphi$. We say that a branch $\mathcal{B}$ agrees with v if the labels of all nodes on $\mathcal{B}$ agree with v. Here we do not stipulate that the tableau T with the branch $\mathcal{B}$ is finished.

Now, let us see what happens. Whenever we have a branch $\mathcal{B}$ that agrees with a valuation v, then $\mathcal{B}$ cannot be closed. So now assume that the branch $\mathcal{B}$ agrees with a valuation v, and that there are nodes on $\mathcal{B}$ that are still marked as undeveloped. Then we select the first such node. It is now clear that as we extend the branch, if we apply one of the negation rules, or true-conjunction or false-disjunction rule, then the resulting branch consists of nodes with labels that agree with v. So we are left with the final case when either the false-conjunction rule or the true-disjunction rule has been applied. But then one of the new branches $\mathcal{B}_i$, $i = 1, 2$, just constructed, has the property that all the nodes on $\mathcal{B}_i$ agree with v. Thus when T_S is finished, there will be an open branch, namely the leftmost one agreeing with v. Thus we proved the following fact.

PROPOSITION 8.4
A set of formulas S is satisfiable if and only if the canonical finished tableau T_S has an open branch.

Finally, let us see that we can use the tableaux technique for testing entailment. This is related to the fact that $S \models \varphi$ is equivalent to the fact that $S \cup \{\neg\varphi\}$ is unsatisfiable. If we attempt to test satisfiability of $S \cup \{\neg\varphi\}$, we put initially on the tableau the nodes labeled $T(\psi)$ for $\psi \in S$ and also $T(\neg\varphi)$. But this really amount of starting our tableau with nodes labeled $T(\psi)$ for $\psi \in S$ and also $F(\varphi)$ (forgetting about $T(\neg\varphi)$, that is). When we do this, develop the tableau until it is finished then, by Proposition 8.4, we get a closed tableau (i.e., one with all branches closed) precisely when $S \models \varphi$. With a slight abuse of notation, let us call this last tableau $T_{S \cup \{\neg\varphi\}}$. Thus we get the following result.

PROPOSITION 8.5
Let S be a set of formulas and φ be a formula. Then $S \models \varphi$ if and only if the canonical finished tableau $T_{S \cup \{\neg\varphi\}}$ is closed.

Tableaux allow for another, quite natural proof of compactness for denumerable theories in propositional logic. Here is a rough draft; we hope the reader fills in the details. Let F be a denumerable infinite set of formulas, $F = \langle \varphi_n \rangle_{n \in N}$. We build a tableau for F as follows. In even steps $2n$ we put on the tableau the formula φ_n. In the odd steps we select on each not-yet-closed branch the first undeveloped formula and apply the appropriate tableaux expansion rule. Because all formulas are finite, every formula will be, eventually, developed on each non-closed branch. Then, like in the proof of completeness for tableaux, we see that the set of formulas on a non-closed (i.e., open) branch generates a Hintikka set. Then such set is satisfiable. So, if F is not satisfiable there cannot be an infinite branch in our tableau and so the tableau itself must be finite! But a finite tableau mentions only a finite number of formulas in F, Thus some finite subset of F is unsatisfiable. Let us observe that the tableau we outlined is not a systematic tableau, but this should not be a problem for the reader.

We conclude this section discussing the connections between tableaux and three-valued logic. Autarkies will also appear in this context. So let a tableau $\mathcal{T}$ be open (i.e., non-closed) but finished. This means that $\mathcal{T}$ has at least one open branch and all signed formulas on that branch are developed. Let $\mathcal{B}$ be such a finished non-closed branch. We assign to $\mathcal{B}$ a partial assignment $v_\mathcal{B}$ as follows:

$$v_\mathcal{B}(p) = \begin{cases} 1 & T(p) \text{ is a label of a node on } \mathcal{B} \\ 0 & F(p) \text{ is a label of a node on } \mathcal{B} \\ u & \text{otherwise.} \end{cases}$$

Here p ranges over Var. The following fact ties the partial assignment $v_\mathcal{B}$ and three-valued evaluation of formulas that are labels of $\mathcal{B}$.

PROPOSITION 8.6
Let S be a set of formulas, $\mathcal{T}$ a tableau for S, and let $\mathcal{B}$ be a finished open branch of $\mathcal{T}$. Let $v = v_\mathcal{B}$ be a partial valuation determined by $\mathcal{B}$ and described above. Then, whenever $T(\varphi)$ is a label of a node of $\mathcal{B}$, $v_3(\varphi) = 1$, and whenever $F(\varphi)$ is a label of a node of $\mathcal{B}$ then $v_3(\varphi) = 0$.

Proof: We prove our assertion by induction on complexity of the formula φ. The case of atomic formulas is obvious; this is how we defined the valuation $v_\mathcal{B}$. So now, let us assume that for all formulas of rank smaller than the rank of φ our assertion is true. Let us assume that $T(\varphi)$ is a label of a node on $\mathcal{B}$. If φ is $\neg\psi$ then, as $\mathcal{B}$ is finished, $F(\psi)$ is a label of a node on $\mathcal{B}$. By inductive assumption, $v_3(\psi) = 0$. But then $v_3(\varphi) = 1$, as desired. Next, let us assume that φ is $\psi_1 \vee \psi_2$. Then again, either $T(\psi_1)$ is on $\mathcal{B}$, or $T(\psi_2)$ is on $\mathcal{B}$. By inductive assumption either $v_3(\psi_1) = 1$ or $v_3(\psi_2) = 1$. But then $v_3(\varphi) = 1$. If φ is $\psi_1 \wedge \psi_2$, then again both $T(\psi_1)$ and $T(\psi_2)$ are labels of nodes on $\mathcal{B}$, thus $v_3(\psi_1) = 1$ and $v_3(\psi_2) = 1$. Thus $v_3(\psi_1 \wedge \psi_2) = 1$, i.e., $v_3(\varphi) = 1$, as desired.

The case of the formulas of the form $F(\varphi)$ is similar and we leave it as an exercise for the interested reader. □

Once we proved Proposition 8.6, we get an alternative proof of Corollary 8.1. Indeed, if $\mathcal{B}$ is open in a finished tableau $\mathcal{T}$ for S, then whenever $T(\varphi)$ is a label of a node of $\mathcal{T}$, then $(v_\mathcal{B})_3(\varphi) = 1$. But for every $\varphi \in S$, $T(\varphi)$ is on $\mathcal{B}$. Thus $v_\mathcal{B}$ evaluates all formulas of S as 1. But then every completion of $v_\mathcal{B}$ to a complete valuation of variables in Var is a model of S. But there are such completions, so S is satisfiable.

Next, we have the following property:

PROPOSITION 8.7
The partial valuation v_B touches every formula ψ such that $T(\psi)$ is a label of a node on B or $F(\psi)$ is a label of a node on B.

Proof: By induction on complexity of ψ. We leave the details to the reader. □
Even though v_B touches all formulas ψ such that either $T(\psi)$ or $F(\psi)$ is on B, it does not assign the value 1 to all such formulas, but only to those that occur on B with the sign T.

PROPOSITION 8.8
Let S be a set of formulas, and let T be a finished tableau for T. Let B be an open branch in T. Then v_B is an autarky for S and $(v_B)_3(S) = 1$.

Proof: The partial assignment v_B touches every formula ψ such that $T(\psi)$ is on B or $F(\psi)$ is on B. Whenever $\varphi \in S$, $T(\varphi)$ is on B. But then v_B touches φ and $(v_B)_3(\varphi) = 1$. □

Now, we can completely characterize autarkies that evaluate the entire set S as 1 by means of tableaux.

In the proof of the next proposition it will be convenient to think about partial valuations as consistent sets of literals. Then the $\preceq_k$ relation is just the inclusion relation.

PROPOSITION 8.9
Let S be a set of formulas, and let T be a finished open tableau for S. Let v be a partial valuation of $Var(S)$. Then $v_3(S) = 1$ if and only if for some open branch B of T, $v_B \preceq_k v$. Consequently autarkies for S that evaluate the entire S as 1 are precisely consistent supersets of sets of the form v_B.

Proof: First, let us assume that a partial valuation v contains a partial valuation of the form v_B for some open branch B. Then, as we have $(v_B)_3(S) = 1$, $v_3(F) = 1$. Thus v touches and satisfies all formulas in S. In particular v is an autarky for F.
Conversely, let us assume that $v_3(S) = 1$. We will construct an open branch B in T so that $v_B \subseteq v$. To this end, we say that a branch B *agrees* with a partial valuation v if for every formula φ, $T(\varphi)$ a label of a node on B implies $v_3(\varphi) = 1$, and $F(\varphi)$ a label of a node on B implies $v_3(\varphi) = 0$. Then all we need to do is to locate an open branch B of T such that B agrees with v. This process of finding B proceeds by induction; we build the desired branch B in stages. The point is that as we build a branch B, we never close the partial branch we construct, and we maintain the condition that the partial branch we constructed agrees with v. When the construction is completed, the branch we find is open, complete, and agrees with v.
We start with the initial part of the putative branch B: a partial branch with nodes labeled $T(\varphi)$ for all formulas $\varphi \in S$. This partial branch through T agrees with v

because v evaluates all formulas in S as 1. Now, let us assume that we constructed a partial branch $\mathcal{B}'$ in $\mathcal{T}$, and that $\mathcal{B}'$ agrees with v, and that there is a formula of the form $T(\varphi)$ or of the form $F(\varphi)$ on $\mathcal{T}'$ which can be developed. There are six cases that need to be considered, depending on the top connective of φ. We will discuss three; the remaining three are similar and left to the reader to complete.

Let us assume that the formula to be developed is $T(\neg \psi)$. Then our inductive assumption is that $v_3(\neg \psi) = 1$. But then it must be the case that $v_3(\psi) = 0$. Then, as we extend $\mathcal{B}$ to $\mathcal{B}'$ by adding a node at the end, and labeling it with $F(\psi)$ we find that $\mathcal{B}'$ agrees with v. The next case we consider is that of the formula $T(\psi_1 \wedge \psi_2)$. Here we extend $\mathcal{B}$ by adding two nodes at the end: one labeled with $T(\psi_1)$, then another labeled with $T(\psi_2)$. But then it is quite clear that the resulting branch $\mathcal{B}'$ agrees with v. Finally let us consider the case of $T(\psi_1 \vee \psi_2)$. Now we have choice; we can either extend $\mathcal{B}$ with $T(\psi_1)$ or with $T(\psi_2)$. But we know that if $v_3(\psi_1 \vee \psi_2) = 1$ then it must be the case that either $v_3(\psi_1) = 1$, or $v_3(\psi_2) = 1$ (both, actually, could be true!). We select one of ψ_i so that $v_3(\psi_i) = 1$ and extend $\mathcal{B}$ to $\mathcal{B}'$ by adding a node at the end and labeling it with $T(\psi_i)$. The resulting branch $\mathcal{B}'$ agrees with v. As mentioned above there are three more cases to consider; those corresponding to the situation where the node to be expanded are labeled with $F(\neg \psi)$, with $F(\psi_1 \wedge \psi_2)$, and with $F(\psi_1 \vee \psi_2)$. We hope the reader completes the construction in each of these cases.

Let us observe that in each of the six cases we did not close the branch. Thus when we complete our construction the branch $\mathcal{B}$ is open. It is finished since all nodes on that branch are developed. It agrees with v by construction. But now, whenever formula $T(p)$ (with p a variable) is on $\mathcal{B}$, then $v(p) = 1$, and when $F(p)$ is on branch $\mathcal{B}$ then $v(p) = 0$. This means that $v_\mathcal{B} \preceq_k v$. But this is precisely what we wanted. Since v was an arbitrary valuation such that $v_3(S) = 1$, we are done. $\square$

8.4 Davis-Putnam algorithm, iterated single-literal resolution

In this section we discuss the algorithm for testing satisfiability of CNF formulas. This is the first of two algorithms for testing the satisfiability of CNF formulas discussed in this book. It is based on the iteration of a process in which occurrences of a single variable (both positive and negative) are eliminated.

Thus, the input formula is assumed to be in the conjunctive normal form, in other words, a set of clauses. Before describing the Davis-Putnam (DP) algorithm[1] [DP60] in detail we need to recall some notation. A literal l occurring in a set of clauses F is *pure* for F if $\bar{l}$ does not occur in clauses of F. Given a set of clauses F and a literal l, we define $F - l$ as the set of those clauses is F that do not contain l. We recall

---
[1] It is sometimes called VER, *variable elimination resolution*.

that $|l|$ is the propositional variable underlying the literal l. We denote by $sgn(l)$ the Boolean 0 if l is a negative literal, and 1 if l is a positive literal.
Let us recall the following fact.

PROPOSITION 8.10
If F is a set of clauses and l is pure for F, then F is satisfiable if and only if $F - l$ is satisfiable.

Note that all we did in our proof of Proposition 8.10 was to show that if l is a pure literal for a set of clauses F, then $\{l\}$ is an autarky for F (we observed it in Chapter 2 when we introduced autarkies).

Next, given a set of clauses F and a variable x, define the new set of clauses $Res(F, x)$ as follows:

$$Res(F, x) = \begin{cases} F - l \text{ if } l \text{ is pure for } F \text{ and } |l| = x \\ \{C_1 \vee C_2 : x \vee C_1 \in F \text{ and } \neg x \vee C_2 \in F \\ \text{and } C_1 \vee C_2 \text{ is not a tautology}\} \quad \cup \\ \{C \in F : x \text{ has no occurrence in } C\}. \end{cases}$$

Here is what is being done in computation of $Res(F, x)$. The first case when l is a pure literal is obvious. In the second case, all clauses in F containing x are resolved against all the clauses in F that contain $\neg x$. The result is maintained, but two types of clauses are eliminated. First, clauses containing either x or $\neg x$ are eliminated. Second, tautologies are eliminated. The clauses that do not contain either x or $\neg x$ are not affected, they pass from F to $Res(F, x)$.

Let us observe that if $l = x$ and l is pure for F, then the result is just $F - l$. Precisely the same thing happens if $l = \neg x$ and l is pure for F. We cannot claim that the size of $Res(F, x)$ is smaller than that of F. But $Res(F, x)$ has at least one variable less, since x is eliminated. Let us observe, though, that more variables may be eliminated in the process. This will happen if, for instance, $l = x$ is pure for F, and some variable y occurs only in clauses containing x.

As observed above, the effect of our computation of $Res(F, x)$ is a set of clauses that do not contain x. But we need to realize that the size of $Res(F, x)$ *may* increase in the process. In fact, if $|F| = n$ then $|Res(F, x)| \leq \frac{n^2}{4}$ and this bound may be the optimal one. Yet, the *number of variables* goes down. Therefore, if we iterate the operation $Res(\cdot, \cdot)$ selecting variables occurring in the formula one by one, then, for a formula F with m variables, we will eliminate all variables in at most m rounds. What are the possible outcomes of such iterations? Actually, two outcomes are possible. One is that we get an empty set of clauses. The other one is that we get a non-empty set of clauses, containing just one clause: the empty clause. In the first case this ultimate result is satisfiable, in the second case it is not satisfiable. Let us observe that the empty clause, if it is computed in some iteration of operation $Res(\cdot, \cdot)$, stays until the elimination of all variables.

Next, we will prove a proposition that forms a basis for the correctness proof of the first of two algorithms that specifically refer to satisfiability of sets of clauses. This is the so-called Davis-Putnam algorithm DP.[2]

First, we need a bit of terminology. Recall that given a non-tautological clause C, $C = l_1 \vee \ldots \vee l_n$, $n \geq 1$, there is exactly one partial assignment v such that

1. $Dom(v) = Var_C$ (i.e., v is defined on variables of C and on no other variables).
2. $v(C) = 0$.

That v is defined as follows:

$$v(x) = \begin{cases} 1 & \text{if } \neg x \text{ is one of } l_1, \ldots, l_n \\ 0 & \text{if } x \text{ is one of } l_1, \ldots, l_n. \end{cases}$$

Since C is not a tautology, v_C is well-defined. We denote that unique v by v_C. Another construction that we will need in the proof that follows is the *reduct* of a set of clauses F by a partial valuation v. When F is a set of clauses and v a partial valuation then the reduct of F by v, in symbols $Red(F, v)$ is computed as follows:

1. If for some l in C, $v(l) = 1$, then the clause C is eliminated.
2. In clauses C surviving the test (1) we eliminate in C all literals l such that $v(l) = 0$.

In particular, the variables of v do not occur in $Red(F, v)$.

Example 8.2
Let F be the following set of clauses

$$\{p \vee q \vee r, \bar{p} \vee \bar{q} \vee s, p \vee t\}$$

and let v be a partial assignment defined on $\{q, t\}$ with values: $v(q) = 0$, $v(t) = 1$. Then the third clause is eliminated, and first two clauses are shortened. Thus $Red(F, v) = \{p \vee \bar{r}, \bar{p} \vee s\}$. □

Now, let F be a set of non-tautological clauses and let x be a propositional variable.

PROPOSITION 8.11

(a) F is satisfiable if and only if $Res(F, x)$ is satisfiable. In fact:

(b) If v is a valuation satisfying F, then v satisfies $Res(F, x)$. Conversely, if v is a partial assignment defined precisely on $Var_{Res(F,x)}$ and satisfying $Res(F, x)$, then v can be extended to a valuation w such that $w \models F$.

---
[2]To confuse matters some, the next algorithm DPLL also has Davis and Putnam as the first two letters.

Proof: Clearly, (b) implies (a), and so we prove (b), only. If the variable x does not occur in F then the only difference between F and $Res(F, x)$ is absence of tautologies. But our assumption was that F does not contain tautologies so in this case $F = Res(F, x)$, and (b) holds.

So let us assume that x actually occurs in F. First we need to show that if a valuation v satisfies F then v satisfies $Res(F, x)$. Indeed, in Corollary 7.5 we established that whenever a valuation v satisfies a set of clauses F, then v also satisfies $Res(F)$. But $Res(F, x) \subseteq Res(F)$, thus $v \models Res(F, x)$, as desired.

Now we go the other way around. We assume that $v \models Res(F, x)$ and that v is defined precisely on variables of $Res(F, x)$. We need to extend v to a valuation w so that $w \models F$.

While it is not strictly necessary, it will be convenient for pedagogical reasons to consider two cases. One, where $Var_{Res(F,x)} = \emptyset$ and another, where $Var_{Res(F,x)} \neq \emptyset$. We will call these cases Case 1 and Case 2.

So let us deal with Case 1, $Var_{Res(F,x)} = \emptyset$. There are three subcases possible. The first one is when x is a pure literal in F. Then $F - x = \emptyset$ and we define valuation w as follows:

$$w(y) = \begin{cases} 1 & y = x \\ 0 & y \in Var_F \setminus \{x\}. \end{cases}$$

What happens in this case is that since $Res(F, x)$ is satisfiable, but $Var_{Res(F,x)} = \emptyset$, it must be the case that $Res(F, x) = \emptyset$ and every clause in F must contain x, so w satisfies F, as we made $w(x) = 1$.

The second subcase of the first case is when $\bar{x}$ is a pure literal in F. Then the valuation equal 0 on all variables of F satisfies F (by the argument similar to the subcase 1).

The third subcase is when neither x nor $\bar{x}$ is pure in F. But $Res(F, x) = \emptyset$. How could it happen? It cannot be that some clause C in F has no x and no $\bar{x}$ in it, because such clause C automatically passes to $Res(F, x)$ (and such clause cannot be a tautology because there is none in F). Thus for *every* clause C in F there is D such that $C = D \vee x$ or there is E such that $C = E \vee \bar{x}$.

It must be the case that at least one of D, E is nonempty for otherwise the resolvent is empty and so $Res(F, x)$ is unsatisfiable.

But in fact *both* D and E must be nonempty. Indeed, let $C_1 = D \vee x$, $C_2 = E \vee \bar{x}$. If D is empty, then the resolvent is E. As C_2 is not a tautology, E is not a tautology. Thus E is a nonempty, non-tautological clause in $Res(F, x)$, which contradicts our assumptions. The argument for $E = \emptyset$ is similar. Thus we established that both D, E are not empty. But $D \vee E$ is not in $Res(F, x)$. This means that $D \vee E$ is a tautology. This in turn implies that whenever $D \vee x \in F$, $E \vee \bar{x} \in F$ then there is a literal l such that l occurs in D, and $\bar{l}$ occurs in E.

We now proceed as follows. We select one clause C in F of the form $D \vee x$. We know that such clause exists as we are in the third subcase. Then we define the following valuation w:

$$w(y) = \begin{cases} 1 & y = x \\ v_D(y) & \text{if } y \text{ occurs in } D \\ 0 & \text{if neither of the previous two cases holds.} \end{cases}$$

We claim that $w \models F$. Clearly, if a clause C' in F contains the literal x, then $w(C') = 1$. So let us assume that C' contains the literal $\bar{x}$ (the only possibility left). Then there must be a literal l different from x and $\bar{x}$ so that l occurs in C, $\bar{l}$ occurs in C'. But $v_D(l) = 0$ by our construction. Therefore $w(l) = 0$, hence $w(\bar{l}) = 1$. Therefore, $w(C') = 1$, as desired. This completes the argument for Case 1.

Once we successfully navigated through Case 1, we are ready for Case 2 (i.e., $Var_{Res(F,x)} \neq \emptyset$). We will follow the same strategy, but there will be four subcases, not three. Our assumption now is that we have a valuation v *defined on variables of $Res(F,x)$ but on nothing else* and we want to extend v to a valuation w such that $w \models F$.

To this end let us consider the collection $G = Red(F, v)$, that is, the reduct of F by v. Let us observe that all clauses purged from F in the step (1) of the reduction process are already satisfied by v thus by any extension of v (by Kleene theorem, Proposition 2.5). Next, we observe that every clause in G is a subclause of a clause of F. Further, for every clause C in F that is not already satisfied by v the reduct of C by v belongs to $Red(F, v)$. So, if we show that G is satisfied by a desired extension w of v, then w satisfies entire F.

The first subcase we consider is when $G = \emptyset$. This means that all clauses of F are already evaluated by v as 1. But we have to extend v to an assignment which is defined on all variables of F. So, we define w as follows:

$$w(y) = \begin{cases} v(y) & \text{if } y \in Dom(v) \\ 0 & \text{if } x \text{ occurs in } F \text{ but not in } G. \end{cases}$$

Clearly, $v \preceq_k w$, so w satisfies all clauses of F.

The second subcase is when x is a pure literal in G. In this case let us define two assignments: w' and w, w' is defined on the domain of v and on x; w is defined on all variables of F. We set

$$w'(y) = \begin{cases} v(y) & \text{if } y \in Dom(v) \\ 1 & \text{if } y = x. \end{cases}$$

We claim that w' already evaluates all clauses in F as 1. Indeed, let $C \in F$. If neither x or $\bar{x}$ occurs in C then $C \in Res(F, x)$, thus v evaluates C as 1, thus w' evaluates C as 1. If the literal x occurs in C then w' evaluates C as 1. So the remaining possibility is that $\bar{x}$ occurs in C. Now, x is a pure literal in $Red(F, v)$. Since x does not belong to the domain of v it must be the case that C did not survive the test (1) of reduction by v. Therefore one of literals of C is evaluated by v as 1, and so w' evaluates C as 1.

Now, we need to extend w' to a valuation w defined on all variables of F. It should be clear that an assignment w

$$w(y) = \begin{cases} w'(y) & \text{if } y \in Dom(w') \\ 0 & \text{otherwise} \end{cases}$$

satisfies F.

The third subcase, when $\bar{x}$ is a pure literal of G, is very similar to the second subcase and we can safely leave it to the reader.

Thus we are left with the fourth case, when neither x nor $\bar{x}$ is pure for G. This means that the literal x occurs in some clause of G whereas the literal $\bar{x}$ occurs in another clause of G (since there are no tautologies in F there are no tautologies in G, so x and $\bar{x}$ do not appear in the same clause of G). We proceed in a manner similar to the third subcase of the first case. We need to define a valuation w which satisfies F. We have at our disposal an assignment v defined on variables of $Res(F, x)$ (and only of them) such that v satisfies all clauses of $Res(F, x)$. To construct w we select one clause containing literal x, say $x \vee C_1$ in G. Then we define w as follows:

$$w(y) = \begin{cases} v(y) & \text{if } y \in Dom(v) \\ 1 & \text{if } y = x \\ v_{C_1}(y) & \text{if } y \in Dom(v_{C_1}) \\ 0 & \text{if none of previous conditions holds.} \end{cases}$$

Our first goal is to show that the assignment w is well-defined. Since x does not belong to the domain of v, the first two cases of our definition do not contradict each other. The third case does not create contradiction since the clause $x \vee C_1$ is in the reduct $Red(F, v)$. Indeed, the variables occurring in C_1 do *not* occur in the domain of v and since $x \vee C_1$ is not a tautology, the variable x does not occur in C_1. The fourth "default" condition does not create problems, of course.

So now, we have to show that w evaluates all clauses in F as 1. So, let us assume that some clause D belongs to F. If v evaluates D as 1, then by Kleene theorem, w evaluates D as 1, as w extends v. If the literal x belongs to D, then w evaluates D as 1, since $w(x) = 1$. The next possibility we have to consider is when literal $\bar{x}$ does *not* occur in D. But then D, which does not contain x and does not contain $\bar{x}$, belongs to $Res(F, x)$ and so v evaluates D as 1, so w evaluates D as 1. Thus we are left with the case when D contains the literal $\bar{x}$ and v evaluates all literals that are in D and are based on variables in $Dom(v)$ as 0. That is:

$$D = \bar{x} \vee D_1 \vee D_2,$$

where D_1 has no occurrence of x, and no occurrence of variables on which v is defined, and moreover, $v_3(D_2) = 0$.

Now, let us look at the clause $x \vee C_1$. This clause, fixed at the beginning of the fourth subcase, belonged to G, that is, the reduct of F by v. This means that this clause $x \vee C_1$ was the effect of reducing a clause C,

$$C = x \vee C_1 \vee C_2,$$

where every literal of C_2 was evaluated by v as 0. The two clauses $x \vee C_1 \vee C_2$ and $\bar{x} \vee D_1 \vee D_2$ can be resolved on x. The result of resolution of these two clauses is $E = C_1 \vee C_2 \vee D_1 \vee D_2$. So, in principle, E should belong to $Res(F, x)$. But it does not belong there. The reason is that if it does, then v evaluates E as 1. But v evaluates C_2 as 0, v evaluates D_2 as 0 so if v evaluates E as 0 or 1 then it assigns to it the same value as it assigns to $C_1 \vee D_1$. But v is *not* defined on any variable occurring in C_1 nor on any variable occurring in D_1. Where does it left us? It must be the case that $C_1 \vee C_2 \vee D_1 \vee D_2$ is a tautology! But v evaluated both C_2 and D_2 as 0. Thus the pair of complementary literals that made E a tautology must be in $C_1 \vee D_1$. This means that for the arbitrary clause D in F, D containing $\bar{x}$, there must be a literal l such that l occurs in C_1 but $\bar{l}$ occurs in D. Now, this l is neither x nor $\bar{x}$ (for it belongs to C_1). As our assignment w extends v_{C_1}, $w_3(l) = 0$. Therefore $w_3(\bar{l}) = 1$, hence w evaluates D as 1. This completes the proof of the fourth subcase of Case 2, and of our proposition as well. $\square$

The key element of Proposition 8.11, to be precise of the implication $\Leftarrow$ of part (b), is that it provides a constructive way to extend *every* satisfying valuation v of $Res(F, x)$ to a satisfying valuation for F. Before we derive algorithms both for testing satisfiability of a set of clauses and for actually finding satisfying valuation, let us recapitulate what happened in the argument. Specifically, we start with an assignment v that satisfies $Res(F, x)$ and is defined on variables of $Res(F, x)$ only. The goal is to extend v to a satisfying valuation w for F (with w defined only on variables of F.) To extend v to w all that is needed is to know the set of those clauses of F that did *not* make it to $Res(F, x)$. The procedure outlined in the proof of Proposition 8.11 will be called below $expand(G, v, x)$. Here, intuitively, G is the set of clauses that were eliminated in passing from F to $Res(F, x)$, x is the variable used in elimination, and v is a satisfying assignment for $Res(F, x)$. Thus, the procedure $expand(G, v, x)$ returns a partial assignment w extending v. When v is a partial assignment defined on variables occurring in a set of clauses of the form $Res(F, x)$ and on no other variable, and G is $F \setminus Res(F, x)$ then $w = expand(G, v)$ satisfies F.

Let us observe that when we pass from F to $Res(F, x)$ and compute the empty clause $\emptyset$ then this clause will be passed to *all* subsequent clause sets (as the empty clause survives the test (1)). And so, at the end when all variables are eliminated, the resulting collection of clauses will be non-empty, and unsatisfiable. Thus any time if an empty clause is computed we need to leave with failure.

We also need a user-provided function *SelectVar* that accepts as an argument a set of clauses F with at least one variable and returns one of variables of F.

Here is a *decision* version of DP algorithm, which we call *DPtest*.

Algorithm DPtest. Input: a CNF F
Output: Decision if F is satisfiable.
if $(F == \emptyset)$
    {**return** ('Input formula satisfiable')}
elseif (F contains an empty clause, $\emptyset$)

{**return** ('Input formula unsatisfiable')}

else

$\{x = SelectVar(F);$
$F = Res(F, x);$
$\{\textbf{return}(DPtest(F))\}$

To see correctness of this algorithm, let us observe that if F is a satisfiable set of clauses then either F is empty and our algorithm returns that F is satisfiable, or F does not contain the empty clause (of course the implication is true, not the equivalence). When F is nonempty but does not contain an empty clause, then it must have at least one variable. Then the correctness follows by induction as the same algorithm is called recursively on a set of clauses with one variable less. Conversely, if our algorithm returns the string 'input formula satisfiable' on input F then it must be the case that F is empty, or our algorithm found that $Res(F, x)$ is satisfiable. In that latter case by Proposition 8.11, F is satisfiable.

Let us observe that in the case of the algorithm *DPtest* all we needed to maintain was the current content of variable F and test if it is empty and also if it contains an empty clause. It should be clear that the size of variable F is exponential in the size of the input formula; it is bound by 3^n where n is the number of variables in F.

In the search version of the DP algorithm we need to maintain more complex data structures. Let us assess the situation, first. In the proof of Proposition 8.11, (b), $\Leftarrow$ in order to compute the extension of the partial valuation v to partial valuation w we had to know the set of clauses that were eliminated when the set $Res(F, x)$ was computed. Those sets are of the form $F \setminus Res(F, x)$ (but both variable x and the set F change as we progress). For that reason we need to maintain them. Then *after* we compute them all (providing we did not find contradiction, i.e., the empty clause), we change direction, and compute the consecutive partial valuations, finding at the end a satisfying valuation for the input formula. We also need to know what variable was used for resolution, so the sequence of these variables also has to be maintained. Let us observe that at the end of the first phase we already know whether the input formula is satisfiable or not. Once we know the input formula is satisfiable, the second part computes the satisfying valuation, going backwards.

The operation that was outlined in the proof of Proposition 8.11 will be denoted by $expand(G, v, x)$ and will have three inputs: a partial valuation v, a variable x which is not in the domain of v, and a set of clauses G such that x or $\bar{x}$ occurs in every clause in G. The intention is that G is the "layer" that was eliminated as we resolved with respect to x, but the definition (implied by the proof of Proposition 8.11) can be given in general. Moreover, let us observe that as we prepare the input for it in Phase I, the operation $expand$ will work correctly on the intended input.

We will define the algorithm *DPsearch* as consisting of two phases. In Phase I (which is similar to the algorithm *DPtest*) we precompute the data which, in Phase II, will be used to compute the satisfying assignment for the input formula. Following the above analysis, in Phase I we compute two arrays: one that we call *layerSeq* will hold layers, the other, called *varSeq*, will hold variables that are used for resolving (i.e., computing sets of the form $Res(F, x)$). We assume that we have user-supplied functions that implement the operation $Res(F, x)$ and a heuristic function *selectVar* (implemented by the user) that on a set of clauses G such that $Var_G \neq \emptyset$ returns one of variables occurring in G.

We need a procedure, which we call *Forward*. This procedure accepts four inputs: a set of clauses G, an array called *varSeq*, an array called *layerSeq*, and an integer i. It then expands both arrays gradually, by computing variables and corresponding layers and putting them into the appropriate place in the arrays.

Algorithm $Forward(G, varSeq, layerSeq, i)$.
if (G contains empty clause, $\emptyset$)
    { **return**($\{p, \bar{p}\}$) }
elseif (G is empty)
    { **return**($varSeq, layerSeq$)}
else   {
    $x = selectVar(G)$;
    $varSeq[i] = x$;
    $layerSeq[i] = G \setminus Res(G, x)$;
    $G = Res(G, x)$;
    $i++$;
    return($Forward(G, varSeq, layerSeq, i)$)}

Once we have the algorithm *Forward*, we iterate it on the input clausal theory F, integer i initialized to 0, and two empty arrays.

Algorithm *DPSearchPhaseOne*
Input: a CNF F with at least one variable. Output: an inconsistent set of literals, or otherwise two arrays, of layers, and of variables.
$G = F$;
$varSeq = \emptyset$;
$layerSeq = \emptyset$;
$i = 0$;
$Forward(G, varSeq, layerSeq, i)$.

After running the algorithm *DPsearchPhaseOne* we either know that the input CNF, F, is unsatisfiable, or we get two arrays, of layers and of variables. These arrays (when computed) serve as an input to Phase II.

In that phase we use the procedure $expand(G, v, x)$ to construct the satisfying assignment for the input formula F. The algorithm *DPsearchPhaseTwo* uses the arrays computed in Phase I, but runs "backwards," processing arrays from the last element to the first one. For that reason we will assume that we have a polymorphic procedure *reverse* that accepts an array, and returns an array with the same items, but listed in reverse order.

Algorithm *DPsearchPhaseTwo*.
Input: two arrays, *varSeq* and *layerSeq* returned by the algorithm

DPsearchPhaseOne.
Output: A satisfying valuation for the input F of Phase I.
$layerSeq = reverse(layerSeq);$
$varSeq = reverse(varSeq);$
$v = \emptyset;$
$j = length(layerSeq);$
    for $(i = 0; i < j; i++)$
    $\{\ G = layerSeq[i];$
    $x = varSeq[i];$
    $v = expand(G, v, x);\ \}$
    $return(v)$

By repeated use of the argument of Proposition 8.11 the assignment returned by the algorithm *DPsearchPhaseTwo* is a satisfying assignment for the input formula F.

Example 8.3
Let F be the following set of clauses.

$\{\ C_1 : p \vee \bar{t},$
$C_2 : p \vee s,$
$C_3 : q \vee r \vee s,$
$C_4 : \bar{q} \vee r \vee s,$
$C_5 : \bar{r} \vee \bar{s},$
$C_6 : u \vee w,$
$C_7 : s \vee \bar{u} \vee \bar{w} \vee x\ \}$

We initially assign to G the input formula F. Let us assume that $p = selectVar(G)$. Then we get $varSeq[0] = p$. Actually, let us observe that selecting p is quite reasonable, since 1 is pure for F and thus the next set of clauses is actually smaller. We also see that $layerSeq[0]$ is $\{C_1, C_2\}$, and G becomes $\{C_3, C_4, C_5, C_6, C_7\}$.
Let us assume that $u = selectVar(G)$. We observe that $selectVar$ is applied to the current content of G, not the original one. We now get $varSeq[1] = u$, and $layerSeq[1]$ equal to $\{C_6, C_7\}$. But now, let us see what becomes of G. When we resolve the previous content of G on u a new clause $s \vee w \vee \bar{w} \vee x$, was created, but did not make it to G because it is a tautology. Thus G becomes $\{C_3, C_4, C_5\}$.
Next, let us assume that $q = selectVar(G)$. We then get $varSeq[2] = q$, $layerSeq[2] = \{C_3, C_4\}$. When we resolved C_3 and C_4 we create a new clause $C_8 : r \vee s$ (repeated literals are eliminated). Thus G becomes $\{C_5, C_8\}$.
Finally, let us assume that $r = selectVar(G)$. We then get $varSeq[3] = r$, $layerSeq[3] = \{C_5, C_8\}$. But now, G becomes empty since the resolvent of C_5 and C_8 is a tautology. As we did not compute the empty clause throughout the process, Phase I outputs the arrays $varSeq$ and $layerSeq$.
We now move to Phase II. Its first step is to reverse the order of the arrays, getting:

$$\langle r, q, u, p \rangle$$

and
$$\langle \{C_5, C_8\}, \{C_3, C_4\}, \{C_6, C_7\}, \{C_1, C_2\} \rangle.$$

The partial valuation h is initialized to $\emptyset$.
Now, using the function *expand* we get the first step: new h is $h(r) = 1$, $h(s) = 0$. Reduction of the next layer gives us the empty reduct. We set $h(q) = 1$.
Next, we reduce clauses of layer $layerSeq[2]$ by h getting two clauses, $u \vee v$ and $\bar{u} \vee \bar{w} \vee x$ (because $h(s) = 0$). We set $h(u) = 1$, $h(w) = 0$ and $h(x) = 0$.
We continue and reduce the layer $layerSeq[3]$ by the current assignment h. The reduced set consists of clauses $p \vee \bar{t}$, and the unit p. As p is pure, we set $h(p) = 1$, $h(t) = 0$. Thus the second phase of the DP algorithm outputs an assignment:

$$\begin{pmatrix} p & q & r & s & t & u & w & x \\ 1 & 1 & 1 & 0 & 0 & 1 & 0 & 0 \end{pmatrix}$$

☐

Let us observe that running of Phase II allows for a lot of leverage. Specifically, at *any* stage of the construction we can "plug-in" another partial valuation (as long as it satisfies the corresponding content of the variable F). This means that as we are running the first phase of the DP algorithm we can break it, solving the current content of F with some other SAT algorithm, get the valuation v, restrict it to variables of the current F and start running the second phase (remember that the second phase is run backwards so at this point we have all layers that we will need). For that reason, the DP algorithm can serve as both pre- and post-processing tool for the DPLL algorithm which we will study in the next section. To make DP work this way, we can, for instance, call DPLL if the size of $Res(F, x)$ increases more than some user-defined ratio. This observation is, actually, stated in [SP05].

8.5 Boolean constraint propagation

In this section we investigate the simplification mechanism used by the DPLL algorithm (described in the next section) for testing satisfiability (and finding satisfying assignment for a set of clauses). This technique computes the set of literals that are entailed *directly* by the input set of clauses. We start with a motivating example.

Example 8.4
Our example consists of two variations on the same theme: some literals can be computed from a set of clauses directly. The second variation is the same one as the first one, except that in addition to a set of clauses we also have an input partial assignment.
(a) Let F be a set of clauses:

$\{C_1 : \bar{p} \vee \bar{q} \vee \bar{s},$
$C_2 : r \vee s \vee \bar{z},$
$C_3 : \bar{z} \vee u,$
$C_4 : z \vee h \vee l,$
$C_5 : p,$
$C_6 : q$
$C_7 : \bar{r}\}$

Then, because the set F contains unit clauses C_5 and C_6, the only way to satisfy these clauses and also C_1 is to assign to s the value 0. But once this is done, the only way to satisfy C_2 is to assign to z the value 0. The assignment to r is also forced on us; it is 0. This, in turn, makes C_3 satisfied, no matter what value we may later assign to the variable u. Finally, for all practical reasons, the clause C_4 (whose value is not known, yet) simplifies to the clause $h \vee l$.

Thus our F forces on us the following partial valuation, necessary for satisfying F: $v(p) = v(q) = 1$, and $v(r) = v(s) = v(z) = 0$.

(b) Now, let us look at a clause set F' consisting of clauses C_1, C_2, C_3, and C_4, and a partial assignment v' specified by $v'(p) = v'(q) = 1$, $v'(z) = 0$. Then together F' and v' force on us the same partial assignment v as in part (a). That is, *any* partial assignment w extending v' and satisfying the formula F must extend v. The argument is exactly the same as in part (a). Also, like in (a), the clause set F' simplifies to the clause set consisting of just one clause, $h \vee l$. □

Thinking algebraically, we computed in our example a fixpoint of a certain operation. To this end let us define an operation bcp_F. This operation accepts as an input a set of literals (if it is consistent, then it is a partial valuation, but we do not limit ourselves to this case) and outputs another set of literals (which may be inconsistent).

$$bcp_F(S) = \{l : \text{There is a clause } C : l_1 \vee \ldots \vee l_{j-1} \vee l, C \in F,$$
$$\bar{l}_1, \ldots, \bar{l}_{j-1} \in S.\}.$$

We then have the following easy fact.

PROPOSITION 8.12
Given a CNF formula F, the operator $bcp_F(\cdot)$ is a monotone operator in the complete lattice $\mathcal{P}(Lit)$.

Let us note that Proposition 8.12, actually, requires a proof which we leave to the reader.

By the Knaster-Tarski fixpoint theorem, for every clause set F, the operation bcp_F possesses a least fixpoint. That least fixpoint is denoted by $\mathrm{BCP}(F)$. It should also be clear that the operator bcp_F is continuous thus, even if F is infinite, the fixpoint is reached in at most ω steps.

But we can also think in terms of logic. The other way to define $\mathrm{BCP}(F)$ is by limiting the form of acceptable resolution proofs. Namely we compute the set of those literals which can be computed from F using resolution where one of parents is always a unit clause. We call such limited form of resolution *unit resolution*. Formally, we define

$$\text{BCP}'(F) = \{l : \{l\} \text{ possesses a unit-resolution proof tree from } F\}.$$

We now prove that both definitions of BCP are equivalent.

PROPOSITION 8.13
For all sets of clauses F, $\text{BCP}(F) = \text{BCP}'(F)$. In other words, both definitions are equivalent.

Proof: For the inclusion $\subseteq$, we need to show that the least fixpoint of bcp_F is included in $\text{BCP}'(F)$. We use induction showing that each iteration $bcp_F^n(\emptyset)$ is included in $\text{BCP}'(F)$. The base case is obvious. Now assume $l \in bcp^{n+1}(\emptyset)$. Then there exists a clause $C : l_1 \vee \ldots \vee l_{j-1} \vee l$ with $\bar{l}_1, \ldots, \bar{l}_{j-1} \in bcp^n(\emptyset)$. But then, by inductive assumption each of $\bar{l}_1, \ldots, \bar{l}_{j-1}$ has a proof using unit resolution. We combine these proofs as in Figure 8.6 (clause C_i in that proof is $l_i \vee \ldots \vee l_j \vee l$). This demonstrates

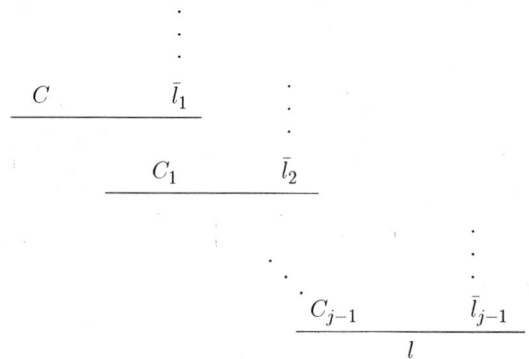

FIGURE 8.6: Derivation of a literal using unit resolution

inclusion $\subseteq$.
Conversely, assume that $l \in \text{BCP}'(F)$. Proceeding by induction on the number of applications of unit resolution we show that $l \in \text{BCP}(F)$. The base case is again obvious. Now assuming that we deal with the resolution tree like we have in Figure 8.6, and $l_1, \ldots, l_{j-1}$ all belong to $bcp^n(\emptyset)$ we see that l itself belongs to $bcp^{n+1}(\emptyset)$. This shows $\supseteq$ and completes the proof. □
But let us keep in mind that $\text{BCP}(F)$ may be inconsistent. Here is an example. Let $F = \{p, \neg p \vee q, \neg p \vee \neg q\}$. Then the unit resolution derives both q and $\neg q$, thus also the empty clause.

Now, let us look at the effect of computing $\mathrm{BCP}(F)$. If we established that the value of a literal l is 1 then every clause C that has a positive occurrence of l is automatically satisfied, regardless of the future assignments of values to other literals occurring in C. Therefore we can eliminate C from further considerations.[3] On the other hand, if the literal $\bar{l}$ occurs in C (that is, l occurs in C negatively) then our assignment of 1 to l cannot contribute to satisfaction of C. Thus $\bar{l}$ can be eliminated from such clause (again, we must be cautious in the case of future backtrack). In other words, we can safely perform the reduct of F with respect to $\mathrm{BCP}(F)$. Moreover, if $\mathrm{BCP}(F)$ is inconsistent, then F itself is inconsistent (the converse implication is false, though) and we have to return a message to the effect that the input formula is unsatisfiable. Otherwise, the reduct R of F with respect to $\mathrm{BCP}(F)$ is satisfiable if and only if F is satisfiable. We will prove this property now.

PROPOSITION 8.14
Let F be a CNF formula. Then F is satisfiable if and only if $\mathrm{BCP}(F)$ is consistent and the reduct of F with respect to $\mathrm{BCP}(F)$ is satisfiable.

Proof: If v is a valuation satisfying F then v also assigns value 1 to all literals in $\mathrm{BCP}(F)$. This follows immediately from the soundness of the resolution, in particular of unit resolution. Next, let R be the reduct of F by $\mathrm{BCP}(F)$. If $D \in R$, then for some literals $l_1, \ldots, l_k \in \mathrm{BCP}(F)$, the clause $C : D \vee \bar{l}_1 \vee \ldots \vee \bar{l}_k$ belongs to F. But $v(\bar{l}_1) = \ldots v(\bar{l}_k) = 0$ so it must be the case that $v(D) = v(C)$. But $v(C) = 1$, so $v(D) = 1$. Since R was an arbitrary clause in the reduct, the implication $\Rightarrow$ is proved.

Conversely, let us assume that $\mathrm{BCP}(F)$ is consistent and the reduct R of F with respect to $\mathrm{BCP}(F)$ satisfiable. Let v be a valuation satisfying R. Let us observe that, by construction, no variable occurring (positively or negatively) in $\mathrm{BCP}(F)$ occurs in R. In the process of reduction from F to R, some variables may have been eliminated altogether; they may occur neither in $\mathrm{BCP}(F)$ nor in R. We need to take care of them as well. Since $\mathrm{BCP}(F)$ is consistent, there is a valuation v' which satisfies $\mathrm{BCP}(F)$. We now define a new valuation w as follows.

$$w(p) = \begin{cases} v(p) & \text{if } p \text{ occurs in } R \\ v'(p) & \text{if } p \text{ occurs in } \mathrm{BCP}(F) \\ 0 & \text{otherwise.} \end{cases}$$

We claim that w satisfies F. Let C be a clause in F. If C is eliminated in the first part of the computation of the reduct R (that is, some literal l of $\mathrm{BCP}(F)$ belongs to C) then $w(C) = 1$ because $w(l) = v'(l) = 1$. Otherwise, $C = D \vee l_1 \vee \ldots \vee l_k$ with $\bar{l}_1, \ldots, \bar{l}_k \in \mathrm{BCP}(F)$, $D \in R$. But $v \models D$, thus for some literal l occurring in R, $v(l) = 1$. But then $w(l) = 1$, and since l occurs in C, $w(C) = 1$, as desired. □

---

[3] That is, until we backtrack, the assignment of 1 to l may be the result of a decision that will have to be rescinded.

In the next section we will see how the operation BCP is used in an algorithm for computation of a satisfying assignment for an input clause set F.

8.6 The Davis-Putnam-Logemann-Loveland (DPLL) Algorithm

Before we state and prove the correctness of the algorithm DPLL, we need a simple proposition that provides the basis of the correctness proof.

PROPOSITION 8.15

Let F be a set of formulas and $p \in At$. Then F is satisfiable if and only if $F \cup \{p\}$ is satisfiable or $F \cup \{\neg p\}$ is satisfiable.

Proof: If $v \models F$, then, since v is defined on all variables in At, $v \models F \cup \{p\}$ or $v \models F \cup \{\neg p\}$. Conversely, if $v \models F \cup \{l\}$ then $v \models F$. Taking $l = p$, or $l = \neg p$ we get a desired implication. □

Proposition 8.15 and another fact that we proved in Section 8.5, Proposition 8.14, imply the correctness of the DPLL algorithm.

Our formulation of DPLL assumes that we have a function BCP which computes BCP of a given input set of clauses F. We assume that we have a function $selectLit(F)$ that returns a literal in Lit_F whenever Var_F is nonempty. We will use this function to find unassigned literal. We also assume that we have a function $reduce(F, v)$ which, on an input set of clauses F and a set of literals v reduces F by v. But now, as the BCP may be inconsistent, we will assume that when this happens, the reduct contains the empty clause $\emptyset$. This is a necessary modification of the reduct, as it is easy to produce an example where BCP is inconsistent but the reduct of the kind we used in our investigations of VER algorithm is consistent. For instance, a clause set $p, \bar{p}, q \vee \neg r$ produces an inconsistent BCP, but a consistent reduct. To distinguish this notion of reduct from the previous one, we will use a different notation: $reduce(F, v)$.

We now introduce the DPLL decision algorithm for SAT.

Algorithm *DPLLdec*. Input: a set of clauses F.
Output: decision if F is satisfiable.
$G = F$;
if (F contains the empty clause $\emptyset$)
    return(*'input formula unsatisfiable'*)
$v = \text{BCP}(G)$;
if (v contains a pair of contradictory literals)
    return(*'input formula unsatisfiable'*)
$G = reduce(G, v)$;
if ($G = \emptyset$)
    return(*'input formula satisfiable'*)
else
    { $l = selectLit(G)$;
    $DPLLdec(G \cup \{l\})$;
    $DPLLdec(G \cup \{\bar{l}\})$};

FIGURE 8.7:   Decision version of DPLL algorithm

PROPOSITION 8.16

If F is a satisfiable set of clauses then DPLLdec(F) *returns* 'input formula satisfiable.' *Otherwise it returns* 'input formula unsatisfiable.'

Proof: We proceed by induction on the number of unassigned variables n of F. Clearly, when n is 0 the output is correct; for $F = \emptyset$ the output is *input formula satisfiable,*' whereas for $F = \{\emptyset\}$ the output is *'input formula unsatisfiable.'*
So now assume that for CNFs with at most n unassigned variables the algorithm answers correctly. We will have to be a bit careful taking into account whether BCP is consistent or not. Now, if $n > 0$, we have two possibilities. When $BCP(F) \neq 0$ and is consistent, then the number of variables in the reduct, and hence in $G \cup \{x\}$ and in $G \cup \{\bar{x}\}$), goes down. We can use inductive assumption directly: DPLL will output the correct message. It is possible that $BCP(G) = \emptyset$. In such case both $BCP(F \cup \{x\})$ and $BCP(F \cup \{\bar{x}\})$ are nonempty. In the reduction process the number of unassigned variables goes down and we can use the inductive assumption. Finally, it is also possible that BCP is inconsistent. In such case F is unsatisfiable, but then we output *'input formula unsatisfiable.'* □

We are now ready to discuss a pseudocode for the search version of DPLL. Here we have to be more careful as we have to maintain the partial valuation that changes as we go up and down the search tree.

The idea is to define an algorithm $DPLLsearch(F, v)$ that on input CNF F and a partial assignment v searches for an extension of the assignment v to a (possibly partial) assignment w such that:

1. $v \leq_k w$.

2. w evaluates all clauses of F as 1.

(The proximity of two figures with algorithms *DPLLdec* and *DPLLsearch* may push the table showing the algorithm *DPLLsearch* to the next page – after all LaTeX- which is what we use - paginates the text automatically.)

Algorithm $DPLLsearch(F, v)$. Inputs: An array of clauses F and partial assignment v. Output: An inconsistent set of literals (when there is no extension w of v such that w evaluates all clauses of F as 1), or a partial assignment w that extends v and evaluates all clauses of F as 1.

Algorithm $DPLLsearch(F, v)$
if $(\emptyset \in F)$
    {**return**$(\{p, \bar{p}\})$}
    $G = Red(F, v)$;
    if $(G = \emptyset)$
    { **return**(v) }
    else
        { $l = selectLit(G)$};
        if (DPLLsearch$(G \cup \{l\}, v)$ noncontradictory)
            { **return** (DPLLsearch$(F, v \cup \{l\})$)}
        else
            { **return** (DPLLsearch$(F, v \cup \{\bar{l}\})$)}

FIGURE 8.8: Pseudocode for DPLLsearch algorithm

Let us observe that the larger the assignment v, the smaller the reduct, $Red(F, v)$. This opens for us the possibility of proving the correctness by induction on the number of *unassigned* variables. Clearly at the base, where all variables are assigned values, the algorithm *DPLLsearch* correctly returns v if v is a satisfying assignment, or contradiction if it is not. Then the correctness follows by induction as in lines 12 and 14 we go down with the number of unassigned variables.

Next, let us observe that while on the first glance running the algorithm $DPLLsearch(F, v)$ on larger input v may seem more expensive, it is, actually, less expensive (as the number of literals with unassigned value goes down). Hence, like in Proposition 8.16 we find the following.

PROPOSITION 8.17

The algorithm DPLLsearch(F, v) finds an extension w of the partial assignment v such that w evaluates all clauses of F as 1, if there is such assignment w. Consequently, for the input $v = \emptyset$, the algorithm DPLLsearch(F, v) finds a partial assignment w which evaluates all clauses of F as 1, if there is such assignment.

We were careful in our Proposition 8.17 not to say that the algorithm outlined in Figure 8.8 returns a satisfying assignment for F. The only (small) issue is that the returned partial assignment may have the domain smaller than the entire set Var. But then, of course, *any* completion of w to an assignment w' defined on all Var is a satisfying assignment for F.

8.7 Improvements to DPLL?

Let us now look at the possibilities of improving the performance of DPLL. We will list four possibilities where the performance can be improved. All these four places were tried with tremendous successes. In fact, it can be said that the progress in the past couple of years is precisely due to the SAT community looking at these places.
(1) Looking at *select* function. A variety of heuristics for choosing a variable and its polarity have been tried. All these proposals attempt to identify a part of the search tree where the solution is more likely to be found. In fact, the choice of selection function amounts to the rule for dynamic construction of the search tree.
(2) Learning. We backtrack to the last decision point each time we encounter a contradiction, that is, we find that the last choice resulted in inconsistent BCP. But what does it mean? It means that the sequence of choices on the branch, $l_1, \ldots, l_k$, together with F entail contradiction. Formally, it means $F \cup \{l_1, \ldots, l_k\} \models \bot$. But this is equivalent to

$$F \models \bar{l}_1 \vee \ldots \vee \bar{l}_k.$$

That is, we just learned a clause that is a consequence of F. This does not help yet. But by analyzing the derivation of that clause $C : \bar{l}_1 \vee \ldots \vee \bar{l}_k$ we may find that some clause subsuming C is a consequence of F. Then we can add such subsumed clause (and possibly more clauses learned from that analysis) to F. This does not change the set of models, or the size of the search tree (because the number of variables does not change), but it does change the behavior of the set of clauses. We may find (and practice confirms this) that some parts of the search tree are pruned. But of course there is no "free lunch." Since every backtrack results in learning a clause, the number of clauses grows, requiring an additional memory for storing clauses. In particular, the learned clauses have to be managed (because their number may grow beyond the available memory). A variety of strategies for managing clauses that have been learned has been proposed in the literature.

(3) Closely related to learning is the issue of *backjumping*. During the analysis of the contradiction (called *conflict analysis*) we may learn that the true reason why we found the contradiction is one of the choices made *before* the last choice. If we learn such a fact, we may prune the entire part of the search tree below such a "bad" choice, avoiding at least one backtrack. Since the size of the search tree is exponential in the number of variables, avoiding visiting a region of the search tree below some choice results in tremendous savings.

(4) It should be clear that most time in running DPLL is spent on computation of BCP. After all, we run BCP after every assignment of truth value at the decision point. So, if we could reduce the number of visits to clauses to see if all but one literal are already falsified by the current partial assignment, we would have a possible significant gain. The idea is, then, to watch *two* literals per clause. As long as neither of these is falsified, the clause cannot be used for contributing to the BCP and the clause does not need to be visited. If one of these is falsified, then we need to visit the clause and see if one of the following happens: new two watched literals can be assigned, or if exactly one of the literals of the clause is not yet falsified (in which case that last literal contributes to BCP), or if all literals in that clause are falsified (forcing the backtrack).

The improvements listed above (and a few others, such as various forms of *lookahead*) as well as better data structures for holding clause sets contributed to the tremendous progress of satisfiability testing in the past number of years.

8.8 Reducing the search for satisfying valuation to testing satisfiability

Our goal is to find a valuation that satisfies a given set of formulas F, if one exists. But let us assume that we have some algorithm $\mathcal{A}$ that only tests if a given input formula is satisfiable, but does not return a satisfying valuation even if one exists. Here is how we can turn $\mathcal{A}$ into an algorithm to actually compute the satisfying valuation. At the very beginning, we run $\mathcal{A}$ to see if a given input formula F is satisfiable. If $\mathcal{A}$ returns '*unsatisfiable*,' we also return '*unsatisfiable*.' We initialize the set U of unassigned variables to Var_F. If $\mathcal{A}$ returns '*satisfiable*,' we select a literal, say l, with underlying variable in U and form a unit clause $\{l\}$ that we add to F. We also delete the underlying variable $|l|$ from U. Then we run $\mathcal{A}$ on this new formula. If $\mathcal{A}$ returns '*satisfiable*,' we substitute for F the clause set $F \cup \{\{l\}\}$, and set $v(l) = 1$. If $\mathcal{A}$ returns '*unsatisfiable*,' we substitute for F the clause set $F \cup \{\{\bar{l}\}\}$, and set $v(l) = 0$. We continue until U becomes empty. Since after the first step the consistency of F is an invariant, and at the end of the computation F contains a complete set of literals, the resulting v is a valuation (it happens to be a subset of F at the end of computation) satisfying F.

We mentioned this technique because closure under resolution (or some of its complete variants) is an example of such decision algorithm which does not directly return the satisfying valuation if one exists. Trivial enhancements, such as simplification after adding l, may improve such algorithms.

8.9 Exercises

1. Here is a formula φ: $(p \Rightarrow q) \wedge r \Rightarrow \neg p$. Build table for φ. Is φ a tautology? Is it satisfiable?

2. In this chapter we defined tableaux expansion rules for connectives $\neg, \wedge$ and $\vee$. We did not formulate such rules for the other De Morgan connectives, $\Rightarrow$ and $\equiv$. But this can be done. Do it.

3. Even more so, one can define tableau expansion rules for *any* connective $c(\cdot, \ldots, \cdot)$. We do not ask you to do this in all generality, but do this for two connectives, the ternary connective ITE and the binary connective $NAND$.

4. We did not discuss the question of satisfiability for formulas represented as DNFs. But the reason why we did not do so is that for DNF testing satisfiability is truly obvious. Here is one procedure: First, we eliminate from a DNF φ all inconsistent terms (i.e., terms that contain some literal and its negation). If the formula shortens to $\emptyset$ we report unsatisfiability. Otherwise we select any term t equal to $l_1 \wedge \ldots \wedge l_k$, define a partial valuation v making all literals $l_1, \ldots, l_k$ true, and extend v to w setting all the remaining variables to 0. Show that the resulting valuation v satisfies φ.

5. We also did not discuss testing satisfiability for formulas defined by Boolean polynomials.

   (a) If we represent a formula φ as a standard polynomial and we do not get the polynomial 0, then φ is satisfiable.

   (b) If we represent a formula φ as a standard polynomial and we do not get the polynomial 1, then $\neg\varphi$ is satisfiable.

   (c) Now, the issue is to find a satisfying assignment for φ. Let $p(x_1, \ldots, x_n)$ be the polynomial representation of φ (the cost of finding such representation may be prohibitive). Once we have a polynomial $p(x_1, \ldots, x_n)$, we can represent the polynomial p as

   $$x_1 \cdot q(x_2, \ldots, x_n) + r(x_2, \ldots, x_n).$$

   Devise an inductive procedure for finding an assignment v that finds the satisfying assignment for p if there is one.

6. Generate any reasonable clause set F (if you do not know how to do this, use this F: $\{p \vee q, \neg p \vee \neg q \vee r, \neg r \vee s \vee t, q \vee r \vee \neg s\}$). Choose any order of variables, and run the DP algorithm on it.

7. Do the same, but now use DPLL (choose some function *select*, for instance, order variables lexicographically, and always select the negative literal first).

8. Derive Quine Theorem (A clause set is satisfiable if and only if its closure under resolution does not contain an empty clause) from the Davis and Putnam Theorem (Proposition 8.11). Hint: select any order of variables. Iterate Davis-Putnam algorithm (Phase I) over the variables. Assuming the input CNF F is unsatisfiable, what will be the result? What does it tells us about the closure of F under resolution?

9. A Gray code of order n is a listing of all assignments on n variables, $x_1, \ldots, x_n$. Often, an additional requirement is that consecutive assignments differ in one place only. Since there are 2^n assignments it will be convenient to list them as columns of a table, not rows. Here are the simplest Gray codes for $n = 1$, and $n = 2$:

$$x_1 \ 0 \ 1$$

and

$$x_1 \ 0 \ 0 \ 1 \ 1$$
$$x_2 \ 0 \ 1 \ 1 \ 0.$$

Here is a general construction:

(Obvious) When M is a listing of a Gray code and M' its reverse, then M' is also a Gray code. In our case above we get:

$$x_1 \ 1 \ 1 \ 0 \ 0$$
$$x_2 \ 0 \ 1 \ 1 \ 0.$$

(a) Let M be a Gray code and M' its reverse listing. We put a row of 0's on top of M and a row of 1's on top of M' and concatenate. Show that the result is again a Gray code. For instance:

$$x_1 \ 0 \ 0 \ 0 \ 0 \ 1 \ 1 \ 1 \ 1$$
$$x_2 \ 0 \ 0 \ 1 \ 1 \ 1 \ 1 \ 0 \ 0$$
$$x_3 \ 0 \ 1 \ 1 \ 0 \ 0 \ 1 \ 1 \ 0.$$

(b) A Gray code is cyclical if, in addition, the last column differs from the first one in one place only. Show that the construction above results in cyclical code.

Chapter 9

Polynomial cases of SAT

9.1 Positive and negative formulas 165
9.2 Horn formulas 167
9.3 Autarkies for Horn theories 176
9.4 Dual Horn formulas 181
9.5 Krom formulas and 2-SAT 185
9.6 Renameable classes of formulas 194
9.7 Affine formulas 199
9.8 Exercises 204

In this chapter we discuss several classes of formulas for which the satisfiability problem can be solved in polynomial time. They include positive formulas, negative formulas, CNF formulas consisting of Horn clauses, CNF formulas consisting of dual Horn clauses, CNF formulas built of of 2-clauses, formulas expressed as collections of linear expressions over the field $\mathbf{Z}_2$ (affine formulas), and DNF formulas. The polynomial time algorithms for these classes vary in difficulty, and some of these cases are very easy.

9.1 Positive formulas and negative formulas

We call a formula φ *positive* if it has no negative occurrences of any variables (that is, its canonical negation normal form has no negative occurrences of any variables). It is, then, easy to check if a formula is positive – it is enough to compute its canonical negative normal form and scan it for presence of negation symbols). For instance, the formula $(p \wedge q) \vee r$ is positive but the formula $\neg(\neg p \vee \neg q) \vee r$ is also positive. It is easy to see that the class of positive formulas contains all variables, and is closed under conjunctions and disjunctions. Let us observe that from the point of view of complexity, our more relaxed definition does not create problems. We can easily decide if a formula is positive or not.

Let us define $\mathbf{1}_{Var}$ as a valuation that assigns to every variable p the value 1. We then have the following fact.

PROPOSITION 9.1
If φ is a positive formula then $\mathbf{1}_{Var} \models \varphi$.

Proof: Since φ is equivalent to its canonical negation-normal form, we can assume that φ has no occurrence of negation symbol. Then, by an easy induction of the complexity of φ, we show that $\mathbf{1}_{Var} \models \varphi$. □

Then we have the following obvious corollary.

COROLLARY 9.1
If S is a set of positive formulas then F is satisfiable.

We also observe the following property of valuations in their "set-of-variables" form.

PROPOSITION 9.2
If φ is a positive formula, M, N are two sets of variables, $M \subseteq N$ and $M \models \varphi$ then $N \models \varphi$.

Proof: By induction on the complexity of formula φ (once again it is enough to consider negation-normal form formulas). □

On analogy with the concept of a positive formula we introduce a *negative formula* as a formula without positive occurrences of variables (in other words one whose canonical negation-normal form has no positive occurrences of variables. For instance $\neg(p \wedge q)$ is a negative formula. In fact let us observe that the negation of a positive formula is negative, and the negation of a negative formula is positive. Like positive formulas, negative formulas are closed under conjunctions and alternatives. We now define $\mathbf{0}_{Var}$ as a valuation that assigns to every variable the Boolean value 0. We then have the following fact.

PROPOSITION 9.3
If φ is a negative formula then $\mathbf{0}_{Var} \models \varphi$.

Corollary 9.1 has an analogue for negative formulas.

COROLLARY 9.2
If F is a set of negative formulas then F is satisfiable.

Likewise Proposition 9.2 has its analogue.

PROPOSITION 9.4
If φ is a negative formula, M, N are two sets of variables, $M \supseteq N$ and $M \models \varphi$ then $N \models \varphi$.

9.2 Horn formulas

Let us recall that a Horn clause is a clause $l_1 \vee \ldots \vee l_k$ such that at most one of literals $l_1, \ldots, l_l$ is positive. In other words a Horn clause is either of the form $p \vee \neg q_1 \vee \ldots \vee \neg q_m$ or of the form $\neg q_1 \vee \ldots \vee \neg q_m$. We further classify the clauses of the form $p \vee \neg q_1 \vee \ldots \vee \neg q_m$ as *program clauses* and clauses of the form $\neg q_1 \vee \ldots \vee \neg q_m$ as *constraints*. A Horn CNF is a CNF consisting of Horn clauses. A Horn clause consisting of program clauses is called a (Horn) *logic program*.

It will sometimes be useful to write the clause $p \vee \neg q_1 \vee \ldots \vee \neg q_m$ as an implication $q_1 \wedge \ldots \wedge q_m \Rightarrow p$. Moreover, it often will be natural in the context of Horn clauses to think about valuations as subsets of the set *Var*.

The fundamental property of Horn clauses is their relationship with families of sets of variables closed under intersections. Specifically, a family $\mathcal{X}$ of subsets of the set *Var* is *closed under intersections* if for every nonempty family $\mathcal{Y} \subseteq \mathcal{X}$, the intersection of all sets in $\mathcal{Y}$, $\bigcap \mathcal{Y}$ belongs to $\mathcal{X}$.

We will now focus on the case of the finite set *Var*. We will comment on the issue of infinite sets of variables later. Recall that for a given set of formulas F, $Mod(F)$ is the set of all valuations v of *Var* such that v satisfies F, i.e., $\{v : v \models F\}$. Under the convention identifying sets of variables with valuations, $Mod(F)$ can be thought of as a family of subsets of *Var*. Here is the fundamental representability result for Horn CNFs.

THEOREM 9.1 (Horn theorem)

Let Var be a finite set of variables and let $\mathcal{S} \subseteq \mathcal{P}(Var)$ be a nonempty family of sets. Then $\mathcal{S}$ is of the form $Mod(H)$ for some collection H of Horn clauses over Var if and only if $\mathcal{S}$ is closed under intersections.

Proof: First, let us assume that H is a collection of Horn clauses. We prove that $Mod(H)$ is closed under intersections. Let $\mathcal{Y}$ be a nonempty family of models of H. We think about the elements of $\mathcal{Y}$ as sets of variables. Let us consider $M = \bigcap \mathcal{Y}$. Let $C \in H$.
Case 1: C is a constraint, that is, $C = \neg q_1 \vee \ldots \vee \neg q_m$. Consider any $N \in \mathcal{Y}$. Since $N \models C$, for some i, $1 \leq i \leq m$, $q_i \notin N$. But $M \subseteq N$, so $q_i \notin M$. Thus $M \models C$, as desired.
Case 2: C is a program clause, $C = p \vee \neg q_1 \vee \ldots \vee \neg q_m$. Take any $N \in \mathcal{Y}$. If for some i, $1 \leq i \leq m$, $q_i \notin N$, then as $M \subseteq N$, $q_i \notin M$, and $M \models C$. So the remaining case is that for all $N \in \mathcal{Y}$ and for all i, $1 \leq i \leq m$, $q_i \in N$. But then, since all sets N in $\mathcal{Y}$ satisfy F, it must be the case that $p \in N$. But then $p \in \bigcap \mathcal{Y}$, that is $p \in M$. Then $M \models C$. This completes the proof of the implication $\Leftarrow$.

Let us observe that the part $\Leftarrow$ of our argument does not depend on the finiteness of the set of variables *Var*.

Now, we assume that a family $\mathcal{S}$ of subsets of a set *Var* is closed under intersections.

We need to exhibit a set of Horn clauses H such that $\mathcal{S} = Mod(H)$. It is quite clear what this set of clauses H should be. Namely, let F be the set of all clauses satisfied in all sets from $\mathcal{S}$ (i.e., $F = Th(\mathcal{S})$) or to eliminate the use of operator Th, $F = \{C : \forall_{M \in \mathcal{S}} M \models C\}$. Then we define $H = F \cap \mathcal{H}$ where $\mathcal{H}$ is the set of all Horn clauses. In simple words, H is the set of all Horn clauses that are satisfied in all sets of $\mathcal{S}$.

Since $H \subseteq F$, every clause $C \in H$ must be satisfied by all sets of $\mathcal{S}$. That is, every set in $\mathcal{S}$ is a model of H. So, all we have to show is the converse inclusion, that is, that every model of H is in $\mathcal{S}$. To this end we prove that if $\mathcal{S}$ is closed under intersections then for every $C \in F$ there is a Horn clause $C' \subseteq C$ (i.e., C' subsumes C) such that $C' \in F$.

Our claim indeed entails the converse inclusion. First, since Var is finite, for every $\mathcal{Y} \subseteq \mathcal{P}(Var)$, $\mathcal{Y}$ is of the form $Mod(F)$ for a set of clauses F (Proposition 3.16). Thus $\mathcal{S}$ is of the form $Mod(F)$ for some F. So now, since $H \subseteq F$, $Mod(F) \subseteq Mod(H)$, and once we prove the subsumption result claimed above, whenever $M \models H$, $M \models F$ as well, thus $M \in \mathcal{S}$. For if $C \in F$, take $C' \subseteq C$, $C' \in H$. Then $M \models C'$ (because $C' \in H$, and so $M \models C$ because C is subsumed by C'). Since $\mathcal{S} = Mod(F)$ and $Mod(F) = Mod(H)$, we will be done.

Thus we need to prove the following: Whenever C is a clause, and C is satisfied in all $M \in \mathcal{S}$ then there is a Horn clause C' such that $C' \subseteq C$ and for all $M \in \mathcal{S}$, $M \models C'$.

Let us take $C \in F$. Since C itself is finite, there must be a clause $C' \subseteq C$ such that C' is inclusion-minimal among clauses in F. This means that $C' \in F$ but any C''' properly contained in C' does not belong to F. We claim that C' is, in fact, a Horn clause. This will be, clearly, enough.

To show that C' is Horn, assume, by way of contradiction, that C' is not Horn. Then it must be the case that for some variables p_1 and p_2, and some clause C'', $C' : p_1 \vee p_2 \vee C''$. But C' is inclusion-minimal in F. This, in turn, means that the clauses $p_1 \vee C''$ and $p_2 \vee C''$ do *not* belong to F. This means that for some $M_1 \in \mathcal{S}$, $M_1 \not\models p_1 \vee C''$, and for some $M_2 \in \mathcal{S}$, $M_2 \not\models p_2 \vee C''$. Now we have to look at C''. Assuming

$$C'' = q_1 \vee \ldots \vee q_i \vee \neg q_{i+1} \vee \ldots \vee \neg q_{i+j}$$

we have

$$p_1 \notin M_1, q_1, \ldots, q_i \notin M_1, q_{i+1}, \ldots, q_{i+j} \in M_1$$

and

$$p_2 \notin M_2, q_1, \ldots, q_i \notin M_2, q_{i+1}, \ldots, q_{i+j} \in M_2.$$

Now we take the intersection of M_1 and M_2, $M_1 \cap M_2$. Then, clearly,

$$p_1 \notin M_1 \cap M_2, p_2 \notin M_1 \cap M_2, q_1, \ldots, q_i \notin M_1 \cap M_2, q_{i+1}, \ldots, q_{i+j} \in M_1 \cap M_2.$$

But this means that $M_1 \cap M_2 \not\models C$, contradicting the fact that $M_1 \cap M_2$ belongs to $\mathcal{S}$. □

Later on, in Chapter 13, when we discuss the Schaefer theorem on the dichotomy of Boolean constraint satisfaction problems, it will be convenient to use Theorem

9.1 in a slightly different form. We will then use the closure properties of sets of valuations (treated as Boolean functions defined on Var). To this end, we introduce *bitwise conjunction* as an operation on Boolean functions. This will be an example of a *polymorphism*.[1] Let $a : (x_1, \ldots, x_n)$, and $b : (y_1, \ldots, y_n)$ be two Boolean vectors of the same length n. The bitwise conjunction of a and b is a vector c of length n such that for all i, $1 \leq i \leq n$, $c_i = a_1 \wedge b_i$. Given a set of variables $Var = \{p_1, \ldots, p_n\}$ with the ordering of variables $p_1 \leq \ldots \leq p_n$, valuations of the set Var are in one-to-one correspondence with Boolean vectors of length n. Thus we can define the bitwise conjunctions on those valuations.

It should be observed that bitwise conjunction (and its close relatives: bitwise negation and bitwise disjunction) are available in most reasonable programming languages.

We then have the following corollary:

COROLLARY 9.3
Let $\mathcal{M}$ be a set of valuations of Var. Then $\mathcal{M}$ is of the form $Mod(H)$ for some Horn theory H if and only if $\mathcal{M}$ is closed under bitwise conjunction.

Proof: We recall (Section 2.1) that valuations are characteristic functions of models. Then we observe that the following holds for arbitrary subsets X, Y of Var (operation $\wedge$ on the left-hand side is the bitwise conjunction):

$$\chi_X \wedge \chi_Y = \chi_{X \cap Y}.$$

Thus, clearly, $Mod(H)$ – treated as a family of sets – is closed under intersections if and only if $Mod(H)$ – treated as a set of valuations – is closed under bitwise conjunctions. □

Looking carefully at the proof of Theorem 9.1 we see that we used two facts. First, every collection of subsets of a finite set Var was of the form $Mod(F)$ for a suitably chosen set of clauses F. Second, we used the fact that $\mathcal{S}$ was closed under the intersection of *two* sets. But in fact, for finite families it entails the stronger property; by an easy induction we show that families closed under intersections of two sets are closed under all intersections of finite nonempty subfamilies.

Therefore, the proof of Theorem 9.1 tells us that for infinite sets Var stronger properties of formulas will be needed to deal with the two obstacles mentioned above (the form of families of the form $Mod(F)$, and the infinitary intersections). In particular it is easy to exhibit families closed under finite intersections but not under infinite intersections. One such family is the family of all cofinite (i.e., subsets $X \subseteq Var$ such that $Var \setminus X$ is finite) subsets of Var.

It should be observed that families of sets closed under intersections are fundamental both in computer science and in mathematics. In the case of mathematics, the

---

[1] The term *polymorphism* has a well-defined meaning in object-oriented programming. This is not the meaning we assign it here. We have no influence on the terminology used in different communities.

families of (say) subgroups of a group or subrings of a ring are closed under intersections. The family of all closed subsets of a topological space is closed under intersections. Theorem 9.1 tells us that, at least in the finite case we can represent such families by means of Horn theories. In view of the results further in this section it is a tremendous advantage.

If the formula H consists of program clauses only, the satisfiability problem for H is particularly simple.

PROPOSITION 9.5
A CNF H consisting of program clauses is satisfiable.

Proof: Since each clause C in H contains a positive literal, $\mathbf{1}_{Var} \models C$. □

For a program H, since the family $Mod(H)$ is closed under *all* intersections, H possesses a *least model*, namely the intersection of all its models. Thus we have the following.

PROPOSITION 9.6
Every Horn CNF consisting of program clauses possesses a least model.

We will denote the least model of a Horn formula H, $lm(H)$, provided it exists. For programs P, Proposition 9.6 tells us that $lm(P)$ exists.

In the case of Horn CNFs consisting of constraints only, the situation is also quite simple. All the constraints are negative formulas, and so, by Proposition 9.3, such CNF is satisfiable; it has a least model, and that model is $\mathbf{0}_{Var}$.

What about the arbitrary Horn CNFs? Here the situation is again quite simple, but we have to be careful: generally, a Horn formula does not have to be satisfiable. Consider the CNF consisting of two unit clauses: p and $\neg p$. It clearly is a Horn formula, and it is inconsistent.

Let us recall that if H is a Horn CNF, then H naturally splits into $H_1 \cup H_2$ where H_1 consists of program clauses in H and H_2 consists of constraints from H.

THEOREM 9.2
Let H be a Horn CNF, and $H_1 \cup H_2$ be its decomposition into program and constraint parts. Then H is consistent if and only if the least model of H_1, $lm(H_1)$, satisfies H_2.

Proof: Clearly, if $lm(H_1) \models H_2$ then $lm(H_1) \models H$ and so H is consistent. Conversely, assume M models H. Then $M \models H_1$, and so $lm(H_1) \subseteq M$. But H_2 consists of negative formulas only and so it is inherited "downwards" (Proposition 9.4). Thus $lm(H_1) \models H$, as desired. □

We then get a corollary which is a fundamental property of Horn theories.

COROLLARY 9.4

If a Horn theory T is satisfiable, then H possesses a least model.

Proof: Let $H = H_1 \cup H_2$ be a decomposition of H into program clauses and constraints. Let M_0 be the least model of H_1. We claim that M_0 is a least model of H. Indeed, if M is a model of H then M is a model of H_1 and so $M_0 \subseteq M$. On the other hand, by Theorem 9.2, M_0 is a model of H_1, thus of entire H. □

Since a consistent Horn theory possesses a least model, we get a useful fact about the derivation of propositional variables in Horn theories.

PROPOSITION 9.7

Let p be a propositional variable, and H be a consistent Horn theory. Then $H \models p$ if and only if $p \in lm(H)$.

Proof: Clearly, if $H \models p$, then p belongs to any set M such that $M \models H$, in particular $p \in lm(H)$.
Conversely, if $H \not\models p$ then there exists a model M of H such that $M \not\models p$. That is, $p \notin M$. But then, as $lm(H) \subseteq M$, $p \notin lm(H)$. □

Combining Proposition 9.7 with the Dowling-Gallier algorithm below we get the following fact.

COROLLARY 9.5

Let p be a propositional variable, and H be a Horn theory. Then we can test if $H \models p$ in linear time.

In view of Theorem 9.2 an algorithm to test satisfiability of a Horn theory could be devised. Namely, we first compute the partition of H into program and constraint parts, H_1 and H_2. Then we compute $lm(H_1)$. Then we test if $lm(H_1) \models H_2$. In this scheme of things, we need to see how complex is it to compute $lm(H_1)$, since the second task is quite simple.

It turns out that the problem of computing the least model of a program H_1 is quite inexpensive. We will see that, actually, computation of that least model reduces to computation of the least fixpoint of a certain monotone operator determined by H_1. Then we shall see how we can compute that least fixpoint in linear time in the size of the program.

Given a Horn *program* H, we shall write the clauses $C : p \vee \neg q_1 \vee \ldots \vee \neg q_k$ of H as $q_1 \wedge \ldots \wedge q_k \Rightarrow p$. This form accentuates the dependence of the variable p (called the *head* of the clause C and often denoted $head(C)$) on the set of variables $\{q_1, \ldots, q_k\}$ (called the *body* of C and denoted $body(C)$).

We now assign to a program H an operator T_H in the complete lattice $\mathcal{P}(Var)$, as follows.

$$T_H(M) = \{head(C) : C \in H, \text{ and } body(C) \subseteq M\}$$

Here is what happens. Let us say that a set M *matches* a Horn clause C if $body(C) \subseteq M$. If we think of a program H as a "machine" producing an output on input M, then the output is produced by matching M with the bodies of clauses C in H. If the matching occurs, the head of C is part of the output. Thus on an input M the output is a *set* of variables – namely of heads of clauses matched by M. Here is the simple fact that has tremendous consequences.

PROPOSITION 9.8
The operator T_H is monotone and continuous.

Proof: Let us assume that M_1 and M_2 are two subsets of Var, and that $M_1 \subseteq M_2$. Then, clearly, each clause $C \in H$ matched by M_1 is matched by M_2. But then the set of heads of clauses matched by M_1 is included in the set of heads of clauses matched by M_2, that is $T_H(M_1) \subseteq T_H(M_2)$.
For the continuity, let us observe that whenever $body(C) \subseteq \bigcup_{n \in N} X_n$, and $\langle X_n \rangle_{n \in N}$ is an increasing family of subsets of Var then, because $body(C)$ is finite, while N is infinite, there must be $m \in N$ such that $body(C) \subseteq \bigcup_{n < m} X_n$. This implies that $T_H(\bigcup_{n \in N} X_n) \subseteq \bigcup_{n \in N} T_H(X_n)$, and so T_H is a continuous operator.
□

By Proposition 9.8 and the Knaster-Tarski theorem, the operator T_H possesses a least fixpoint. Then we need to ask what is it. There is a natural candidate for this, and so we have the following result.

PROPOSITION 9.9 (van Emden and Kowalski theorem)
Let H be a Horn program. Then the least model of H coincides with the least fixpoint of the operator T_H.

Proof: Since the operator T_H is monotone and continuous, the least fixpoint of T_H is $\bigcup_{n \in N} T_H^n(\emptyset)$. Let us call this set N_H.
First, let us show that the least model of H, $lm(H)$ is included in N_H. To this end it suffices to show that N_H satisfies H. But indeed, if $C \in H$, $C = p \vee \neg q_1 \vee \ldots \vee \neg q_k$, then either some of q_j, $1 \leq j \leq k$ does not belong to N_H and so N_H satisfies C, or all q_j, $1 \leq j \leq k$, do belong to N_H. In this latter case, $p \in T_H(N_H)$. But N_H is a fixpoint of T_H, thus $p \in N_H$, and so N_H satisfies C.
The converse inclusion ($N_H \subseteq lm(H)$) requires induction. By Proposition 1.3 the set N_H is $\bigcup_{n \in N} T_H^n(\emptyset)$. By induction on n we show that $T_H^n(\emptyset) \subseteq lm(H)$. This is certainly true for $m = 0$, as $T_H^0(\emptyset) = \emptyset$. Assuming $T_H^n(\emptyset) \subseteq lm(H)$, we see that every clause of H that is matched by $T_H^n(\emptyset)$ is matched by $lm(H)$. But $lm(H)$ satisfies H. Then it is easy to see that for each clause C such that $lm(H)$ matches C, the head of C must belong to $lm(H)$. Thus $T_H^{n+1}(\emptyset) \subseteq lm(H)$.
Now we have established that for all $n \in N$, $T_H^n(\emptyset) \subseteq lm(H)$. Therefore

$$\bigcup_{n \in N} T_H^n(\emptyset) \subseteq lm(H),$$

i.e., $N_H \subseteq lm(H)$, as desired. □

In the case of the finite set *Var* of variables, all monotone operators in $\mathcal{P}(Var)$ are of the form T_H for a suitably chosen Horn program H.

PROPOSITION 9.10

Let Var be a finite set of variables. Then for every monotone operator $O : \mathcal{P}(Var) \to \mathcal{P}(Var)$, there is a Horn CNF H such that $T_H = O$.

Proof: Here is the construction of the desired H. Since *Var* is finite, so is every subset $M \subseteq Var$. Now, for a given $M \subseteq Var$, $M = \{q_1, \ldots, q_k\}$, define

$$H_M = \{q_1 \wedge \ldots \wedge q_k \Rightarrow p : p \in O(M)\}.$$

Then define $H = \bigcup_{M \subseteq Var} H_M$. Clearly, H is a Horn CNF, by construction. We will now show that for all $M \subseteq Var$, $T_H(M) = O(M)$.

The inclusion $\supseteq$ is obvious. Indeed, if $p \in O(M)$, $M = \{q_1, \ldots, q_k\}$, then the clause $q_1 \wedge \ldots \wedge q_k \Rightarrow p$ belongs to H and so p belongs to $T_H(M)$.

The converse implication uses the monotonicity of the operator O. Let us assume $p \in T_H(M)$. Then, for some clause $C : s_1 \wedge \ldots \wedge s_r \Rightarrow p$, M matches C. That is,

$$\{s_1, \ldots, s_r\} \subseteq \{q_1, \ldots, q_k\}.$$

Since $C \in T_H$, by the construction of the formula H, it must be the case that $p \in O(\{s_1, \ldots, s_r\})$. But then, by monotonicity of O, $p \in Q(\{q_1, \ldots, q_k\})$, i.e., $p \in O(M)$, as desired. □

Let us observe that, since the operator T_H is monotone, by the Knaster-Tarski fixpoint theorem T_H has a largest fixpoint as well. This largest fixpoint can be computed by iteration of T_H starting with the entire set of variables, *Var*. When *Var* is finite, this largest fixpoint will be reached in finitely many steps. But if H is infinite then, unlike in the case of the least fixpoint, the number of steps needed to reach that fixpoint may be larger than ω.

The representation of the least model $lm(H)$ of a Horn program H as the least fixpoint of the operator T_H immediately entails a polynomial-time algorithm for computation of $lm(H)$. Here is how this can be done. We start with the empty set. Then, in each round of the algorithm we scan the program H for those unused clauses which are matched by the currently computed set of variables. Those clauses are marked as used, and their heads are added to the currently computed set of variables. If the currently computed set of variables does not increase in a round, we halt and output the currently computed set of variables, as the least model of H.

Since in each round the size of the set of unused clauses decreases, while the size of the set of currently computed variables increases, it is clear that we will halt after the number of rounds bounded by the minimum of the number of clauses in the program H and the number of variables occurring in the program H, thus, clearly, by the size of the program H. Within each round, the work performed by our algorithm is bound by the size of the program. Therefore this algorithm computes the least model of a program H in at most $|H|^2$ steps. But it turns out that we can do better. This is the essence of the Dowling-Gallier algorithm [DG84] for computation of the least model of program H.

The idea of the Dowling-Gallier algorithm is to introduce the *counter* for the body of each clause. Each variable is pointing to each clause in whose body it occurs. Each clause points to the variable that constitutes its head. We will assume that the variables occur in the body of each clause at most once. Then, as a variable q is added to the set of currently computed variables, the counters of the clauses that have that variable q in the body are decremented by one. When the counter associated with a clause C is decremented to 0, the body of the clause C is entirely included in the set of currently computed variables. Then the head p of the clause C is added to the set of currently computed variables (if it is not already computed!), and we continue to decrement counters, this time of clauses containing p. But now, in this computation, we see that each variable will be processed at most once, and in fact the variables that do not end up in the least model will not be visited at all. The effect of this is that we will compute the least model of H in time bound by a constant times the size of H.

We certainly can assign a counter to a constraint as well. We will decrement that counter by 1 each time a variable occurring in a constraint is included in the least model. But now our limitation is that we *cannot* get such counter equal 0. If this happens, the constraint in question is not satisfied and we fail. Let us look at an example.

Example 9.1

Let H consist of the following four clauses:

$C_1 : p.$
$C_2 : \neg p \vee q.$
$C_3 : \neg p \vee \neg q \vee r.$
$C_4 : \neg p \vee \neg r.$

The counters associated with these clauses initially hold, respectively: 0, 1, 2, and 2. In the first round we derive variable p. The remaining counters hold 0, 1, and 1. In the second round we derive q. Now the counters associated with C_2 and C_3 contain 0 and 1. In the next round we derive r. The counter associated with C_4 goes down to 0 and we fail.

But now let us alter the clause C_4 to $C_4' : \neg p \vee \neg s$. At the end of the computation we have the counter associated with C_4; down to 1, but not to 0. Thus the CNF $H' : \{C_1, C_2, C_3, C_4'\}$ is consistent. □

Let us observe that the collection of Horn clauses is closed under resolution. That is, if two Horn clauses are resolvable, then the resolvent is also a Horn clause. But an interesting twist on resolution in Horn case is that a limited form of resolution can be used for *computation*. A *unit resolution* is a restriction of the resolution proof rule to the situation where at least one of the inputs is a unit clause. The unit resolution proof of a clause C is a resolution proof of C where all the applications of resolution are unit resolutions. As before we can represent unit resolution proofs as trees. In those proofs, each application of resolution must have at least one input being a unit clause (i.e., a literal). Here is an appealing property of Horn programs.

LEMMA 9.1
Let H be a Horn program and p be a propositional variable. Then the variable p belongs to the least model $lm(H)$ if and only if there is a unit resolution proof of p from H. Moreover, in that proof, every unit input is a variable.

Proof: First, assume that we have a unit resolution proof of p from H. Then $H \models p$, and so $p \in lm(P)$.

Conversely, for each variable $p \in lm(P)$ we shall construct a unit resolution proof D for p. We will use the fixpoint characterization of $lm(H)$ proved in Proposition 9.9. Namely, the model $lm(H)$ is the union $\bigcup_{n \in N} T_h^n(\emptyset)$. Thus, by induction on n we will prove that for each $n \in N$, for each $p \in T_H^n(\emptyset)$, p possesses a unit resolution proof. The base case is obvious since $T_H^0(\emptyset) = \emptyset$. Now, assuming that the desired property is true for all $q \in T_H^n(\emptyset)$, consider $p \in T_H^{n+1}(\emptyset)$. Then there is a clause

$$C : p \vee \neg q_1 \vee \ldots \vee \neg q_k$$

in H such that $q_1, \ldots, q_k$ all belong to $T_H^n(\emptyset)$. By inductive assumption, each q_i, $1 \leq i \leq k$ possesses a unit resolution proof D_j. We now combine these proofs, as in Figure 9.1. The resulting derivation is a unit derivation of p. □

Let us define $Res^1(H)$ as the closure of H under the *unit* resolution rule of proof. Clearly $Res^1(H) \subseteq Res(H)$, in particular the unit resolution is *sound*.

We will now use Lemma 9.1 to get a completeness result for Horn clauses and unit resolution. Let us observe that since Horn programs are always consistent, the resolution proofs, and in particular unit resolution proofs cannot produce the empty clause $\emptyset$. The situation changes when constraints are permitted.

PROPOSITION 9.11
Let H be a Horn CNF. Then H is inconsistent if and only if $\emptyset \in Res^1(H)$. In other words unit resolution is all that is needed to establish inconsistency of Horn CNFs.

Proof: If $\emptyset \in Res^1(H)$, then, in particular $\emptyset \in Res(H)$ and so H is inconsistent.

Conversely, assume that H is inconsistent. Then, by Theorem 9.2, the least model $lm(H_1)$ of program part H_1 of H does not satisfy some constraint $C : \neg q_1 \vee \ldots \vee \neg q_k$ of H_2. This means that all q_j, $1 \leq j \leq k$, belong to $lm(H_1)$. By Lemma 9.1 all these variables q_j possess a unit resolution proof D_j from H_1, thus from H. We will now combine these proofs and the clause C to a derivation of a contradiction. This combination is shown in Figure 9.2. □

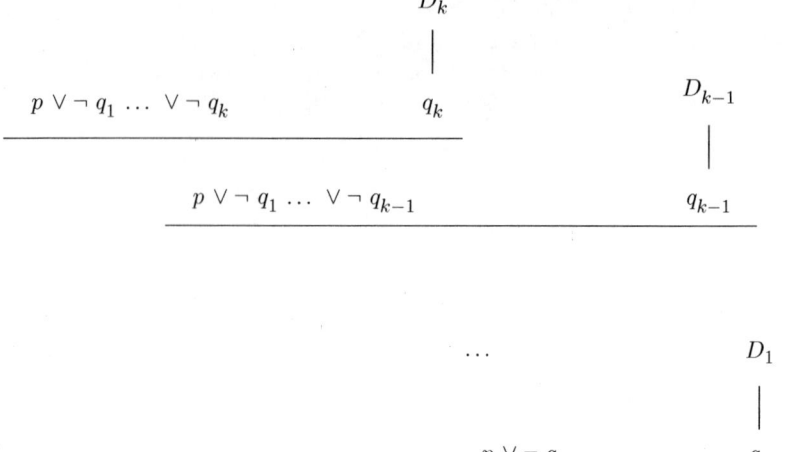

FIGURE 9.1: Combining unit derivations

9.3 Autarkies for Horn theories

We will study nonempty autarkies for Horn theories, that is, collections of Horn clauses. Recall that a Horn clause is a clause with at most one positive literal. Such clause C is a *program clause* if C contains exactly one positive literal and is a *constraint* if it contains no positive literals. Thus a Horn theory H splits into the union $H_1 \cup H_2$ of the set of its program clauses and the set of its constraints.

First we discuss general results for autarkies of Horn theories. When v is a partial valuation that is, a consistent set of literals, then v splits into its positive part v_P and negative part v_N, according to the polarity (i.e., sign) of literals. v_P contains positive literals from v and v_N negative literals from v. Then it is natural to say that v is *positive* if $v = v_P$, and v is *negative* if $v = v_N$.

We start with a useful property of autarkies for Horn theories.

PROPOSITION 9.12
Let H be a Horn theory, and v a partial valuation containing at least one negative literal. If v is an autarky for H, then v_N is an autarky for H as well.

Proof: Let us assume that v is an autarky for H. Assume that v_N touches a clause $C \in H$.
Case 1: If C is a constraint, then v_N touches C on a negative literal, and so v_N being negative itself must contain that literal. Thus $(v_N)_3(C) = 1$.

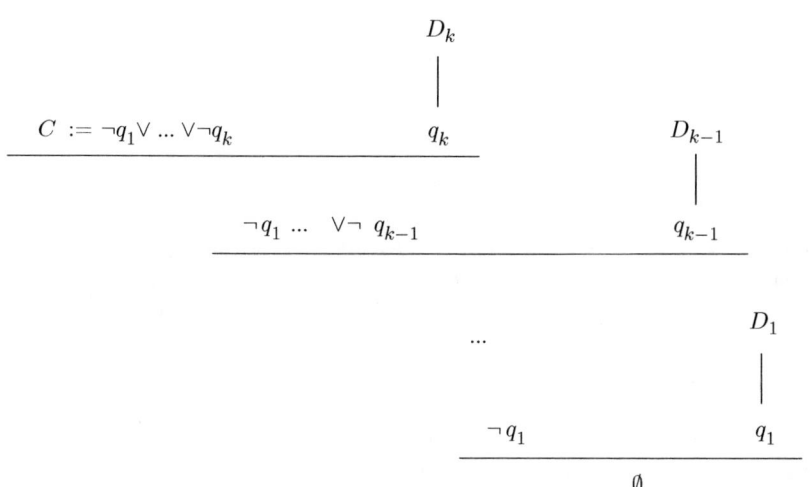

FIGURE 9.2: Getting contradiction using unit derivations and unsatisfied constraint

Case 2: Now, let C be a program clause and v_N touches C. If one of the literals of v occurs in C we are done. So now assume that

$$C : p \vee \neg q_1 \vee \ldots \vee \neg q_m$$

and that $\neg p \in v_N$ (that is v_N happens to touch C on p). Now, since $v_N \subseteq v$, v also touches C. Then $v_3(C) = 1$ because v is an autarky for H, and so one of literals $p, \neg q_1, \ldots, q_m$ must belong to v. But it is not p because we assumed $\neg p \in v$, and v is consistent. Therefore for some i, $1 \leq i \leq m$, $\neg q_i \in v$, and so that $\neg q_i$ belongs to v_N. But then $(v_N)_3(C) = 1$, as desired. □

Now we find that Horn theories have a surprising property.

PROPOSITION 9.13 (Truszczynski)
Let H be a Horn theory. If H possesses an autarky, then H possesses a positive autarky, or H possesses a negative autarky.

Proof: Let us assume that H possesses an autarky, but not a positive autarky. Let v be an autarky for H. Then v, being nonempty, contains at least one negative literal. But then v_N is an autarky for H, and, of course, v_N is negative. □

Now, let us observe that not all Horn theories possess autarkies, for instance a theory consisting of two unit clauses: p and $\neg p$ is inconsistent and also has no nonempty autarky. Then there are clausal theories (of course non-Horn) that have no positive and no negative autarkies (but possess autarkies). One such theory is $\{p \vee q, \neg p \vee \neg q\}$. It has two autarkies (which happen to be satisfying valuations) $\{p, \neg q\}$ and $\{q, \neg p\}$.

Now we get the following corollary.

COROLLARY 9.6
Let H be a Horn theory possessing autarkies. Then every inclusion-minimal autarky for H is positive or negative.

Proof: If v is a positive autarky for H then every $\subseteq$-smaller autarky for H is positive. Similarly for negative autarkies. But if v is an autarky for H containing both positive and negative literals, then by Proposition 9.13, the assignment v cannot be inclusion-minimal autarky for H. □

Unfortunately, neither of the results above tells us how to *test* if a Horn theory possesses an autarky. We will investigate this problem now. To this end we discuss positive and negative autarkies of arbitrary clausal theories.

Hence our next step concerns properties of positive autarkies for *arbitrary* clausal theories. Subsequently, we will use this result and a related one to study autarkies for Horn theories. Let us recall the formulas φ_V where φ is a formula, and V is a set of variables. φ_V was the effect of substituting the positive occurrences of variables from V in F by $\bot$ and negative occurrences of variables from V by $\top$.

Since we are dealing with clauses, let us get a better feel of the form of clause C_V. Let C be the clause

$$p_1 \vee \ldots \vee p_k \vee p_{k+1} \vee \ldots \vee p_{k+r} \vee \neg q_1 \vee \ldots \vee \neg q_l \vee \neg q_{l+1} \vee \ldots \vee \neg q_{l+s},$$

where $p_1, \ldots, p_k, q_1, \ldots, q_l \notin V$ and $p_{k+1} \ldots, p_{k+r}, q_{l+1}, \ldots, q_{l+s} \in V$. When we compute C_V we get:

$$C_V : p_1 \vee \ldots \vee p_k \vee \bot \vee \ldots \vee \bot \vee \neg q_1 \vee \ldots \vee \neg q_l \vee \neg \top \vee \ldots \vee \neg \top.$$

Surely all the $\bot$ and $\neg \top$ can be eliminated from C_V without changing semantics. Thus C_V amounts to simply eliminating from C all literals l where $|l|$ belongs to V.

PROPOSITION 9.14
Let F be a set of clauses, $\mathcal{C}_F$ its set of constraints, and let $V = Var(\mathcal{C}_F)$. If v is a set of positive literals disjoint from V, then v is an autarky for F if and only if v is an autarky for F_V.

Proof: First, let us assume that v is an autarky for F. Then since v consists of positive literals only, it cannot satisfy any constraint. Therefore it must be the case that $Var(v) \cap V = \emptyset$. So let us assume that v touches some clause $C \in F_V$. Then $C = D_V$ for some $D \in F$. Since v touches C, v touches D. Since v is an autarky for F, $v_3(D) = 1$. But then by Proposition 3.2 $v_3(D_V) = 1$, thus $v_3(C) = 1$, as desired.

Conversely, let us assume that v is an autarky for F_V. Let $C \in F$. We know that C looks like the following:

$$p_1 \vee \ldots \vee p_k \vee p_{k+1} \vee \ldots \vee p_{k+r} \vee \neg q_1 \vee \ldots \vee \neg q_l \vee \neg q_{l+1} \vee \ldots \vee \neg q_{l+s},$$

where $p_1, \ldots, p_k, q_1, \ldots, q_l \notin V$ and $p_{k+1} \ldots, p_{k+r}, q_{l+1}, \ldots, q_{l+s} \in V$. Let $D = C_V$, i.e.,
$$D : p_1 \vee \ldots \vee p_k \vee \neg q_1 \vee \ldots \vee \neg q_l.$$

Since v touches C one of propositional variables $p_1, \ldots, p_{k+r}, q_1, \ldots q_{l+s}$ belongs to v (because v is positive). But this variable cannot be $p_{k+1}, \ldots, p_{k+r}, q_{l+1}, \ldots, q_{l+s}$ because these latter variables belong to V. Thus one of $p_1, \ldots, p_k, q_1, \ldots, q_l$ occurs in v. That is, v touches D. But v is an autarky for F_V. Thus $v_3(D) = 1$. But all literals of D occur in C, and so $v_3(C) = 1$. Since C is an arbitrary clause in F touched by v, v is an autarky for F. □

We now introduce the notion of a *dual-constraint*. A *dual-constraint* is a clause of the form $p_1 \vee \ldots \vee p_k$ where $p_1, \ldots, p_k$ are variables. Let $\mathcal{D}_F$ be the set of dual constraints of F. Here is a fact similar to Proposition 9.14.

PROPOSITION 9.15
Let F be a set of clauses, $\mathcal{D}_F$ its set of dual-constraints, and let $V = Var(\mathcal{D}_F)$. If v is a set of negative literals disjoint from V, then v is an autarky for F if and only if v is an autarky for F_V.

One proof of Proposition 9.15 follows the line of the proof of Proposition 9.14. Amateurs of algebra will follow an alternative line. They will apply the permutation π of literals such that for all l, $\pi(l) = \bar{l}$. This permutation moves positive partial valuations to negative partial valuations, constraints to dual constraints and vice-versa. Then, after applying π to F we can use directly Proposition 9.14, and then apply π again. Small technical details (like showing that $\pi(C_V) = \pi(C)_V$) can be easily filled by the reader.

Propositions 9.14 and 9.15 imply the following corollary.

COROLLARY 9.7
It is possible to establish if an arbitrary set of clauses F possesses a positive autarky, or a negative autarky in linear time in the size of F, $|F|$.

Our final goal is to *compute* autarkies of Horn theories. Let us look for a moment for the case of positive autarkies, and limit ourselves to the case of Horn theories. Let us start with a finite set of Horn clauses H and let us define two sequences $\langle H_i \rangle$ and $\langle V_i \rangle$ inductively as follows. $H_0 = H$, $V_0 = Var(\mathcal{C}_H)$. $H_{i+1} = H_{V_i}$, $V_{i+1} = Var(\mathcal{C}_{H_i})$. Here is what happens. We run the following loop: we compute the set of variables occurring in constraints of the current theory. Then we reduce the current theory by the computed set of variables, get a new theory, and continue.

What are possible final outcomes? Either the current theory becomes empty, or it is not empty but the set of its constraints becomes empty and there is nothing to reduce by. Let us look at each of these possibilities.

Outcome 1: The current theory becomes empty. Then this current theory has no autarkies, in particular positive autarkies. Chaining back, we report that the original

Horn theory H had no positive autarkies.

Outcome 2: The current theory is not empty, but it has no constraints. Then it means that the last theory H_k in our sequence is a Horn program. Now, a Horn program definitely possesses a positive autarky, namely the collection M of those variables p so that p occurs positively in some clause of H_k is a positive autarky for H_k. But by repeated application of Proposition 9.14 all H_is have exactly the same positive autarkies! Thus we return the set M as an autarky for H.

Let us call the algorithm outlined above A_+. Our previous discussion results in the following.

PROPOSITION 9.16

The algorithm A_+ is correct. If a Horn theory H possesses a positive autarky, then the algorithm A_+ returns one such autarky. If H does not possess a positive autarky, then this will be established by the algorithm A_+. Moreover, the number of runs of the loop in the run of A_+ is bound by $|Var(H)|$ and the algorithm A_+ runs in linear time in the size of H.

We are now ready to discuss the computation of negative autarkies for Horn theories. We proceed as above, in the case of positive autarkies, but now we use Proposition 9.15. Recall that in that proposition we have established that if v is a set of negative literals, and the set of variables V occurring in dual constraints of F is disjoint from the variables occurring in v then v is an autarky for F if and only if v is an autarky for F_V. There are several things that we need to take into account in this case when we deal with a Horn CNF, H. First, let us observe that since H is Horn, the reduction H_V is again Horn. Second, we observe that since H is Horn, the dual constraints are necessarily positive units (for these are the only dual-constraints that are Horn clauses). But now, we observe that if we proceed as before and accumulate the bigger and bigger collection of positive units by which we reduce (notice that after we reduce by some units we may create more units, etc.) we compute *precisely* the least fixpoint of the operator T_P where P is the program part of H! We can either reduce by means of positive units and then iterate this operation, or simply compute the fixpoint of T_P and then reduce by it. Because the fixpoint of T_P is computed, it follows that the resulting reduct has the property that each of its clauses (if any still exists after reduction) possesses at least one negative literal. Two outcomes are possible:

Outcome 1: The reduct by this fixpoint M becomes empty. Then the resulting theory has no nonempty negative autarkies, and so the original theory H has no nonempty autarky.

Outcome 2: The reduct is not empty. Then the set of all negative autarkies of that reduct H' has precisely the same negative autarkies as H. It is also clear that H' possesses at least one negative autarky: it is the set of all negative literals occurring in H'. We call the simple algorithm we described (computing the least fixpoint of T_P, where P is the program part of H, eliminating that set of variables from H, reporting '*no negative autarkies*' if that reduction is empty, or the set of all negative literals in the reduction, otherwise) A_-. We then have the following.

PROPOSITION 9.17
The algorithm A_- is correct. If a Horn theory H possesses a negative autarky, then the algorithm A_- returns one such autarky. If H does not possess a negative autarky, then this will be established by the algorithm A_-. Moreover, the algorithm A_- runs in linear time in the size of H.

What is the effect of Propositions 9.16 and 9.17? Since a Horn theory that possesses an autarky must possess a positive autarky or a negative autarky (Proposition 9.13), if both our algorithms A_+ and A_- fail to find one, it must be the case that H has no autarkies at all.

We thus have the following corollary.

COROLLARY 9.8 (Truszczynski)
It is possible to establish if a Horn theory H possesses an autarky in linear time in the size of H.

9.4 Dual Horn formulas

A *dual Horn* clause is a clause $C : l_1 \vee \ldots \vee l_k$ where at most one of literals l_j, $1 \le j \le k$ is a negative literal. So, for instance, the clause $p \vee q \vee \neg r$ is a dual Horn clause, while $\neg p \vee \neg q$ is not a dual Horn clause. A dual Horn CNF is a CNF consisting of dual Horn clauses.

Let us introduce the following permutation Inv of the set of literals.

$$Inv(l) = \begin{cases} \neg p & \text{if } l = p \\ p & \text{if } l = \neg p. \end{cases}$$

In other words, $inv(l) = \bar{l}$. It should be clear that the permutation Inv is its own inverse; $Inv^{-1} = Inv$.

The permutation Inv, as any permutation, extends its action to clauses. That is, when $C = l_1 \vee \ldots \vee l_k$, we set $Inv(C) = Inv(l_1) \vee \ldots \vee Inv(l_k)$. The following is an easy observation that relates Horn and dual Horn CNFs.

PROPOSITION 9.18
Let F be a CNF formula. Then F is a dual Horn if and only if $Inv(F)$ is Horn. Likewise, F is Horn if and only if $Inv(F)$ is dual Horn.

If instead of looking at valuations we look at sets of variables, we see that under the identification of valuations and sets, $Inv(M) = Var \setminus M$. Then, Proposition 2.23, implies the following.

PROPOSITION 9.19
Let M be a set of variables, and let F be a CNF formula. Then $M \models F$ if and only if $Var \setminus M \models Inv(F)$.

The permutation Inv, when thought about in terms of sets (instead of valuations) reverses inclusion. Since Inv is one-to-one operation on sets (different sets have different complements), we can extend the action of Inv to the next type (from sets, to families of sets). That is, we can make Inv act on sets of sets of variables by letting
$$Inv(\mathcal{X}) = \{Inv(X) : X \in \mathcal{X}\}.$$
The following property of families of sets is an obvious consequence of De Morgan laws for sets.

PROPOSITION 9.20
Let $\mathcal{X}$ be a family of subsets of the set of variables Var. Then $\mathcal{X}$ is closed under unions if and only if $Inv(\mathcal{X})$ is closed under intersections. Likewise, $\mathcal{X}$ is closed under intersections if and only if $Inv(\mathcal{X})$ is closed under unions.

Proposition 9.20 allows us to find the following counterpart of Theorem 9.1.

THEOREM 9.3
Let Var be a finite set of variables and let $\mathcal{S} \subseteq \mathcal{P}(Var)$ be a nonempty family of sets. Then $\mathcal{S}$ is of the form $Mod(dH)$ for some collection dH of dual Horn clauses over Var if and only if $\mathcal{S}$ is closed under unions.

As in Section 9.2 we can think in terms of valuations instead in terms of sets. There, we introduced the operation of bitwise conjunction. Here we consider the operation of *bitwise disjunction*.

Let $a : (x_1, \ldots, x_n)$, and $b : (y_1, \ldots, y_n)$ be two boolean vectors of the same length n. The bitwise disjunction of a and b is a vector c of length n such that for all i, $1 \leq i \leq n$, $c_i = a_1 \vee b_i$. As before, given a set of variables $Var = \{p_1, \ldots, p_n\}$ with the ordering of variables $p_1 \leq \ldots \leq p_n$, valuations of the set Var are in one-to-one correspondence with Boolean vectors of length n. Thus, on analogy with bitwise conjunctions, we can define the bitwise disjunctions on those valuations.
We then have the following corollary.

COROLLARY 9.9
Let $\mathcal{S}$ be a family of valuations of a finite set Var. Then $\mathcal{S}$ is of the form

$Mod(dH)$ for a set of dual Horn clauses dH if and only if S is closed under bitwise disjunction.

It should be clear that all the results of Section 9.2 will have counterparts for the case of dual Horn CNFs. Moreover, the proofs can always be found by first applying permutation Inv, then doing some work on the Horn side, then coming back by means of another application of Inv. This technique will always work as long as the argument on the Horn side is semantic in nature. For instance, let us define a dual Horn program clause as one that has exactly one negative literal, and a dual Horn program as a CNF consisting of dual Horn program clauses. Then based on Propositions 9.5 and 9.6 and on the fact that inclusion is reversed by Inv we get the following.

PROPOSITION 9.21

1. A CNF dH consisting of dual Horn program clauses is satisfiable.
2. Every dual Horn CNF consisting of dual program clauses possesses a largest model.

The technique of monotone operators can, equally, be applied to dual Horn programs. Let dH be a dual Horn program. We have two options. The first is to apply permutation Inv, getting a Horn program H. The program H determines its operator O_H. The least fixpoint of that operator generates, via application of inversion (i.e., complement), the largest model of dH. The second possibility is to define the operator associated with dH directly. Here is how we do this:

$$O_{dH}(M) = Var \setminus \{q : \exists_{C \in dH} \ C = p_1 \vee \ldots p_k \vee \neg q \ \& \ p_1, \ldots, p_k \notin M\}.$$

This operator O_{dH} is monotone and continuous in the lattice $\mathcal{P}(Var)$ but with the ordering relation reversed. The 'least' fixpoint of O_{dH} is inclusion-largest among the models of dH. Perhaps a bit surprisingly, the operator O_{dH} is also monotone in the lattice $\langle \mathcal{P}(Var), \subseteq \rangle$, but not, in general, continuous.

We also observe that further results of Section 9.2 also generalize, as long as we are careful and see which lattice is actually being used. In the case of Lemma 9.1 concerned with unit resolution the reformulation requires some care.

LEMMA 9.2
Let dH be a dual Horn program and p be a variable. Then the variable p is false in the largest model of pH if and only if there is a unit resolution proof of $\neg p$ from H. Moreover, in that proof, every unit input is a negated variable.

On the other hand, Proposition 9.11 lifts to the present context (dual Horn CNFs) verbatim.

PROPOSITION 9.22

Let dH be a dual Horn CNF. Then H is inconsistent if and only if $\emptyset \in Res^1(H)$. In other words unit resolution is all that is needed to establish inconsistency of dual Horn CNFs.

The algorithmic aspects of computing the largest model of a dual Horn program, and of testing consistency of dual Horn CNFs are very similar to the case of Horn programs. The reason is that the operation Inv is inexpensive both for valuations and for formulas. In either case the cost is linear in the size of the formula. The net effect is that the Dowling-Gallier algorithm can be applied (either directly with minimal modifications, or by going through the transformation Inv, executing the Dowling-Gallier algorithm and coming back, by another application of Inv). In either case the complexity of the corresponding algorithms does not increase.

Finally, let us look at the issue of autarkies for the dual Horn CNFs. Here, everything about computation of autarkies lifts from the Horn case verbatim. The reason is that the permutation Inv transforms Horn theories to dual Horn theories and vice versa. Permutation Inv preserves autarkies; $Inv(v)$ is an autarky for $Inv(F)$ is and only if v is an autarky for F. Moreover, if v is a set of positive literals then $Inv(v)$ is a set of negative literals, and vice versa.

Taking all this into account, we collect the properties demonstrated in our considerations on autarkies for Horn theories and get a complete picture for dual Horn theories. Let us recall that for a set of literals v, v_P is its positive part, while v_N is its negative part.

PROPOSITION 9.23

Let dH be a dual Horn CNF, v a set of literals containing at least one positive literal.

1. If v is an autarky for dH, then v_P is an autarky for dH.

2. If dH possesses an autarky, then dH possesses a positive autarky or dH possesses a negative autarky.

3. All inclusion-minimal autarkies for dH are positive or negative.

The algorithms A_+ and A_- work for us now exactly as they did in the case of Horn theories. Then, obviously, Corollary 9.8 lifts verbatim to the dual Horn case. We thus have the following corollary.

COROLLARY 9.10

It is possible to establish if a dual Horn theory dH possesses an autarky in linear time in the size of dH.

9.5 Krom formulas and 2-SAT

When a clause has at most two literals, that is empty, a unit clause l or of the form $l_1 \vee l_2$, we call it a *Krom* clause, or 2-clause. A *Krom CNF* (often called in literature 2CNF) is a collection of Krom clauses. Here is a property of Krom clauses that makes it distinct. The set of Krom clauses is closed under the application of resolution rule of proof. Indeed, if $C_1 : l_1 \vee l_2$ and $C_2 : \bar{l}_1 \vee l_3$ are resolvable Krom clauses then the resolvent is $l_2 \vee l_3$, again a Krom clause.

When $|Var| = n$, there is $\binom{2n}{2} + 2 \cdot n + 1$ Krom clauses, thus $4 \cdot n^2 + 1$ Krom clauses altogether. This bound on the number of Krom clauses shows that the DP algorithm will never take more than $O(n^2)$ space (recall that we have to store not only the resolvents, but also the sets of clauses containing a given variable). This, in turn, shows that there is a polynomial $d(n)$ so that the number of steps executed in the first phase of the DP algorithm is bound by $d(n)$. The second phase is even simpler. Thus we get the following fact.

PROPOSITION 9.24
The DP algorithm solves the satisfiability problem for collections of Krom clauses in polynomial time.

It is quite clear that the space requirements are still quite high, of order $O(n^3)$. The space requirements, in general, for DPLL are simpler (of the order $O(m)$, where $m = |F|$, with F the input formula, thus in Krom case $O(n^2)$). But we will see that the DPLL algorithm behaves much simpler on Krom inputs.

Let us recall that DPLL does different things at initialization and at the splitting. At the initialization we compute the BCP of input formula. This collection of literals, if inconsistent, forces unconditional failure, not the backtrack. Specifically, let us recall the following fact previously stated (and proven) as Proposition 8.14.

LEMMA 9.3
Let F be a CNF and let v be $\mathrm{BCF}(v)$. Then F is consistent if and only if v is consistent, and the set of clauses $reduct(F, v)$ is consistent. Moreover, the set of literals $\mathrm{BCP}(reduct(F, v))$ is empty.

Now we have a fact that shows an advantage of Krom clauses.

LEMMA 9.4
Let K be a Krom CNF, and let l be a literal such that $l, \bar{l} \notin \mathrm{BCP}(K)$. Let $v = \mathrm{BCP}(K \cup \{l\})$, and let $K_1 = reduct(K, v)$. Then:

1. $K_1 \subseteq K$.

2. *If v is consistent then: K is consistent if and only if K_1 is consistent.*

3. If v is consistent then a satisfying valuation for K can be computed from v and a satisfying valuation for K_1 in linear time.

Proof: (1) Let us observe that in the reduction process, once we shorten a Krom clause, the shortened clause becomes a unit. Therefore all clauses are, actually, subsumed by $\mathrm{BCP}(K \cup \{l\})$ or not touched by it at all. Thus the reduct, K_1, is a subset of K.
(2) If v is consistent then, since $K_1 \subseteq K$, every valuation satisfying K also satisfies K_1. The other direction is simple: Given a valuation w satisfying K_1 and the partial valuation $\mathrm{BCP}(K \cup \{l\})$, we can combine them as follows:

$$u(p) = \begin{cases} v(p) & \text{if } v(p) \text{ defined} \\ w(p) & \text{otherwise.} \end{cases}$$

The resulting valuation u satisfies K.
(3) Follows from the construction in (2). □
A careful inspection of Lemma 9.4 shows that, in fact, we are dealing with an autarky (cf. Section 2.3). Specifically, we have the following fact.

PROPOSITION 9.25
Let K be a set of Krom formulas. For every set of literals v, if $\mathrm{BCP}(K \cup v)$ is consistent, then it is an autarky for K.

Proof: Let $w = \mathrm{BCP}(K \cup v)$. Let $l \in w$. If $\{\bar{l}\}$ is a clause of K then w is inconsistent. If $C \in K, l \in C$ then $w_3(C) = 1$. So let us assume (the last possibility) that $\bar{l} \in C$, but C is not a unit. Then $C = \bar{l} \vee m$ with $l \in w$. But then $m \in w$, thus $w_3(C) = 1$, as desired. □
Let us observe that Lemma 9.4 does not lift to 3-clauses. Here is the example. Let F consist of clauses: $\neg p \vee q \vee r, \neg p \vee \neg q \vee r, \neg p \vee q \vee \neg r, \neg p \vee \neg q \vee \neg r$.
Then F is satisfiable (set $p = 0$), adding p to F results in a consistent BCP (namely p), but the resulting reduct is inconsistent.
Here is a consequence of Lemma 9.4, namely that once we run BCP on $K \cup \{l\}$ and find the resulting set of literals v consistent, there is no need *ever* to visit $K \cup \{\bar{l}\}$. Either the reduct of K by v is inconsistent and then K is unsatisfiable, or that reduct is satisfiable, and then K is satisfiable. Thus, backtracking disappears from the picture!
This does *not* mean that we will only visit one literal in a decision node: if $\mathrm{BCP}(K \cup \{l\})$ is inconsistent, we will have to go to that other branch.
To handle this situation we will now introduce a new variable *visited* with Boolean values. If we are at the first case of computing BCP at a decision point, then we set up this variable to 0. If $\mathrm{BCP}(K \cup \{l\})$ is inconsistent and we have to move to $\mathrm{BCP}(K \cup \{\bar{l}\})$ then this variable *visited* is changed to 1, to indicate that we cannot backtrack.
We now present the "Krom" version of the DPLL algorithm.

Polynomial cases of SAT

At the beginning we preprocess, by computing BCP. If we get a contradictory set of literals we leave with failure. Otherwise we have a partial valuation v and reduce K by v.

Now we proceed recursively. We have an initial partial valuation v, and K_1. If K_1 is empty, we return any valuation extending v (v was a partial valuation; it did not have to evaluate all variables).

If K_1 is nonempty, we select a literal l occurring in K_1 and set the variable *visited* to 0. If $\text{BCP}(K \cup \{l\})$ is consistent (it is definitely nonempty!), we merge v with $\text{BCP}(K_1 \cup \{l\})$, getting the new v, reduce K_1 by v getting the new K_1, and continue. If, however $\text{BCP}(K_1 \cup \{l\})$ is inconsistent, we change the variable *visited* to 1. Now we run BCP on $K_1 \cup \{\bar{l}\}$. If the resulting w is inconsistent, we leave with failure. Otherwise, we merge v with w getting the new v, reduce K_1 by w getting the new K_1, and continue.

It should be clear from our discussion that the algorithm above correctly tests satisfiability of Krom theories (i.e., sets of Krom clauses) K and if such theory K is satisfiable returns a satisfying valuation for K.

We will now look at some graph-theoretic issues associated with Krom CNFs and use some results from the graph theory to get a better algorithm for testing satisfiability of Krom formulas.

Before we study the graph-theoretic issues we need a simple lemma relating satisfiability of Krom formulas and BCP.

LEMMA 9.5
Let K be a Krom formula. Then K is satisfiable if and only if, for every variable p, at least one of $\text{BCP}(K \cup \{p\})$, $\text{BCP}(K \cup \{\neg p\})$ is consistent.

Proof: We prove our assertion by induction on the number n of variables in K. When n is 1 there are four clauses over $Var = \{p_1\}$. We can disregard the tautology $p_1 \vee \bar{p}_1$. There are 8 subsets of the remaining three clauses. Five of these are inconsistent and have inconsistent BCP. In the remaining three cases the assertion holds. This establishes the base of induction.

Now let us look at the inductive step. If a clause set K is satisfiable and v is a satisfying assignment for K then, without loss of generality, we can assume $v(p_1) = 1$. Then v satisfies $BCP(K \cup \{p\})$ and so this latter set of literals is consistent. Conversely, let us assume that for every variable $p \in Var$, at least one of $BCP(K \cup \{p\})$, $BCP(K \cup \{\bar{p}\})$ is consistent. We assume that $BCP(K \cup \{p_1\})$ is consistent. Let $K_1 = Red(K, BCP(K, \{p_1\}))$. Then $K_1 \subseteq K$ (Lemma 9.4). If K_1 is empty, then every clause of K is subsumed by $BCP(K, \{p_1\})$ and since $BCP(K, \{p_1\})$ is consistent, K is satisfiable. Otherwise, for every variable q of K_1,

$$BCP(K_1 \cup \{q\}) \subseteq BCP(K \cup \{q\})$$

and

$$BCP(K_1 \cup \{\bar{q}\}) \subseteq BCP(K \cup \{\bar{q}\})$$

Since at least one of the right-hand sides is consistent, at least one of the left-hand sides is consistent. But K_1 has at least one variable less. By inductive assumption, there is an assignment w of variables of K_1 such that w satisfies K_1. We extend w to a valuation w' setting

$$w'(x) = \begin{cases} w(x) & \text{if } x \in Var_{K_1} \\ 1 & \text{if } x \in BCP(K \cup \{p_1\}) \\ 0 & \text{if } \bar{x} \in BCP(K \cup \{p_1\}) \end{cases}$$

Then, clearly, w' satisfies K_1 and also w' satisfies $BCP(K \cup \{p_1\}$. But every clause in $K \setminus K_1$ is subsumed by a literal from $BCP(K \cup \{p_1\})$. Thus w' satisfies K, as desired. □

Let us recall that in the initialization phase of DPLL, if we do not catch the inconsistency outright, we get as an output a consistent set of literals (which we called v), and a Krom formula K_1. This last formula K_1 has the property that all clauses in K_1 have exactly two literals.

We will now assign to the Krom CNF, consisting of clauses with exactly two literals, a graph G_K as follows. The vertices of G_K are all literals of the set of variables occurring in K. Thus not only the literals actually occurring in K are nodes of G_K, but their duals as well. Now, for edges, whenever a clause $l \vee m$ belongs to K we define two edges of G_K: one from $\bar{l}$ to m and another from $\bar{m}$ to l.

The intuition is that the clause $l \vee m$ means, procedurally, two implications: $\bar{l} \Rightarrow m$ and $\bar{m} \Rightarrow l$. Each of this implications can be used for chaining in the computation of BCP.

Now, in the directed graph G_K we will write $l_1 \mapsto l_2$ if there is a directed path starting at l_1 and ending in l_2, or if $l_1 = l_2$. We write $l_1 \sim l_2$ if $l_1 \mapsto l_2$ and $l_2 \mapsto l_1$. The relation $\sim$ is an equivalence relation. Its cosets are called *strong connected components* of G_K. The following property of the graph G_K is easily proved by induction on the length of paths: if $l \mapsto m$ then also $\bar{m} \mapsto \bar{l}$.

Before we go any further, let us observe that if $l \sim m$, then $BCP(K \cup \{l\}) = BCP(K \cup \{m\})$.

Here is the basic connection of G_K and satisfiability.

PROPOSITION 9.26

Let K be a Krom CNF consisting of clauses with exactly two literals. Then K is satisfiable if and only if no strong connected component of G_K contains a pair of contradictory literals.

Proof: Let us assume that K is satisfiable, but G_K has a connected component containing p and $\neg p$. Then $BCP(K \cup \{p\}) = BCP(K \cup \{\neg p\})$. But then any valuation satisfying K must satisfy both p and $\neg p$, a contradiction.

Conversely, let us assume that no strongly connected component of G_K contains a pair of contradictory literals. We want to show that K is satisfiable. Let us assume it is not satisfiable. Then, by Lemma 9.5, it must be the case that for some variable p, both $\mathrm{BCP}(K \cup \{p\})$ and $\mathrm{BCP}(K \cup \{\neg p\})$ are contradictory. In particular, in graph G_K, for some r, $p \mapsto r$ and $p \mapsto \bar{r}$, and also $\bar{p} \mapsto s$ and $\bar{p} \mapsto \bar{s}$. But then, it is the case that $r \mapsto \bar{p}$, and $s \mapsto p$. Now the contradiction is imminent:

$$p \mapsto r \mapsto \bar{p} \mapsto s \mapsto p.$$

Thus the composition of these four paths creates a cycle with p and $\neg p$ on it, a contradiction. □

Now it is very easy to test satisfiability of Krom CNFs in linear time. All we need to do is to be able to compute strong connected components and test those for the absence of a pair of complementary literals, but it is quite clear that this can be done in linear time in the size of the graph, thus of K.

The satisfiability test outlined above does not produce a satisfying valuation directly. We will now show how such computation can be found in linear time. That is, given an input Krom CNF (a set of 2-clauses), K, we will compute a satisfying valuation for K if K is satisfiable. We will use the graph G_K defined above. The construction bears a lot of similarity to the second phase of the DP algorithm.

So, let G_K be the graph assigned to K above. We can certainly compute G_K in linear time, and also compute and test strongly connected components of G_K for consistency in linear time. So, let us assume that we found that no strongly connected component of G_K is inconsistent. We proceed as follows. We form the quotient graph of G_K whose nodes are strongly connected components of G_K. Let us call this graph H_K. The edges of H_K are formed as follows: a strongly connected component c points to the strongly connected component c' if c is different from c' and for some edge (x, y) of G_F, x belongs to c, and y belongs to c'. Then the graph H_K is acyclic and so we can topologically sort H_K. Let the resulting list of strongly connected components be $\langle c_1, \ldots, c_n \rangle$. This list has the property that whenever (l, m) is an edge of G_K and l, m are in different strongly connected components of G_K then the strongly connected component of l occurs in the list earlier than the component of m.

Our goals now are the following:

1. We need to read off the list $\langle c_1, \ldots, c_n \rangle$ a valuation v of variables of K.

2. Then we need to show that v satisfies K.

As mentioned above, our construction is quite similar to the second phase of the DP algorithm. We construct the desired valuation going backwards, dealing first with the last of the element of the list, c_n, then the previous one, c_{n-1}, etc. We construct the desired valuation v by induction. To start, since no connected component of G_K contains a pair of dual literals, the component c_n, treated as a set of literals, is consistent. Thus it is a partial valuation, which we call v_n. Now assume that we constructed a partial valuation v_{n-r} and that

$$v_n \preceq_k v_{n-1} \preceq_k \ldots \preceq v_{n-r}.$$

Here is how we construct v_{n-r-1}. It is an extension of v_{n-r} by these literals in c_{n-r-1} which are not contradicted by v_{n-r}. Formally

$$v_{n-r-1} = v_{n-r} \cup \{l \in c_{n-r-1} : \bar{l} \notin v_{n-r}\}.$$

That is, in the terminology of Chapter 2, $v_{n-r-1} = v_{n-r} \oplus c_{n-r-1}$. In particular, $v_{n-r} \preceq_k v_{n-r-1}$. At the end of the construction we have

$$v_n \preceq_k v_{n-1} \preceq_k \cdots \preceq_k v_1.$$

Now we define as v the assignment v_1. By construction all variables of K are evaluated by v.

So, all we need to do is to check that the valuation v satisfies K. Let $C : l \vee m$ be a clause in K. That is, the graph G_K has both edges $(\bar{l}, m)$ and $(\bar{m}, l)$. We want to show that v satisfies C. To this end, let us consider the *last* strongly connected component containing any of $\bar{l}, \bar{m}, l$, or m. If this last strongly connected component contains $\bar{l}$ then as there is an edge from $\bar{l}$ to m, m must be in the same, or later, strongly connected component. But under our assumption it means that m is in the last component containing any of $\bar{l}, \bar{m}, l$ or m. The reasoning in the case of $\bar{m}$ is similar. Therefore that last component must contain either l or m (or both). But then no later strongly connected component contains $\bar{l}$ or $\bar{m}$ so either l or m (or both) are evaluated by v as 1, Thus $v \models C$, as desired.

We will now show a characterization of families of models of Krom CNFs analogous to one we found in Section 9.2. There, in Theorem 9.1, we proved that a family $\mathcal{X}$ of subsets of Var is of the form $Th(H)$ for a CNF H consisting of Horn clauses if and only if $\mathcal{X}$ was closed under intersections. Here we will prove a similar result, but related to closure under a different operation. It will be natural to use this time 2-valued interpretations rather than sets, but, as usual, it is mainly a matter of taste.

The ternary operation *maj* assigns to three arguments p, q, and r the value taken by at least two of them. Next, we look at the "polymorphic" version of *maj*.

To define an operation on interpretations that takes an interpretation as its value, we need to define the value at every propositional variable. We define a ternary operation *maj* on interpretations by the following condition

$$maj(v_1, v_2, v_3)(p) = \begin{cases} 1 & \text{if } |\{i : v_i(p) = 1\}| \geq 2 \\ 0 & \text{otherwise.} \end{cases}$$

Here is what happens: either at least two of v_1, v_2, v_3 take value 1 on p, and then the result of *maj* also takes value 1 on p or at least two of v_1, v_2, v_3 take value 0 on p, and then the result of *maj* also takes value 0 on p. Thus $maj(v_1, v_2, v_3)$ "goes with majority" on every variable p. This means that for every literal l we have the following equivalence:

$$maj(v_1, v_2, v_3) \models l \quad \text{if and only if} \quad |\{i : i \in \{1,2,3\} \wedge v_i \models l\}| \geq 2.$$

Now we are able to formulate the desired characterization of families of valuations definable by Krom (2SAT) CNFs. This characterization is due to Schaefer, and later, in Chapter 13, when we prove the Schaefer characterization theorem for Boolean constraint satisfaction problems (Theorem 13.2), we will need it.

PROPOSITION 9.27
Let Var be a finite set of propositional variables. Let $\mathcal{F}$ be a set of valuations of Var. Then there exists a set of Krom clauses S such that $\mathcal{F} = Mod(S)$ if and only if $\mathcal{F}$ is closed under the operation maj.

Proof: We need to prove two implications. First, given a family $\mathcal{F}$ definable by a set of Krom clauses, we need to prove that $\mathcal{F}$ is closed under maj. Then, we will have to prove that given a family $\mathcal{F}$ closed under maj we can exhibit a family S of 2-clauses (Krom clauses) such that $\mathcal{F} = Mod(S)$.

So, let us assume that $\mathcal{F}$ is of the form $Mod(S)$ where S consists of 2-clauses. The case of a unit clause is obvious. So let us assume that a clause $l_1 \vee l_2$ belongs to S, and that v_1, v_2, v_3 belong to $\mathcal{F}$. All we need to show is that $maj(v_1, v_2, v_3)$ also satisfies $l_1 \vee l_2$. Because if we show this, then as $l_1 \vee l_2$ is an arbitrary clause in S, $maj(v_1, v_2, v_3)$ also satisfies S and so belongs to $\mathcal{F}$. So let $l_1 \vee l_2$ be satisfied by each of v_1, v_2, v_3. Then it cannot be the case that $|\{i : v_i \models l_1\}| \leq 1$ and $|\{i : v_i \models l_2\}| \leq 1$ because then we have

$$3 = |\{i : v_i \models l_1 \vee l_2\}| =$$
$$|\{i : v_i \models l_1\} \cup \{i : v_i \models l_2\}| \leq |\{i : v_i \models l_1\}| + |\{i : v_i \models l_2\}| \leq 2,$$

which is a contradiction. Thus for some $j \in \{1, 2\}$

$$|\{i : v_i \models l_j\}| \geq 2.$$

But then $maj(v_1, v_2, v_3) \models l_j$, and so $maj(v_1, v_2, v_3) \models l_1 \vee l_2$. Thus the implication $\Rightarrow$ is proved.

Now, let $\mathcal{F}$ be a family of valuations such that $\mathcal{F}$ is closed under maj. We need to find S consisting of Krom clauses and such that $\mathcal{F} = Mod(S)$. It is quite clear what S should be, namely,

$$S = \{C : |C| \leq 2 \text{ and for all } v \in \mathcal{F}, v \models C\}.$$

Clearly, all valuations in $\mathcal{F}$ belong to $Mod(S)$. All we need to show is that valuations in $Mod(S)$ belong to $\mathcal{F}$. To this end, let $v \in Mod(S)$. We show that if all valuations $w \in \mathcal{F}$ satisfy some clause D, then it is also the case that $v \models D$. We prove this fact by induction on the length of the clause D. This is certainly true for all clauses D of length at most 2. This is how we defined S, and v belongs to $Mod(S)$. So let us fix a valuation v in $Mod(S)$. Let us assume that D is of length $k + 1$ with $k \geq 2$ and that for clauses D' of length at most k our assertion holds. We show that our assertion is true for D as well. For otherwise we have

$$D = l_1 \vee \ldots \vee l_k \vee l_{k+1},$$

which is true at all valuations w that are in $\mathcal{F}$ but D is not true at v. Could our clause D be subsumed by a shorter clause D' that is true in all valuations in $\mathcal{F}$? No, because such shorter clause D' being true in all valuations of $\mathcal{F}$ would be true, by inductive assumption, in v. But then D, being weaker, would be satisfied by v as well. So all clauses strictly stronger than C must be falsified by *some* valuation in $\mathcal{F}$. In particular there are valuations v_1, v_2, and v_3 such that

$$v_1 \not\models l_1 \vee \ldots \vee l_k,$$
$$v_2 \not\models l_1 \vee l_3 \vee \ldots \vee l_k \vee l_{k+1}, \text{ and}$$
$$v_3 \not\models l_2 \vee \ldots \vee l_{k+1}.$$

But this means that

$$v_1 \models \neg l_1 \wedge \ldots \wedge \neg l_k,$$
$$v_2 \models \neg l_1 \wedge \neg l_3 \wedge \ldots \wedge \neg l_k \wedge \neg l_{k+1},$$
$$v_3 \models \neg l_2 \wedge \ldots \wedge \neg l_{k+1}.$$

But now, let $w = maj(v_1, v_2, v_3)$. Then, clearly,

$$w \models \neg l_1 \wedge \neg l_2 \ldots \wedge \neg l_k \wedge \neg l_{k+1}.$$

This means that $w \not\models C$. But this contradicts that $\mathcal{F}$ is closed under maj, as all valuations in $\mathcal{F}$ satisfy C.

So now we know that v satisfies all clauses satisfied by all valuations from $\mathcal{F}$. Now, Proposition 3.16 tells us that there is a set of clauses S such that $\mathcal{F} = Mod(S)$. Then, every clause C from S is satisfied by all valuations from $\mathcal{F}$. Therefore every such clause C is satisfied by v. But this means that $v \in Mod(S)$, that is, $v \in \mathcal{F}$, as desired. □

One may complain that our construction is non-effective, we did not actually *compute* S in polynomial time. But in fact we could do so. We need to be a bit careful about the parameters because the size of the family $\mathcal{F}$ may be exponential in the size of Var. So let k be the number of valuations in $\mathcal{F}$ and let n be the number of variables. There are at most $O(n^2)$ clauses over the set of variables of size n. There is only $O(n^2)$ of such clauses satisfied by any valuation v. We can find the intersection of all the sets of 2-clauses satisfied by all these valuations in $O(n^2 k)$ steps, in particular in $O(m^2)$ where m is the total size of the family $\mathcal{F}$.

We formalize our discussion above into a proposition.

PROPOSITION 9.28

Let S be a set of all satisfying valuations for some Krom theory K. Then we can find one such K in polynomial time in the size of S.

We will now use Proposition 9.27 to obtain an alternative characterization of the set of valuations of the form $Mod(T)$ where T is a set of 2-clauses (i.e., Krom clauses).

The binary Boolean operation $+$ was defined (actually more than once) above. Specifically, $+$ is the complement of equivalence. The operation $+$ is a group operation in $Bool$.

Similarly, as we did before in our considerations of the sets of models of Horn formulas, and sets of models of dual Horn formulas, we can treat $+$ as a polymorphism on the set of functions from Var to $Bool$ (i.e., valuations). Thus we define a binary operation $+$ on valuations. Let f_i, $i = 1, 2$ be valuations (i.e., the Boolean functions defined on Var). We define $f_1 + f_2$ by the equation

$$(f_1 + f_2)(p) = f_1(p) + f_2(p)$$

for all $p \in Var$. We also remark that $+$ is a group operation in the set of valuations. This fact, actually, requires a proof, which we leave to the reader.

Now, let $\mathcal{F}$ be a set of valuations, and let $v \in \mathcal{F}$. Then we say that $X \subseteq Var$ is a *change set* for v and $\mathcal{F}$ if $v + \chi_X$ belongs to $\mathcal{F}$. The intuitive meaning of a change set for v and $\mathcal{F}$ is this: we "flip" the values of v on the set X (and leave the values unchanged outside of X). The resulting valuation should still be in $\mathcal{F}$.

PROPOSITION 9.29
Let $\mathcal{F}$ be a set of valuations of Var. Then $\mathcal{F}$ is $Mod(F)$ for some set of 2-clauses F if and only if for all $v \in \mathcal{F}$, for all X_1, X_2 which are change sets for v and $\mathcal{F}$, the intersection $X_1 \cap X_2$ is also a change set for v and $\mathcal{F}$.

Proof: First, let us observe that for any pair of valuations v_1, v_2 there is a set X such that $v_1 + \chi_X = v_2$. Namely, the set

$$X = \{p : v_1(p) \neq v_2(p)\}$$

is such a set. Actually, such a set X is unique.

Next, we have the following key identity which is of independent interest. Let v be a valuation, and let X_1, X_2 be subsets of Var. Then:

$$maj(v, (v + \chi_{X_1}), (v + \chi_{X_2})) = v + \chi_{X_1 \cap X_2}.$$

To see that this identity (that postulates the equality of two valuations, that on the left-hand side, and that on the right-hand side) holds, we need to see that both functions take the same value on the entire Var. Let $p \in Var$. Four cases are possible. First, let $p \in X_1 \cap X_2$. Then:

$$v(p) \neq (v + \chi_{X_1})(p) = (v + \chi_{X_2})(p).$$

Thus
$$maj(v, (v + \chi_{X_1}), (v + \chi_{X_2}))(p) = 1 + v(p).$$

But $p \in X_1 \cap X_2$ and so $v + \chi_{X_1 \cap X_2}(p) = 1 + v(p)$ and so the desired equality holds.

The second case is when $p \in (Var \setminus (X_1 \cup X_2))$. In this case

$$v(p) = (v + \chi_{X_1})(p) = (v + \chi_{X_2})(p).$$

Thus
$$maj(v, (v + \chi_{X_1}), (v + \chi_{X_2}))(p) = v(p).$$

But also $v(p) = v + \chi_{X_1 \cap X_2}(p)$, and so the desired equality holds. The third case is when $p \in X_1 \setminus X_2$

$$v(p) = (v + \chi_{X_2})(p) \neq (v + \chi_{X_1})(p).$$

Thus
$$maj(v, (v + \chi_{X_1}), (v + \chi_{X_2}))(p) = v(p).$$

But $p \notin X_1 \cap X_2$ and so $v(p) = (v + \chi_{X_1 \cap X_2})(p)$, and thus the desired equality holds.

The fourth case, $p \in X_2 \setminus X_1$, is very similar to the third one, and we leave it to the reader to prove.

With the identity above proven, we are now ready to prove our proposition. First, let us assume that $\mathcal{F}$ is $Mod(F)$ where F is a set of 2-clauses. Then, by Proposition 9.27, the family $\mathcal{F}$ is closed under the polymorphism maj. Let $v \in \mathcal{F}$ and let X_1, X_2 be two change sets for v and $\mathcal{F}$. Let $v' = v + \chi_{X_1}$, and $v'' = v + \chi_{X_2}$. Then both v', v'' belong to $\mathcal{F}$ since X_1, X_2 are change sets for v and $\mathcal{F}$. Thus $maj(v, v', v'')$ belongs to $\mathcal{F}$. But then $maj(v, (v + \chi_{X_1}), (v + \chi_{X_2}))$ belongs to $\mathcal{F}$ and by the identity proven above, $v + \chi_{X_1 \cap X_2}$ belongs to $\mathcal{F}$, i.e., $X_1 \cap X_2$ is a change set for v and $\mathcal{F}$.

Conversely, let us assume that for all $v \in \mathcal{F}$, for all change sets X_1, X_2 for v and $\mathcal{F}$, $X_1 \cap X_2$ is also a change set for v and $\mathcal{F}$. Let us select arbitrary valuations w, w', and w'' in $\mathcal{F}$. Then, as we observed before, there are sets X_1 and X_2 such that $w' = w + \chi_{X_1}$, and $w'' = w + \chi_{X_2}$. Then X_1, X_2 are change sets for w and $\mathcal{F}$. But then $X_1 \cap X_2$ is a change set for w and $\mathcal{F}$. But our identity says that

$$w + \chi_{X_1 \cap X_2} = maj(w, (w + \chi_{X_1}), (w + \chi_{X_2})),$$

and so $maj(w, (w + \chi_{X_1}), (w + \chi_{X_2})) \in \mathcal{F}$, that is $maj(w, w', w'') \in \mathcal{F}$. Since w, w', w'' were arbitrary valuations in $\mathcal{F}$, $\mathcal{F}$ is closed under the polymorphism maj, that is, $\mathcal{F}$ is of the form $Mod(F)$ for a set of 2-clauses F, as desired. $\square$

9.6 SAT as a tool for manipulation of formulas, renameable variants of classes of clauses

It turns out that formulas can be used to manipulate formulas, or to be precise, to compute manipulations. The idea is to assign to formulas some other formulas which express the task of transforming the input formulas. It may sound strange at first glance, but as we will see, we can sometimes use satisfiability to compute operations on formulas.

In Section 9.2 we found that it is quite inexpensive to test satisfiability of a set of clauses H, provided that H was a set of Horn clauses. Not only were we able to easily test if H was satisfiable, but we also had an inexpensive algorithm to compute a least model of H if H was satisfiable. In fact, we have a candidate for a valuation v satisfying H. If v does not satisfy H (a matter easy to check), H is unsatisfiable. Now suppose that someone gave us a CNF F and a consistent permutation of literals π such that $\pi(F)$ happens to be a Horn CNF. Here is how we then test satisfiability of F and if F is satisfiable, compute a satisfying valuation for F. First we form an auxiliary formula $H = \{\pi(C) : C \in F\}$. Next, since H is Horn, we test if H is satisfiable. Since satisfiability of F is equivalent to satisfiability of $\pi(F)$, that is, H, therefore if we find that H is unsatisfiable, we report that F is unsatisfiable. If H is satisfiable, and $v \models H$, then we compute $\pi^{-1}(v)$ and return it as a satisfying valuation for F.

The only problem with this approach is that unless someone gave us π we do not know (at least for now) if such π exists at all. It is our goal now to show that, actually, it is quite inexpensive to test if such permutation exists, and if so, find it. First we need a bit of terminology. A consistent permutation π of literals is called a *permutation of variables* if π does not change signs of literals: variables are moved to variables and negative literals to negative literals. Then we call a consistent permutation π of literals a *shift permutation* if π does not change underlying variables, that is, for every variable p, $\pi(p) = p$, or $\pi(p) = \neg p$.

We first have the following fact.

PROPOSITION 9.30

1. *If π is a consistent permutation of literals, then π uniquely decomposes into a product (composition) $\pi_1 \circ \pi_2$ where π_1 is a permutation of variables, and π_2 is a shift permutation*
2. *Permutations of variables and shift permutations commute.*

Proof: (a) Consistent permutations of literals satisfy the condition $\pi(\bar{l}) = \overline{\pi(l)}$. Thus $\pi(p)$ determines $\pi(\neg p)$ and the underlying variable of $\pi(p)$ and of $\pi(\neg p)$ is the same. Let us strip $\pi(p)$ of its polarity; we obtain a permutation π_1 of variables. Now for π_2, let us define

$$\pi_2(p) = \begin{cases} p & \text{if } p = \pi_1(q) \text{ and } \pi(q) = p \\ \neg p & \text{if } p = \pi_1(q) \text{ and } \pi(q) = \neg p. \end{cases}$$

Then π_2 is a shift permutation and $\pi = \pi_1 \circ \pi_2$ because all the π_2 does is to restore the correct (with respect to π) polarity of the value of $\pi_1(p)$. We leave the proof of the uniqueness of decomposition to the reader.

(b) It should be clear that it does not matter if we first move the variable to the proper underlying variable and then change the sign (if needed) or conversely. □

The next step of our considerations is to observe that the permutations of variables (as opposed to general consistent permutations of literals) preserve the property of "being Horn." That is, the image of a Horn CNF under a permutation of variables is again a Horn CNF.

PROPOSITION 9.31
Let π be a permutation of variables. Then a CNF is Horn if and only if $\pi(F)$ is Horn.

Proof: Since π is a permutation of variables it preserves signs, and so the number of positive and negative literals in a clause does not change. □

But now Proposition 9.31 has a consequence.

PROPOSITION 9.32
Let F be a CNF. Then there exists a consistent permutation π such that $\pi(F)$ is Horn if and only if there exists a shift σ such that $\sigma(F)$ is Horn.

Proof: The implication $\Leftarrow$ is obvious. Now assume that for some consistent permutation of literals π, $\pi(F)$ is Horn. Let us decompose the permutation π into $\pi_1 \circ \pi_2$, where π_1 is a permutation of variables, and π_2 is a shift. We claim that $\pi_2(F)$ is Horn. Indeed, $\pi(F) = \pi_1 \circ \pi_2(F) = \pi_1(\pi_2(F))$. Thus $\pi_2(F)$ is Horn, by Proposition 9.31. □

Proposition 9.32 tells us that if we want to test if a given input formula can be moved by some renaming to a Horn formula, all we need to do is to check if some shift does the job. Let us now introduce a suitable terminology. Let $\mathcal{K}$ be a class of formulas. We say that a formula ψ is *renameable-$\mathcal{K}$* if for some consistent permutation of literals π, $\pi(\psi)$ belongs to $\mathcal{K}$. With this terminology we are investigating renameable-Horn formulas. We also found, that in case of renameable-Horn, all we need are shift permutations. Now, let us look at an example to see what conditions on such shift permutation can be derived from a given clause.

Example 9.2
Let C be the clause:

$$p \vee q \vee r \vee \neg s \vee \neg t \vee \neg u.$$

Here are the constraints on a putative shift π. First, after the shift, at most one of p, q, and r may remain positive. Second, we may shift one (but no more) of literals s, t, or u but if we do so, then *all* of p, q, and r must be shifted. Further refining these requirements we see that: (1) At least one variable of each pair from $\{p, q, r\}$ must be shifted. (2) From each pair out of $\{s, t, u\}$ at most one may be shifted. (3) If any of $\{r, s, t\}$ has been shifted, then each of p, q, and r must be shifted. Let us move to a new set of variables. This new set has a variable $shift(x)$ for each $x \in Var$. The intuitive meaning of the variable $shift(x)$ is 'x needs to be shifted.' Using the fact that the implication $l \Rightarrow m$ is just a 2-clause: $\bar{l} \vee m$ we formalize the constraints listed above as a collection of 2-clauses:

1. $shift(p) \vee shift(q), shift(p) \vee shift(r), shift(q) \vee shift(r)$
2. $\neg shift(s) \vee \neg shift(t), \neg shift(s) \vee \neg shift(u), \neg shift(t) \vee \neg shift(u)$
3. $\neg shift(s) \vee shift(p), \neg shift(s) \vee shift(q), \neg shift(s) \vee shift(r), \neg shift(t) \vee shift(p), \neg shift(t) \vee shift(q), \neg shift(t) \vee shift(r), \neg shift(u) \vee shift(p), \neg shift(u) \vee shift(q), \neg shift(u) \vee shift(r)$.

Let us call the CNF composed of these 15 clauses S_C. How can we recover a shift π out of a valuation satisfying S_C? Here it is: whenever $v(shift(x)) = 1$, assign $\pi(x) = \neg x$. Otherwise, $\pi(x) = x$. For instance the valuation v such that $shift(p) = 1, shift(q) = 1, shift(r) = 1, shift(s) = 1, shift(t) = 0, shift(u) = 0$ generates a shift π_v such that

$$\pi(p \vee q \vee r \vee \neg s \vee \neg t \vee \neg u) = \neg p \vee \neg q \vee \neg r \vee s \vee \neg t \vee \neg u.$$

□

Let us generalize the construction from Example 9.2. Given a clause

$$C : p_1 \vee \ldots \vee p_k \vee \neg q_1 \vee \ldots \vee \neg q_l,$$

define a collection of 2-clauses consisting of three groups:

Group 1. $shift(p_i) \vee shift(p_j), 1 \leq i < j \leq k$.

Group 2. $\neg shift(q_i) \vee \neg shift(q_j), 1 \leq i < j \leq l$.

Group 3. $\neg shift(q_i) \vee shift(p_j), 1 \leq i \leq l, 1 \leq j \leq k$.

What is important here is that the size of $S(C)$ is bound by the square of the size of C.

Now, let us define $S_F = \bigcup_{C \in F} S_C$. Here is a fact that justifies our construction.

PROPOSITION 9.33 (Lewis theorem)
There is a bijection between valuations satisfying the Krom CNF S_F and shift permutations of literals transforming F to a Horn CNF.

Proof: First, let us assume that v is a valuation satisfying S_F. We define a permutation π_v by setting:

$$\pi_v(x) = \begin{cases} x & \text{if } v(shift(x)) = 0 \\ \neg x & \text{if } v(shift(x)) = 1. \end{cases}$$

Then, clearly, π_v is a shift permutation. Moreover, π_v changes polarity of x precisely if $v(shift(x)) = 1$. But now it is quite clear that the clause $\pi(C)$ must be Horn, for every $C \in F$.

Conversely, if π is a shift permutation moving F to a Horn clause, then let us define v_π by

$$v(shift(x)) = \begin{cases} 1 & \text{if } \pi(x) = \neg x \\ 0 & \text{otherwise.} \end{cases}$$

Now, clearly, $v \models S_C$ for each $C \in F$ and so $v \models S_C$. It is also clear that the correspondence between vs and πs is a bijection. □

Now, the size of S_F is bound by the polynomial in the number of variables in F, and S_F is a Krom CNF. This implies the following fact.

PROPOSITION 9.34
The problem of testing if a formula F is renameable Horn can be done in polynomial time in the size of F. If it is, we can compute a shift permutation turning formula F into a Horn formula in linear time in the number of variables of F out of a valuation satisfying S_F.

It is easy to construct an example of a non-renameable Horn CNF. Here is one: $\{p \vee q \vee \neg r \vee \neg s, \neg p \vee \neg q \vee r \vee s\}$.

If the reader has started to suspect that a slightly different Krom CNF can be used to test if F is renameable dual Horn, she is right. Let us call the collection of Krom clauses generated to describe the shift into dual Horn CNF, T_F (we did not compute T_F, we hope the reader does). The procedure and discussion above allow us to obtain the following fact.

PROPOSITION 9.35
The problem of testing if a formula F is renameable dual Horn can be done in polynomial time in size of F. If actually F is renameable dual Horn, we can compute a shift permutation turning formula F into a dual Horn formula in a linear time in the number of variables of F out of a valuation satisfying T_F.

Generally, if a collection of formulas $\mathcal{K}$ is invariant with respect to permutation of variables, then the issue of testing renameability of a given F to a formula in $\mathcal{K}$ reduces to shifts.

Here is one such class. Let us call a clause C an "either-or" clause (in short EO) if all literals in C are positive, or if all literals in C are negative. An EO CNF is one consisting of EO clauses (we may have some purely positive and some purely negative). In Chapter 11 we will see that the problem of testing if an EO CNF formula is satisfiable is NP-complete.

It turns out that the technique of shifts discussed above allows us to test if a formula F is renameable-EO. Just for the fun of it, we suggest that the reader devises an example of non-renameable-EO CNF.

9.7 Affine formulas

We consider yet another case when the decision problem is "easy." This is the case of *affine formulas*. Recall that we discussed a binary operation in *Bool* which we denoted by $+$. Recall that the relational structure $\langle Bool, +, \wedge, 0, 1 \rangle$ is an algebraic structure called a *field*. In fact the readers with a bit of experience with abstract algebra will recognize that this is the familiar field $\mathbf{Z}_2$. We used this fact (and additional properties of field operations in Z_2) when we studied the polynomial representation (Zhegalkin polynomials) of Boolean functions in Chapter 2. Here we will not use the full power of the fact that we deal with a field, but it will turn out that procedure, called *Gaussian elimination* allows us to test collections of formulas involving *only* the functor $+$ and the negations of such formulas for satisfiability.

It will be convenient to think about such formulas in algebraic terms, namely as *linear equations*. Specifically, a valuation v satisfies a formula $x_{i_1} + \ldots + x_{i_k}$ if and only if the field $\mathbf{Z}_2$ satisfies the equation

$$v(x_{i_1}) + \ldots + v(x_{i_k}) = 1,$$

whereas v satisfies the formula $\neg(x_{i_1} + \ldots + x_{i_k})$ if and only if

$$v(x_{i_1}) + \ldots + v(x_{i_k}) = 0.$$

But now, a set of affine formulas is equivalent to the set of the corresponding linear equations over $\mathbf{Z}_2$. A technique called *Gaussian elimination* is used to solve such systems of equations. The idea is to use *row reduction* to make sure that a given variable occurs in just one equation. This is based on tautologies: $(p + p) \equiv \bot$, and $(p + q) \wedge (p + r) \Rightarrow ((q + r) \equiv \bot))$.

Exotic as it may seem, all that happens here is that we enforce the fact that the characteristic of the field $\mathbf{Z}_2$ is 2. Thus $x + x = 0$, and the linear combination rule holds (see also Chapter 10.) The difference here is that we are dealing with a field of characteristic 2, and so if two equations share a variable, they will not share it after addition. The point here is that addition of equations is a reversible operation. Adding two equations and getting a third one does not change the set of satisfying valuations (i.e., solutions), and in fact, if we eliminate one of the original equations, we can get it back.

With the appropriate renaming, every set of linear equations over $\mathbf{Z}_2$ can be taken into *diagonal form* (we are not in the business of teaching algebra here, but it should be clear what happens in the current context). Two outcomes are possible. The first is that we get an inconsistent set of equations that is one that includes an equation $0 = 1$ (in the context of logic we derive after a series of rewritings the formula $\bot \equiv \top$). Or we do not get an inconsistent set of equations, but rather a set of equations presented in Table 9.1. Trivial equations ($0 = 0$ and $1 = 1$) are eliminated. Now, all we need to do is to select any way we want the values of variables $x_{n+1}, \ldots, x_{n+m}$ and we get the solution for all variables. If an expression f_j does *not* contain variables at all,

TABLE 9.1: Reduced set of equations, consistent case

x_1	$+$	$\ldots$	$=$	$f_1(x_{n+1}, \ldots, x_{n+m})$
	x_2	$+ \ldots$	$=$	$f_2(x_{n+1}, \ldots, x_{n+m})$
		$\ldots$	$=$	$f_j(x_{n+1}, \ldots, x_{n+m})$
		$\ldots \quad +x_n$	$=$	$f_n(x_{n+1}, \ldots, x_{n+m})$

it is $\bot$, or $\top$. In such case the variable x_j is completely constrained, regardless of the values of variables x_{n+1}, x_{n+k}.

The algorithm for solving the sets of affine formulas (that is, systems of linear equations over $\mathbf{Z}_2$) are implemented either as the so-called Gauss algorithm, or as the Gauss-Jordan[2] algorithm.

We will now show a characterization of sets S of Boolean vectors such that $S = Th(F)$ for some set of affine formulas F. To this end, we will tailor the usual definitions of vector space and affine sets to our current situation. We say that a set of vectors $S \subseteq \{0,1\}^n$ is a vector space if it is closed under the sums of two vectors. Here the sum of vectors $\vec{x}$ and $\vec{y}$, $\vec{x} + \vec{y}$ is a vector $\vec{z}$ such that for all i, $1 \leq i \leq n$, $\vec{z}(i) = \vec{x}(i) + \vec{y}(i)$. The well-known fact from linear algebra is that a vector space (remember, we are in Boolean case!) is of the form $S = \{\vec{x} : M \cdot \vec{x} = \vec{0}\}$. Now, a set of vectors S is called *affine* if there is a vector $\vec{y}$ and a matrix M such that $S = \{\vec{x} : M\vec{x} = \vec{y}\}$.

We will consider sets S of vectors closed under sums of three vectors. This is a property *weaker* than closure under sums of two vectors. Nevertheless we have the following fact.

LEMMA 9.6
Let S be a set of Boolean vectors, $S \subseteq \{0,1\}^n$. Let us assume that $\vec{0} \in S$. Then S is closed under sums of two vectors if and only if S is closed under sums of three vectors.

Proof: If S is closed under sums of two vectors then, obviously, S is closed under sums of three vectors. Now, if S is closed under sums of three vectors, but we can take one of those $\vec{0}$, then we can see that S is closed under sums of two vectors as well. □

We now characterize affine sets of vectors in a manner analogous to Theorems 9.1 and 9.3, and Proposition 9.27.

PROPOSITION 9.36
Let S be a set of Boolean vectors, $S \subseteq \{0,1\}^n$. Then S is affine if and only

---
[2] Not the topologist Camille Jordan of the Jordan curve theorem fame, but the engineer Wilhelm Jordan.

if S is closed under sums of three vectors.

Proof: First, let us assume that S is affine. Let us fix the vector $\vec{y}$ and the matrix M so that $S = \{\vec{x} : M\vec{x} = \vec{y}\}$. Now, if $\vec{x}_1, \vec{x}_2, \vec{x}_3 \in S$ then we have:

$$\vec{y} = \vec{y} + \vec{y} + \vec{y} = (M \cdot \vec{x}_1) + (M \cdot \vec{x}_2) + (M \cdot \vec{x}_3) = M \cdot (\vec{x}_1 + \vec{x}_2 + \vec{x}_3),$$

thus $\vec{x}_1 + \vec{x}_2 + \vec{x}_3$ belongs to S.

Conversely, let us assume that S is closed under sums of three vectors. We have two cases to consider.

Case 1: $\vec{0} \in S$. Then, since S is closed under sums of three vectors, S is closed under sums of two vectors (Lemma 9.6). But then S is a linear space, there is a matrix M such that $S = \{\vec{x} : M \cdot \vec{x} = \vec{0}\}$, and all we need to do is to set $\vec{y} = \vec{0}$.

Case 2: $\vec{0} \notin S$. Let us fix a vector $x_0 \in S$. We now define

$$S' = \{\vec{x} + \vec{x}_0 : x \in S\}.$$

We claim that S' is closed under sums of three vectors. Indeed, let $\vec{x}_1 + \vec{x}_0$, $\vec{x}_2 + \vec{x}_0$, and $\vec{x}_3 + \vec{x}_0$ all be in S'. Then by commutativity and the fact that $\vec{x}_0 + \vec{x}_0 = \vec{0}$ we get

$$(\vec{x}_1 + \vec{x}_0) + (\vec{x}_2 + \vec{x}_0) + (\vec{x}_3 + \vec{x}_0) = (\vec{x}_1 + \vec{x}_2 + \vec{x}_3) + \vec{x}_0.$$

But $\vec{x}_1 + \vec{x}_2 + \vec{x}_3 \in S$ so $(\vec{x}_1 + \vec{x}_2 + \vec{x}_3) + \vec{x}_0 \in S'$.

But now, there is a matrix M such that $S' = \{\vec{x} : M \cdot \vec{x} = \vec{0}\}$, and let us define $\vec{y} = M \cdot \vec{x}_0$. We claim that

$$S = \{\vec{x} : M \cdot \vec{x} = \vec{y}\}.$$

For inclusion $\subseteq$, let us observe that if $\vec{x} \in s$, then $\vec{z} = \vec{x} + \vec{x}_0 \in S'$, so

$$\vec{0} = M \cdot \vec{z} = (M \cdot \vec{x}) + (M \cdot \vec{x}_0) = M \cdot (\vec{x} + \vec{y}).$$

For inclusion $\supseteq$, let us assume that $M \cdot \vec{x} = \vec{y}$. Since $M \cdot \vec{x}_0 = \vec{y}$, we get $M \cdot (\vec{x} + \vec{x}_0) = \vec{0}$. But then $\vec{x} + \vec{x}_0 \in S'$, and so $\vec{x} \in S$. □

Thus testing if a set of vectors S is affine can be done in polynomial time in the size of S (just checking if all sums of three vectors from S are in S.)

We can also find an explicit form of a theory F consisting of affine formulas such that $S = Th(F)$ using Gaussian elimination. The technique we present comes from studies of error-correcting codes.

Let us assume that we have already established that a table T describes an affine set of vectors, $\{\vec{r}_1, \ldots, \vec{r}_k\}$, each vector r_i, $1 \leq i \leq k$ of length n. In algebraic notation our task is to find a matrix M and a vector $\vec{b}$ such that

$$T = \{\vec{r} : A \cdot \vec{r} = \vec{b}\}.$$

We are a bit imprecise; on the one hand we talk about T as a table (where the order of rows is fixed) and on the other as a set of vectors. In fact, the order in which we list the rows of T does not matter. The issue here is if we can find such matrix A and a vector $\vec{b}$ in polynomial time in the total size of the problem (i.e., in the size of T). Let us observe that the size of T is $k \cdot n$.

First, let us observe that the problem easily reduces to the case when the rows of T form a linear space. Indeed, an affine set T is a linear space if and only if the zero-vector $\vec{0}$ belongs to T. So after we test if T is affine we first check if $\vec{0}$ belongs to T. If it is, T is a linear space. If not, then the set

$$T' = \{\vec{r} + \vec{r}_1 : \vec{r} \in T\}$$

is a linear space. So we find a matrix A such that

$$T' = \{\vec{s} : A \cdot \vec{s} = \vec{0}\}.$$

Then we set $\vec{b} = A \cdot \vec{r}_1$. Then, if $\vec{r} \in T$ then $\vec{r} + \vec{r}_1 \in T'$ so

$$A \cdot (\vec{r} + \vec{r}_1) = \vec{0} = A \cdot \vec{r} + A \cdot \vec{r}_1 = A \cdot \vec{r} + \vec{b}.$$

Thus $A \cdot \vec{r} = \vec{b}$, as desired. Conversely, if $A \cdot \vec{s} = \vec{b}$ then by setting $\vec{t} = \vec{s} + \vec{r}_1$ we check that $A \cdot \vec{t} = \vec{b} + \vec{b} = \vec{0}$. Thus $\vec{t} \in T'$ and since $\vec{s} = \vec{t} + \vec{r}_1$, $\vec{s} \in T$.

So now we have reduced our problem to one in which we can assume that T is a linear space. Using Gaussian elimination with respect to rows of the matrix T we find a maximal set of independent vectors in T. When we find m such vectors, T has precisely 2^m vectors (because with $|\mathbf{Z}_2| = 2$ every m-dimensional subspace of $\{0,1\}^n$ has 2^m vectors). So now, possibly with renaming the variables, we can find a matrix B so that we can write T as

$$(I\ B),$$

where I is an $m \times m$ identity matrix (1 on the main diagonal, 0 outside main diagonal), and B is an $m \times (n-m)$ matrix. We then produce a new matrix A by setting

$$\begin{pmatrix} B \\ J \end{pmatrix}.$$

Here J is an $(n-m) \times (n-m)$ identity matrix. This matrix A is known in the theory of error-correcting codes as a *parity-check* matrix and it generates the space orthogonal to the space T. This actually requires a proof (which we only indicate because after all it is part of a different domain: error-correcting codes). First we need to prove that rows of $(I_m\ B)$ are orthogonal to columns of $\begin{pmatrix} B \\ I_{n-m} \end{pmatrix}$. There are all sorts of proof for this. One such proof uses induction on the number of 1s in matrix B. Then, using dimension reasoning we observe that the dimension of the space spanned by $\begin{pmatrix} B \\ I_{n-m} \end{pmatrix}$ is $n-m$ and so T consists *precisely* of vectors orthogonal to the matrix $\begin{pmatrix} B \\ I_{n-m} \end{pmatrix}$. But this means that T consists of the solutions of the system of equations described by $A \cdot \vec{r} = \vec{0}$. This, together with the reduction described above, completes the argument of the correctness of our construction.

Polynomial cases of SAT

The described procedure is polynomial because Gaussian elimination can be done in polynomial time. We summarize our discussion above in a proposition which will be used when we discuss the Schaefer theorem.

PROPOSITION 9.37
Let S be a set of all satisfying valuations for some affine theory A. Then we can find one such A in polynomial time in the size of S.

We will illustrate the procedure described above with an example.

Example 9.3
Let T be the following table:
$$\begin{pmatrix} 0 & 0 & 1 \\ 0 & 1 & 0 \\ 1 & 0 & 0 \\ 1 & 1 & 1 \end{pmatrix}.$$

The table T is affine, but not a linear space. Selecting the first row we get out of T the space:
$$\begin{pmatrix} 0 & 0 & 0 \\ 0 & 1 & 1 \\ 1 & 0 & 1 \\ 1 & 1 & 0 \end{pmatrix}.$$

Using row elimination we find the basis
$$\begin{pmatrix} 0 & 1 & 1 \\ 1 & 0 & 1 \end{pmatrix}.$$

Transposing the first and second column we get
$$\begin{pmatrix} 1 & 0 & 1 \\ 0 & 1 & 1 \end{pmatrix}.$$

The matrix B is
$$\begin{pmatrix} 1 \\ 1 \end{pmatrix}.$$

The number $n - m$ is 1. Then the matrix A is
$$\begin{pmatrix} 1 \\ 1 \\ 1 \end{pmatrix}.$$

Thus T' is
$$\{\vec{r} : A \cdot \vec{r} = \vec{0}\}.$$

Consequently T' is determined by a single equation,

$$x_1 + x_2 + x_3 = 0.$$

This equation characterizes T', not T. But all we need to do now is to compute $A \cdot \vec{r}_1$, that is, $A \cdot (0, 0, 1)$. This value is 1 and so:

$$T = \{(x_1, x_2, x_3) : x_1 + x_2 + x_3 = 1\}.$$

☐

9.8 Exercises

1. We noted that every consistent Horn theory H possesses a least model. Using the Dowling-Gallier algorithm we can find this model, $lm(H)$, in linear time. Now, use this algorithm to test (in quadratic time) if H possesses a unique model.
2. If H consists of program clauses, the problem of multiple models is completely trivial. Why?
3. Of course the previous two problems can be lifted to the dual Horn case. Do it.
4. Write the truth table for the ternary Boolean function maj.
5. Let us assume that a class of theories $\mathcal{T}$ has the property that there is an algorithm that tests satisfiability of theories in $\mathcal{T}$ in polynomial time. Use this algorithm to test if theories in $\mathcal{T}$ have multiple models (also in polynomial time).
6. Use problem (5) to check if a Krom theory possesses multiple models.
7. Use problem (5) to check if an affine theory possesses multiple models.
8. We are given two Horn theories, H_1 and H_2. Design an algorithm to test if H_1, H_2 have the same models.
9. Design a similar technique for Krom theories.
10. We observed that testing if a formula is renameable Horn can be decided in polynomial time. It is known that testing if a formula is renameable to a program is, actually, an NP-complete problem. This is unpleasant news. However, testing if a Krom formula can be renamed into a program can be done in polynomial time. Design an appropriate algorithm.
11. Of course, the same can be done for dual Horn programs. Use the technique used in this chapter to do so.
12. Write the truth table for the ternary Boolean function $sum3$.

Chapter 10

Embedding SAT into integer programming and into matrix algebra

10.1 Representing clauses by inequalities 206
10.2 Resolution and other rules of proof 207
10.3 Pigeon-hole principle and the cutting plane rule 209
10.4 Satisfiability and $\{-1, 1\}$-integer programming 214
10.5 Embedding SAT into matrix algebra 216
10.6 Exercises 225

In this chapter we look at the effects of changing representations of clauses and of sets of clauses. We will explore (but this is, literally, "the tip of the iceberg") the results of representing clauses and sets of clauses in two formalisms: linear programming (but really in integer programming) and matrix algebra. In the first of these areas, we represent each clause as an inequality. As the result, the sets of clauses are represented by systems of inequalities. The basic idea here is to identify Boolean values 0 and 1 with integers 0 and 1 (nothing strange in itself; something similar happens in some programming languages). This representation is *faithful* in the following sense: the solutions to the resulting sets of inequalities, *as long as they are in the desired form, taking only values* 0 *and* 1, can be "pulled back," resulting in satisfying valuations for the original CNF. In fact, we will see that there is more than one representation of clausal logic in the integer programming. The second representation, closer in spirit to Kleene three-valued logic, requires solutions taking only values -1 and 1. The simple act of conversion brings, however, advantages: techniques applicable in integer programming can be used to locate satisfying valuations.

Yet another translation that we will explore assigns to CNFs matrices over some ordered ring (in reality integers are sufficient). This is the subject of Section 10.5. Translation of CNFs into matrices allows us to reduce the search for satisfying valuations and autarkies to suitably formulated algebraic questions.

10.1 Representing clauses by inequalities

Assume that we have a clause, say

$$C := l_1 \vee \ldots \vee l_k.$$

Let us interpret the atom p as an *integer* variable but ranging over the set $\{0, 1\}$. In other words, we interpret Boolean 0 and 1 as integers 0 and 1. Now, let us substitute for each negative literal $\bar{p}$ in C the algebraic expression $1-p$, substitute the operation symbol $+$ for each occurrence of $\vee$ in C, and append at the end the inequality symbol ≥ 1. The effect of this syntactic transformation is the following integer inequality:

$$l_1 + \ldots + l_k \geq 1. \tag{10.1}$$

We will denote by I_C the result of this transformation of the clause C, and for a CNF (i.e., a set of clauses) F, we write I_F for $\{I_C : C \in F\}$. We also call the inequality (10.1) a *pseudo-Boolean* (or *pseudo-Boolean inequality*) [ARSM07].

Continuing our identification of Boolean 0 and 1 with integer 0 and 1, we can lift the notion of a valuation to the integer context. Specifically, a valuation v is now a function from Var to integers with the restriction that the range of v is included in $\{0, 1\}$. As was the case in logic, the valuation v uniquely extends its action to all polynomials. In particular, $v(\bar{p}) = 1 - v(p)$, for all variables p. We now have the following simple observation.

PROPOSITION 10.1

1. Let C be a clause and let v be a valuation. Then $v \models C$ if and only if v is a solution to I_C. Consequently:

2. Let F be a CNF and let v be a valuation. Then $v \models F$ if and only if v is a solution to I_F.

Proof: (1) Let v be a valuation, $v \models C$. Then for some i, $1 \leq i \leq k$, $v(l_i) = 1$. Also, for all j, $1 \leq j \leq k$, $v(l_j) \geq 0$. Therefore $\sum_{i=1}^{k} v(l_i) \geq 1$, and so v is a solution to the inequality I_C.

Conversely, let v be a solution to the inequality I_C, i.e., $\sum_{i=1}^{k} v(l_i) \geq 1$. Since all values of v are either 0 or 1, at least one $v(l_i)$ must be 1. But then $v \models C$, as desired. (2) follows directly from (1). □

The effect of Proposition 10.1 is that the system of integer inequalities faithfully represents SAT as long as we limit ourselves to the solutions in $\{0, 1\}$.

Let us look at an example.

Example 10.1
Let F consist of the clauses: $p \vee q \vee \neg r, p \vee \neg q$, and $q \vee r$. We translate F to inequalities: $p + q + (1 - r) \geq 1, p + (1 - q) \geq 1$, and $q + r \geq 1$. But now we can apply rules of elementary algebra and get the following system of inequalities:

$p + q - r \geq 0$
$p - q \geq 0$
$q + r \geq 1.$

There are many solutions to this system of inequalities. Of these, for instance the assignment v specified by $p = 1, q = 0, r = 1$ is a solution taking only values 0 and 1, and so we have a satisfying assignment for the CNF F. □

Now, integer programming, and its special case 0-1 integer programming, where we care only about solutions that take the values 0 and 1, is a domain with a long history. Surely, they found something we did not find in SAT.

10.2 Resolution and other rules of proof

The most natural rule of proof in integer programming is the *linear combination rule* (LCR). Let r, s be non-negative integers (actually, non-negative reals work as well). The instance of LCR determined by two inequalities in the premise, and two non-negative integers r and s is:

$$\frac{a_1 x_1 + \ldots + a_n x_n \geq a \quad b_1 x_1 + \ldots + b_n x_n \geq b}{(ra_1 + sb_1)x_1 + \ldots + (ra_n + sb_n)x_n \geq ra + sb}$$

We were taught in elementary school that the linear combination rule is sound. That is, whenever $(c_1, \ldots, c_n)$ is a sequence of numbers satisfying both premises, then $(c_1, \ldots, c_n)$ satisfies the conclusion as well.

Surprisingly, the linear combination rule generalizes resolution. To this end, let us observe that $l + \bar{l} = 1$ regardless what the literal l is. Next, let $p \vee C_1$ and $\neg p \vee C_2$ be two clauses which are resolved on variable p. The clause $p \vee C_1$ translates (with slight abuse of notation) to $p + C_1 \geq 1$, while $\neg p \vee C_2$ translates to $(1 - p) + C_2 \geq 1$. Then with $r = s = 1$ we get

$$p + C_1 + (1 - p) + C_2 \geq 2.$$

This expression rewrites to $C_1 + C_2 \geq 1$, as desired.

If we substitute $\bar{a} = 1 - a$ then the inequality is no longer the sum of literals, it is the weighted sum of variables, and the right-hand side no longer has to be 1. There is a delicate issue of repeated occurrences of variables in clauses, and also of tautologies. We will deal with these issues in the Section 11.6 Exercises. Clauses without repeated occurrences of variables translate to inequalities where all coefficients on the left-hand side are either 1 or -1. When we resolve, there is a possibility of getting a coefficient 2 (or -2) of a variable. We will see soon, however, that this case can be eliminated.

If the left-hand side of our inequality admits only coefficients 1 and -1, then we can transform that left-hand side to an expression which is a sum of literals. With such transformation it is very easy to see when such inequality is valid, satisfiable, or unsatisfiable in the $0 - 1$ world. The reason is that if an inequality $I = f(p_1, \ldots, p_k) \geq r$ (with a linear polynomial f) has t negative literals on the left-hand side, then adding t to both sides we can transform it to an inequality of the form

$$l_1 + \ldots + l_k \geq s.$$

But such inequality is satisfiable (in the $0 - 1$ world) if $s \leq k$, a tautology if $s \leq 0$ and unsatisfiable if $s > k$.

Example 10.2

1. The inequality $-p_1 - p_2 - p_3 \geq -4$ is a tautology. Indeed it is equivalent to $\bar{p}_1 + \bar{p}_2 + \bar{p}_3 \geq -1$ which is always true.

2. The inequality $-p_1 - p_2 - p_3 \geq 1$ is antitautology (always false). Indeed it is equivalent to $\bar{p}_1 + \bar{p}_2 + \bar{p}_3 \geq 4$ which is always false.

3. The inequality $-p_1 + p_2 \geq 1$ is, of course, satisfiable.

□

We observed that it seems that the combination rule is *different* from *resolution* if a literal appears in both C_1 and C_2. But as we will see we can handle this as well, for there is a sound rule for deriving additional inequalities, a rule that simplifies the coefficients. This rule is called the *cutting plane rule* and it is valid for all integer inequalities, not only when we care about 0–1 solutions. Here it is: when r is a positive integer then:

$$\frac{a_1 x_1 + \ldots + a_n x_n \geq b}{\lceil \frac{a_1}{r} \rceil x_1 + \ldots + \lceil \frac{a_n}{r} \rceil x_n \geq \lceil \frac{b}{r} \rceil}.$$

(Recall that $\lceil \cdot \rceil$ is the familiar *ceiling* function which assigns to the real number x the least integer bigger than or equal to x.)

PROPOSITION 10.2
The cutting plane rule is sound.

Proof: Clearly, for a positive integer r, $\lceil \frac{a}{r} \rceil \cdot r \geq \frac{a}{r} \cdot r = a$. Therefore if $a_1 x_1 + \ldots + a_n x_n \geq b$ then

$$\lceil \frac{a_1}{r} \rceil r x_1 + \ldots + \lceil \frac{a_n}{r} \rceil r x_n \geq a_1 x_1 + \ldots + a_n x_n \geq b$$

Thus
$$\lceil\frac{a_1}{r}\rceil x_1 + \ldots + \lceil\frac{a_n}{r}\rceil x_n \geq \frac{b}{r}.$$
But the left-hand side is an integer, thus we get
$$\lceil\frac{a_1}{r}\rceil x_1 + \ldots + \lceil\frac{a_n}{r}\rceil x_n \geq \lceil\frac{b}{r}\rceil.$$

Now, with this additional rule we can get resolution in the integer programming setting. After we combine the inequalities, either we get a tautology (when there is a pair of complementary literals after resolving), or a contradiction (if we have a pair of contradictory literals), or an inequality with all coefficients equal to 1 (remember, we combine literals), or we have a final case which we analyze now. This is the case when we combine two clauses, where $C_1 \cap C_2 \neq \emptyset$. Then, after the combination we get an inequality where coefficient(s) of some literal is equal to 2 (there may be more than one). But now, we apply the cutting plane rule to this inequality with $r = 2$. What we get is the inequality corresponding to the resolvent.

10.3 Proving the pigeon-hole principle using the cutting plane rule

To illustrate the possible uses of techniques originating in the integer programming community, we will use both the change of the representation (to the inequalities instead of clauses) and the linear combination and cutting plane rules to get a "short" proof of the *Dirichlet pigeon hole principle*. By short we mean an argument that will use a number of applications of these rules bound by a polynomial in the parameter determining the problem.

There are various forms of the Dirichlet principle. One says that if you have m pigeons and n holes and $m > n > 0$ then there is no assignment of pigeons to holes so that each pigeon gets her own hole. But it can also be interpreted as follows: any assignment of m pigeons to n holes with $m > n$ has to have at least two different pigeons in some hole. This principle, surprisingly, is nothing else but the induction principle. The derivation is sort of circular, using the least element in every nonempty set of integers. This fact goes beyond the scope of this book.

It is known that the CNF formula PH_n, expressing the fact that there is no assignment of $n + 1$ pigeons to n holes so that no two pigeons share a hole, is provable using resolution *without* subsumption. Specifically, in a moment we will produce a CNF PH_n such that PH_n possesses this property. It will be clear that this formula is unsatisfiable. One shows that the closure of PH_n under resolution generates the empty clause, $\emptyset$. But the argument using the resolution rule alone must have $O(c^n)$ steps (n is a variable here). This is a classical result by Haken [Ha85]. But we will see (because it is so much easier) that we can do this in a polynomial number of steps if we have at our disposal the cutting plane rule.

We will introduce the SAT representation of PH_n ($n+1$ pigeons, n holes). Our language has an atom $p(i,j)$ for each i, $1 \leq i \leq n+1$, and each j, $1 \leq j \leq n$. The CNF PH_n consists of two groups of clauses:

1. For each i, $1 \leq i \leq n+1$, a clause

$$p(i,1) \vee \ldots \vee p(i,n)$$

(each pigeon has its hole).

2. For each pair j_1, j_2 such that $j_1 \neq j_2$, $1 \leq j_1 \leq n+1$, $1 \leq j_2 \leq n+1$ and for each k, $1 \leq k \leq n$, a clause

$$\neg p(j_1, k) \vee \neg p(j_2, k)$$

(no two different pigeons are in the same hole).

When we encode our collection of clauses as integer inequalities we get (with the range of parameters as above)

$$\sum_{j=1}^{n} p(i,j) \geq 1 \tag{10.2}$$

and

$$\overline{p(j_1, k)} + \overline{p(j_2, k)} \geq 1 \tag{10.3}$$

whenever $1 \leq j_1 < j_2 \leq n+1$.

Let us observe that altogether, given n, we have only $n+1$ clauses each of length n in (10.2), and $n \cdot (n+1) \cdot n$ clauses in (10.3). Altogether there is $O(n^3)$ clauses and the size of the entire formula is also $O(n^3)$.

By a *structural rule* we mean a rule

$$\frac{L + l + \bar{l} \geq j}{L \geq j - 1},$$

where L is an algebraic expression.

PROPOSITION 10.3
There is a polynomial-length proof of contradiction out of PH_n in the proof system consisting of linear combination rules, cutting plane rules, and structural rules.

Proof: We will get a contradiction from the set of inequalities (10.2) and (10.3). As the first step, we will prove, by induction on j that for every k, $1 \leq k \leq n$,

$$\sum_{i=1}^{j} \overline{p(i,k)} \geq j - 1. \tag{10.4}$$

(It is a curious formula. What it says is that if you try to pack into the kth hole first j pigeons then at least $j-1$ will not be there).

Proof of the inequality (10.4). The case of $j = 1$ is obvious, and the case of $j = 2$ is one of our input inequalities. Now, assume that for each k, $1 \leq k \leq n$, $\sum_{i=1}^{j} \overline{p(i,k)} \geq j-1$. We can multiply both sides by $j-1$ (this is nothing but adding $j-1$ copies of our inductive assumption, and so it is allowed with our rules of proof). We then have:

$$(j-1)(\sum_{i=1}^{j} \overline{p(i,k)}) \geq (j-1)^2. \tag{10.5}$$

We will now add to this inequality j inequalities

$$\overline{p(1,k)}+ \qquad \overline{p(j+1,k)} \geq \qquad 1 \tag{10.6}$$
$$\ldots + \qquad \ldots \geq \qquad 1 \tag{10.7}$$
$$\overline{p(j,k)}+ \qquad \overline{p(j+1,k)} \geq \qquad 1 \tag{10.8}$$

What happens when we sum up these inequalities? The summands within the inequality (10.5) have now not $j-1$ but j occurrences. Each of them; so this amounts to incrementing the factor $j-1$ outside by 1, that is, to j. But this is not all. Let us observe that in the subsequent k inequalities the second term was identical and was $\overline{p(j+1,k)}$. Therefore adding all our inequalities produces the left-hand side:

$$j \cdot \sum_{i=1}^{j+1} \overline{p(i,k)}.$$

What about the right-hand side? It is $(j-1)^2 + j$, that is $j^2 - j + 1$. In other words, we got the inequality

$$j(\sum_{i=1}^{j+1} \overline{p(i,k)}) \geq j^2 - j + 1. \tag{10.9}$$

Let us apply to this inequality the cutting plane rule with $r = j$. We then get the following inequality:

$$\sum_{i=1}^{j+1} \overline{p(i,k)} \geq \lceil \frac{j^2 - j + 1}{j} \rceil. \tag{10.10}$$

Inequality (10.10) is rewritten to:

$$\sum_{i=1}^{j+1} \overline{p(i,k)} \geq \lceil j - 1 + \frac{1}{j} \rceil. \tag{10.11}$$

But now the right-hand side of the inequality (10.11) is, clearly, j. Indeed, j is the least integer bigger than $j - 1 + \frac{1}{j}$. Therefore the proof of the inductive step, and thus of the statement, is completed.

So now we look at the case of $j = n + 1$, and get for every k, $1 \leq k \leq n$ the inequality

$$\sum_{i=1}^{n+1} \overline{p(i,k)} \geq n. \tag{10.12}$$

Let us sum up all the inequalities from our input of the form (10.2). We are summing $n + 1$ of these inequalities and so get a double sum on the left-hand side and $n + 1$ on the right-hand side:

$$\sum_{\substack{1 \leq i \leq n+1 \\ 1 \leq j \leq n}} p(i,j) \geq n+1. \tag{10.13}$$

Let us sum up all the inequalities of the form (10.12). There were n of them, the right-hand side in each was n, so we get the following inequality:

$$\sum_{\substack{1 \leq i \leq n+1 \\ 1 \leq j \leq n}} \overline{p(i,j)} \geq n^2. \tag{10.14}$$

Now, let us sum up the inequalities (10.13) and (10.14). Let us look at the left-hand side first. We get, after obvious change of the order of summation,

$$\sum_{\substack{1 \leq i \leq n+1 \\ 1 \leq j \leq n}} (p(i,j) + \overline{p(i,j)}). \tag{10.15}$$

Since the inner sum $p(i,j) + \overline{p(i,j)}$ is *always* 1, the left-hand side rewrites to $n(n+1)$. But the right-hand side is $n^2 + n + 1$, that is $n(n+1) + 1$. This is a desired contradiction.

But it is not the end of our argument; we need to see how many rule applications the entire process required. There was an inductive process. Each of these inductive processes required summation of $2k$ inequalities, then one application of the cutting plane rule. We realize that summing up inequalities is an application of the linear combination rule. Altogether, clearly, $O(n^2)$ proof steps were needed at this point. Then, we summed up two sets of inequalities (one had $n + 1$ inequalities, the other one n). Then we used the structural rule ($l + \bar{l} = 1$) $O(n^2)$ times. Therefore we used $O(n^2)$ steps. This completes the argument. □

We discussed the inequalities of the form $l_1 + \ldots + l_k \geq 1$. What about the dual inequalities $l_1 + \ldots + l_k \leq 1$? Clearly, since at most one of l_i, $1 \leq i \leq k$ can be true, at least $k - 1$ of dual literals $\bar{l}_i$ must be true. That is the inequality $\bar{l}_1 + \ldots + \bar{l}_k \geq k - 1$ can be derived. This is not the *negation* of the original inequality (which is $l_1 + \ldots + l_k \leq 0$, that is $l_1 = 0$, $l_2 = 0$, etc.). It is a dual inequality. This also tells us that language of inequalities allows for expressing succinctly various other properties of sets of propositional variables.

Let us look again at the inequalities (where we look for 0–1 solutions)

$$a_1 l_1 + \ldots + a_n l_n \geq a$$

with positive coefficients. Such inequalities can have all sorts of interpretations. For instance, we may think about *weight* function wt on the set of literals and then this inequality requires that

$$wt(l_1)l_1 + \ldots + wt(l_n)l_n \geq a.$$

We end this section by asking *can we somehow handle Boolean constraint propagation* for the inequalities? In order to deal with this question, we first generalize the inequalities to *cardinality atoms*. Those will be expressions of the form rXs where X is a set of literals and $0 \leq r \leq s \leq |X|$ are two integers. The meaning of such cardinality atom is: "Out of literals of X at least r but no more than s are true." This is a generalization of inequalities discussed above, except that we require that all coefficients be equal to 1. The inequality $l_1 + l_2 + \ldots + l_m \geq r$ is written rX where $X = \{l_1, \ldots, l_m\}$. Likewise $l_1 + l_2 + \ldots + l_m \leq s$ is written Xs. The expression rXs denotes the conjunction of these two inequalities. We will look in a later chapter at other, more complex inequalities.

What would be the analogue of unit resolution for such inequalities? Let $l \in X$. The rule looks like this:

$$\frac{rXs \quad \bar{l}}{r(X \setminus \{l\})s}.$$

Moreover, we have the following rules

$$\frac{rX}{l},$$

when $|X| = r$, and $l \in X$. Let us observe that the expression rX with $r = |X|$ is a generalization of the unit clause. It asserts that all literals (thus units) in X are true. When $l \in X$, then we have another propagation rule (which reduces to elimination of literal in the clausal case):

$$\frac{rXs \quad l}{(r-1)(X \setminus \{l\})(s-1)}.$$

But we can go even further. Assume that we treat the expressions rXs as a new kind of *propositional variables* (i.e., atoms). Then we can form clauses from such atoms. Let us observe that not only such atoms generalize ordinary atoms via $a \equiv 1\{a\}1$, but also negative atoms: $\neg a \equiv 0\{a\}0$. All sorts of normal forms can be shown easily for formulas formed of cardinality atoms. One such normal form is this:

$$r_1X_1s_1 \wedge \ldots \wedge r_mX_ms_m \Rightarrow t_1Y_1u_1 \vee \ldots \vee t_sY_su_s.$$

This normal form generalizes the usual clausal form. The point here is that one can use the syntactic properties of cardinality atoms for speeding up processing. Again, we will look at such clauses in a later chapter.

10.4 Satisfiability and $\{-1,1\}$-integer programming

It is natural to ask if the faithful embedding of clauses into integer programming described above is unique, in the sense that there are no others. Actually, there are other embeddings of clausal logic into integer programming. For the sake of demonstrating that there is much to be discovered here, we will look at one such embedding (there are surely other embeddings as well!).

Given a valuation v, we assign to it a new valuation, but not into $\{0,1\}$ but rather into $\{-1,1\}$, as follows. Namely, for a valuation v we assign to it a function $w_v : Var \to \{-1,1\}$ as follows.

$$w_v(p) = \begin{cases} 1 & \text{if } v(p) = 1, \\ -1 & \text{if } v(p) = 0. \end{cases}$$

Clearly, the assignment $v \mapsto w_v$ is a bijection between the set of valuations of Var into $\{0,1\}$ and the set of valuations of Var into $\{-1,1\}$.

Next, given a clause

$$C := p_1 \vee \ldots \vee p_m \vee \neg q_1 \vee \ldots \vee \neg q_n$$

let us assign to C an integer inequality

$$i_C := p_1 + \ldots + p_m - q_1 - \ldots - q_n \geq 2 - (m+n).$$

Again, this is a bijection between clauses and very special integer inequalities involving all variables where all coefficients of those variables are equal to $-1, 0,$ or 1 (except that the terms with coefficient 0 are not listed for obvious reasons). Next, given a CNF F, we can assign to it a system of inequalities $i_F = \{i_C : C \in F\}$. First, we need to formally say what it means that an assignment $w : Var \to \{-1,1\}$ is a solution to an inequality (yes, we know that this is taught in elementary school). To this end, we first formally define $w(p_1 + \ldots + p_m - q_1 - \ldots - q_n)$. We define inductively: $w(p_1 + \ldots + p_m) = \sum_{j=1}^{m} w(p_j)$, and then: $w(-q_j) = -w(q_j)$, and $w(p_1 + \ldots + p_m - q_1 \ldots - q_{n-1} - q_n) = w(p_1 + \ldots + p_m - q_1 \ldots - q_{n-1}) - w(q_n)$. Then, w is a solution to the inequality $p_1 + \ldots + p_m - q_1 \ldots - q_n \geq r$ if $w(p_1 + \ldots + p_m - q_1 \ldots - q_n) \geq r$. After this ocean of formalism we are ready to formulate and prove a result on the embedding of clausal logic into $\{-1,1\}$-integer programming.

PROPOSITION 10.4

1. *For every valuation v and for every clause C, $v \models C$ if and only if w_v is a solution to i_C and therefore,*

SAT, integer programming, and matrix algebra

2. For every valuation v and a CNF F, $v \models F$ if and only if w_v is a solution of i_F.

Proof: By induction on the sum $m + n$.
Base case, subcase $m = 1, n = 0$. Then the clause C is the unit clause p_1, and i_C is the inequality $p \geq 1$. But as w_v takes the values in $\{-1, 1\}$, the inequality $p \geq 1$ is in fact an equality, $p = 1$. Thus satisfaction of C by v, and satisfaction of i_C by w_v are equivalent in our case.
Base case, subcase $m = 0, n = 1$. Then the clause C is unit $\neg q_1$, and i_C is $-q \geq 1$, i.e., $q \leq -1$. But again, $q \leq -1$ is equivalent to $q = -1$, and the equivalence holds.
Induction step. First, let us consider the case when $n = 0$. Then C is $p_1 \vee \ldots \vee p_m$, and i_C is $p_1 + \ldots + p_m \geq 2 - m$. First, let us suppose that $v \models C$. Then for some j, $v(p_j) = 1$. We can assume $j = m$, and we proceed as follows:

$$w_v(p_1 + \ldots + p_{m-1} + p_m) = w_v(p_1 + \ldots + p_{m-1}) + w_v(p_m) =$$
$$w_v(p_1 + \ldots + p_{m-1}) + 1.$$

By inductive assumption $w_v(p_1 + \ldots + p_{m-1}) \geq 2 - (m - 1)$. Adding one to both sides and taking into account the above equations, we get

$$w_v(p_1 + \ldots + p_{m-1} + p_m) \geq 2 - (m - 1) + 1 = 2 - m.$$

Next, we have to show that if w_v is a solution to $p_1 + \ldots + p_m \geq 2 - m$ then $v \models p_1 \vee \ldots \vee p_m$. Indeed, if $v \not\models p_1 \vee \ldots \vee p_m$, then $v(p_1) = \ldots = v(p_m) = 0$, $w_v(p_1) = \ldots = w_v(p_m) = -1$, and so $w_v(p_1 + \ldots + p_m) = -m < 2 - m$. Thus for at least one of p_j, $1 \leq j \leq m$, $w_v(p_j) = 1$, and so $v(p_j) = 1$, thus $v \models C$.
Next, consider the subcase when $n > 0$. Then C is $p_1 \vee \ldots \vee p_m \vee \neg q_1 \vee \ldots \vee \neg q_{n-1} \vee \neg q_n$. Let D be $p_1 \vee \ldots \vee p_m \vee \neg q_1 \vee \ldots \vee \neg q_{n-1}$. First, let us assume $v \models D$. Then by inductive assumption,

$$w_v(p_1 + \ldots + p_{m-1} + p_m - q_1 - \ldots - q_{n-1}) \geq 2 - (m + n - 1).$$

Therefore

$$w_v(p_1 + \ldots + p_{m-1} + p_m - q_1 - \ldots - q_{n-1} - q_n) \geq 2 - (m + n - 1) - w_v(q_n).$$

But this last term $-w_v(q_n)$ is definitely bigger or equal to -1. Thus we get

$$w_v(p_1 + \ldots + p_{n-1} + p_n - q_1 - \ldots - q_{n-1} - q_n) \geq 2 - (m+n-1) - 1 = 2 - (m+n),$$

as desired.
Second, let us assume $v \not\models D$ (but $v \models C$). Then it must be the case that: $v(p_1) = \ldots = v(p_m) = 0$, $v(q_1) = \ldots = v(q_{n-1}) = 1$, and $v(q_n) = 0$. The last equality follows from the fact that under the assumption $v \not\models D$, the only way to satisfy C is with q_n set to 0. Then

$$w_v(p_1 + \ldots + p_{m-1} + p_m - q_1 - \ldots - q_{n-1} - q_n) = (m + (n-1)) \cdot (-1) + 1 = 2 - (m+n).$$

To complete our argument we now have to prove that if w_v satisfies the inequality

$$p_1 + \ldots + p_{m-1} + p_m - q_1 - \ldots - q_{n-1} - q_n \geq 2 - (m+n),$$

then $v \models C$.
If $w_v(q_n) = -1$, then $v(q_n) = 0$, and $v \models C$. So let us assume $w_v(q_n) = 1$. Then as

$$w_v(p_1 + \ldots + p_{m-1} + p_m - q_1 - \ldots - q_{n-1} - q_n) \geq 2 - (m+n),$$

we have

$$w_v(p_1 + \ldots + p_{m-1} + p_m - q_1 - \ldots - q_{n-1}) \geq 2 - (m+n) + w_v(q_n)$$
$$= 2 - (m+n) + 1 = 2 - (m + (n-1)).$$

But by the inductive assumption this means that $v \models D$, and since D subsumes C we are done. □

It is also clear that our Proposition 10.4 implies a faithful representation of inequalities $p_1 + \ldots + p_m + (1 - q_1) + \ldots (1 - q_n) \geq 1$ (with 0–1 solutions) by means of inequalities $p_1 + \ldots + p_m - q_1 - \ldots - q_n \geq 2 \geq 2 - (n+m)$ (with $\{-1, 1\}$ solutions).

10.5 Embedding SAT into matrix algebra

When we discussed the minimal unsatisfiability (Section 7.1) and proved that for a minimally unsatisfiable set of clauses F, $|Var(F)| \leq |F|$, we introduced certain data structure, namely a bipartite graph that was associated with F, where one part consisted of F, the other of $Var(F)$ and an edge connected a clause C with a variable p if p occurred in C (regardless of whether it was a positive or negative occurrence). A closer look at such a graph reveals some problems. Namely, the graph G_F considered in Section 7.1 does not uniquely determine F. Here is an example: the CNF F consisting of clauses $\{p \vee q, \neg p \vee q, \neg q\}$ determines precisely the same graph as the CNF F' consisting of clauses $\{p \vee q, \neg p \vee q, q\}$. From our perspective it is not acceptable: the formula F is unsatisfiable, but the formula F' is satisfiable. Hence, to represent formulas faithfully we need an additional element. We modify the graph G_F by introducing *weights* of edges. When p occurs in C positively, we assign to that edge the weight 1; if it occurs negatively, we assign the weight -1.

Such representation is faithful; we can recompute F from such a weighted bipartite graph G_F. It is also clear that this representation is much more in the spirit of the second representation of satisfiability in integer programming (Section 10.4). One way of thinking about the weighted bipartite graph representing a set of clauses F is that it is a full bipartite graph with edges weighted with a weight taking values $-1, 1, 0$. It is just that we do not list edges weighted 0. If we think in those terms, then we are again somewhere close to Kleene's three-valued logic, with 1 represented by 1, 0 represented by -1, and u represented by 0. But now, if we have a weighted complete bipartite graph $G = \langle X, Y, E, wt(\cdot) \rangle$ with one part $X = \{a_1, \ldots, a_m\}$ and the other part $Y = \{b_1, \ldots, b_n\}$, with all possible edges in E and with a real-valued weight function wt, then we can assign to the graph G an $(n \times m)$-real matrix M by setting its entries as follows:

$$M_{i,j} = wt(e(i,j)),$$

where $e(i, j)$ is the edge connecting the vertex a_i with the vertex b_j. For a weighted bipartite graph which is not necessarily complete, we can extend this to the following definition:

$$M_{i,j} = \begin{cases} wt(e(i,j)) & \text{if there is an edge incident with } a_i \text{ and } b_j, \\ 0 & \text{otherwise.} \end{cases}$$

When we compose these two representations (first, sets of clauses represented by a weighted bipartite graph, then a weighted bipartite graph represented by a matrix) we get a *matrix representation of a* CNF F. Assuming $F = \{C_1, \ldots, C_m\}$ and $Var(F) = \{p_1, \ldots, p_n\}$ the matrix $M(F)$ is defined by the following conditions:

$$M(F)_{i,j} = \begin{cases} 1 & \text{if } p_j \text{ occurs in } C_i \text{ positively,} \\ -1 & \text{if } p_j \text{ occurs in } C_i \text{ negatively,} \\ 0 & \text{if } p_j \text{ does not occur in } C_i. \end{cases}$$

The assignment $F \mapsto M(F)$ is one to one (assuming a fixed ordering of clauses and fixed ordering of variables), as long as F contains no tautological clauses. The reason is that when F contains tautologies, $M(F)$ is not well-defined.

There are several operations on matrices that preserve satisfiability. First, any permutation of rows in $M(F)$ is just another listing of the same F. Next, a permutation of columns corresponds to a permutation of variables. Furthermore, permutation with a possible complement (change of sign of the entire column) corresponds to consistent permutation of literals and does not change satisfiability. All these facts can be formally expressed, and we will look at them in Section 10.6 Exercises.

Now, in addition to the above, there are two operations on matrices worth mentioning. The first one is the elimination of duplicate rows. This does not change satisfiability. All that happens is that a duplicate clause is eliminated. Another is related to duplicate columns. If two columns corresponding to variables p_k and p_l are duplicates of each other, then it is quite easy to see that the partial assignment $\{p_k, \neg p_l\}$ forms an autarky for F. In such case we can simplify matrix $M(F)$ without changing the satisfiability status: namely, we eliminate all *rows* that have a non-zero entry in the kth column as well as both columns. The resulting matrix represents a formula G which is satisfiable if and only if F is satisfiable. A similar situation occurs when a column corresponding to p_i is equal to a column corresponding to p_j multiplied by -1. In such case $\{p_i, p_j\}$ is an autarky, and we can proceed as in the previous simplification. These simplification procedures justify our introduction of a new syntax. It will turn out that there are more benefits.

Not every matrix with $-1, 1, 0$ entries is a matrix of a formula. But we are close enough. At least when a matrix M represents a CNF then (up to limitations discussed above, that is, order of rows and renamings), it represents a unique formula.

Now, unlike the previous two interpretations we are going to interpret satisfiability by linear algebraic means. The matrices need to be over any *ordered ring*, but for simplicity we will discuss matrices over real numbers. Integers or rationals would do equally well.

First, we will assign to every real matrix M its *sign pattern*. This is a matrix M' defined as follows:

$$M'_{i,j} = \begin{cases} 1 & \text{if } M_{i,j} > 0, \\ -1 & \text{if } M_{i,j} < 0, \\ 0 & \text{if } M_{i,j} = 0. \end{cases}$$

Now, let us fix F, and thus the parameters m (the size of F) and n (the size of $Var(F)$) and the orders of clauses and variables (thus fixing the meaning of rows and columns). The sign patterns of $n \times n$ matrices induces a relation. The relation $\sim$ is defined by the formula:

$$M \sim N \text{ if } M' = N'.$$

PROPOSITION 10.5
The relation $\sim$ is an equivalence relation.

The matrices M that interest us are those for which $M \sim M(F)$. We need to have in mind that some matrices M are not $\sim$-equivalent to any $M(F)$ for a CNF formula F. But if they are, the formula F for which $M \sim M(F)$ (remember that we fixed the orders of parameters) is unique.

Next, let us look at matrices with a single row, that is, vectors with real values. If we take M' for such a single-row matrix we get a vector composed of -1s, 1s, and 0s. We now have two different interpretations to such vector. The first one is a CNF consisting of a single clause. But there is another interpretation as well. It is an interpretation where the vector $\vec{x}$ is interpreted as a *set of literals*, namely, the set of literals $\{p_j : \vec{x}(j) = 1\} \cup \{\neg p_j : \vec{x}(j) = -1\}$. This set of literals can be thought of (as we often do) as a partial interpretation of $Var(F)$. This is the meaning we assign to those vectors. Here is one example: let $\langle p, q, r \rangle$ be a list of 3 propositional variables, and let $\vec{x}$ be the vector $(-3, 2, 0)$. The vector $\vec{x}$ determines the partial assignment $(-1, 1, 0)$ which corresponds to the following consistent set of literals: $\{\neg p, q\}$.

This duality of interpretations (either by a single-row matrix or by a vector) corresponds to the analogous phenomenon in logic: a collection of literals v can either be interpreted by its disjunction (and then we talk about a clause) or by its conjunction (and then we talk about partial valuation.)

Now, if a matrix M is $\sim$-equivalent to the matrix $M(F)$ for a propositional formula F we say that M *has the same sign pattern* as $M(F)$ and abusing the language, the same sign pattern as F. Testing if M has the same sign pattern as F is easy and can

Example 10.3
The matrix M
$$\begin{pmatrix} 2 & 2 & -1 \\ 3 & 5 & -2 \end{pmatrix}$$
has the sign matrix M'
$$\begin{pmatrix} 1 & 1 & -1 \\ 1 & 1 & -1 \end{pmatrix}$$
which has duplicate rows. Such matrices M are not considered here. On the other hand the matrix M
$$\begin{pmatrix} 2 & 2 & -1 \\ -3 & 0 & -2 \end{pmatrix}$$
has the sign matrix M'
$$\begin{pmatrix} 1 & 1 & -1 \\ -1 & 0 & -1 \end{pmatrix},$$
which represents the CNF $F = \{p \vee q \vee \neg r, \neg p \vee \neg r\}$. □

Now, we define the following set of vectors associated with a CNF F:
$$H_F = \{\vec{x} \in \mathbf{R}^n : \exists_M (M \sim M(F) \text{ and } M\vec{x} > \vec{0})\}.$$

So H_F consists of n-dimensional vectors of real values (any ordered ring would do, though) such that the dot product of *some* matrix M with the same sign pattern as the sign pattern of F results in a vector with all positive entries.

First, let us observe the following property of the set H_F.

PROPOSITION 10.6
The set H_F is a cone in $\mathbf{R}^n$; that is, H_F is closed under vector sum and the product by positive reals.

Proof: Let us observe that the set of matrices $\sim$-equivalent to $M(F)$ is closed under sums and it is also closed with respect to multiplication by positive reals. This, clearly, implies the assertion. □

But now, we can compute out of H_F the set of partial valuations generated out of elements of H_F. Let us formalize it.[1] So formally (we know that, in effect, we repeat

---
[1] We juggle with two different formalisms to handle three-valued interpretations, one using values $1, 0$, and u, the other using $1, -1$, and 0. It is horrible, but those are used in two different communities and we will not force them to use other people's formalism.

the definition given above, but we still need it formally!) we have:

$$v_{\vec{x}}(i) = \begin{cases} 1 & \text{if } x_i > 0, \\ 0 & \text{if } x_i < 0, \\ u & \text{if } x_i = 0. \end{cases}$$

($\vec{x}$ here is an n-dimensional vector, and x_i is the ith coordinate of $\vec{x}$.)
Now, we define:

$$P_F = \{v_{\vec{x}} : \vec{x} \in H_F\}.$$

So what is P_F? It is the set of partial assignments (so there will be a connection to Kleene three-valued logic) which can be read off vectors that left-multiplied by some matrix M, M sign-equivalent to F, resulting in a vector with all positive entries. Now, we have the following fact.

PROPOSITION 10.7
Let F be a CNF. Then

$$P_F = \{v : v \text{ is a partial valuation of Var and } v_3(F) = 1\}.$$

Proof: Let $m = |F|$ and $n = |Var(F)|$. We first prove the inclusion $\subseteq$. Let us assume that a partial valuation v belongs to the set P_F. Then there exists a matrix M, $M \sim M(F)$, and a real vector $\vec{x}$ such that $v_{\vec{x}} = v$ and $M\vec{x} > \vec{0}$. Now, $M\vec{x} > \vec{0}$ means that for every row M_i of M, the dot product of M_i and $\vec{x}$ is positive. Formally,

$$M_i \cdot \vec{x} > \vec{0},$$

that is, $\sum_{j=1}^{n} M_{i,j} \cdot x_j$ is positive. Let us fix i and look at $\sum_{j=1}^{n} M_{i,j} \cdot x_j$. It must be the case that for at least one j, $M_{i,j} \cdot x_j > 0$. But this means that both $M_{i,j}$ and x_j are non-zero, and that they are of the same sign. In other words either $M_{i,j}$ is positive and x_j is positive or $M_{i,j}$ is negative and x_j is negative. Now, recall that $M \sim M(F)$. Therefore, looking at the clause C_i and the partial valuation $v_{\vec{x}}$ we find that one of two cases must happen: either p_j occurs in C_i positively and $v_{\vec{x}}(p_j) = 1$, or p_j occurs in C_i negatively and $v_{\vec{x}}(p_j) = 0$. In either case, $(v_{\vec{x}})_3(C_i) = 1$. Since i was an arbitrary index of the row (i.e., C_i was an arbitrary clause in F) the inclusion $\subseteq$ is proved.
Conversely, let us assume that v is a partial valuation evaluating F as 1. We can compute out of v a real vector $\vec{x}$ such that $v_{\vec{x}} = v$. There are many such vectors. We will consider the following one:

$$\vec{x}(i) = \begin{cases} 1 & \text{if } v(p_i) = 1, \\ -1 & \text{if } v(p_i) = 0, \\ 0 & \text{if } v(p_i) = u. \end{cases}$$

Now, we need to find a matrix M such that $M \sim M(F)$ and $M\vec{x} > \vec{0}$. There are many such matrices. We show one. The matrix $M(F)$ may be no good, because the interaction of entries of the ith row with $\vec{x}$ may result in a negative result. So we need to modify $M(F)$ as follows. First, since $v_3(F) = 1$, for every clause C_i there is a literal $l_{j(i)}$, $1 \leq j(i) \leq n$ with $l_{j(i)}$ occurring in C_i such that $v(l_{j(i)}) = 1$. This means that $l_{j(i)}$ occurs in C_i and it is evaluated by v as 1. So here is how we define a desired matrix M.

$$M_{i,j} = \begin{cases} M(F)_{i,j} & \text{if } j \neq j(i), \\ n \cdot M(F)_{i,j} & \text{if } j = j(i). \end{cases}$$

Thus in the row i we modify $M(F)$ in a single place $j(i)$ by multiplying the value by the length of the vector $\vec{x}$. The first observation is that each row of the matrix M is non-zero. Indeed, $l_{j(i)}$ occurs in C_j, thus the entry $M(F)_{i,j}$ is non-zero, thus $M_{i,j(i)}$ is non-zero.

Now, let us look at the dot product of M and $\vec{x}$, $M\vec{x}$. This product can be expressed as a sum of two terms:

$$(\sum_{j=1, j \neq j(i)}^{n} M_{i,j} \cdot x_j) + (M_{i,j(i)} \cdot x_{i,j(i)}).$$

The ith row of M differs from the ith row of $M(F)$ in just one place, $j(i)$. Therefore all the terms $M_{i,j}$ in the first part of our sum are $1, -1$, or 0. Likewise all the terms x_j in the first part of our sum are $1, -1$, or 0. The effect is that we can estimate the first term of our sum:

$$|\sum_{j=1, j \neq j(i)}^{n} M_{i,j} \cdot x_j| \leq n - 1.$$

But the second term in our sum is $M_{i,j(i)} \cdot x_{j(i)}$ which is, by construction of M, $n \cdot M(F)_{i,j(i)} \cdot x_{j(i)}$. But $M(F)_{i,j(i)} \cdot x_{j(i)} = 1$. Thus

$$M_{i,j(i)} \cdot x_{j(i)} = n.$$

Therefore we can estimate the sum $\sum_{i=1}^{n} M_{i,j} \cdot x_j$; it is certainly bigger than or equal to 1. Thus we conclude

$$\sum_{i=1}^{n} M_{i,j} \cdot x_j \geq 1 > 0.$$

As i was arbitrary, we conclude that the inclusion $\supseteq$ is also true, and our assertion is proved. □

Before we get a corollary expressing the satisfiability in linear algebraic terms, let us look at an example showing that moving to matrices with the same sign pattern is really necessary.

Example 10.4
Let F be the following set of clauses: $\{p \vee q, p \vee \neg q, \neg p \vee \neg q\}$. The matrix $M(F)$ is:

$$\begin{pmatrix} 1 & 1 \\ 1 & -1 \\ -1 & -1 \end{pmatrix}.$$

The inequality $M(F)\vec{x} > \vec{0}$ has no solution for it requires both $x_1 + x_2 > 0$ (first row) and $-x_1 - x_2 > 0$ (third row). But here is a matrix M with the same sign pattern:
$$\begin{pmatrix} 1 & 1 \\ 1 & -1 \\ -1 & -5 \end{pmatrix}.$$
This matrix imposes different constraints on the vector $\vec{x}$. Those are three inequalities: $x_1 + x_2 > 0$, $x_1 - x_2 > 0$, and $-x_1 - 5x_2 > 0$. This system of inequalities has solutions, for instance $x_1 = 1, x_2 = -.5$ is a solution, generating satisfying valuation $\{p, \neg q\}$. If we execute the construction of the second part of Proposition 10.7 we get the following matrix M_1:
$$\begin{pmatrix} 2 & 1 \\ 2 & -1 \\ -1 & -2 \end{pmatrix}.$$
The matrix M_1 has the property that $M_1\vec{x} > \vec{0}$ has a solution and one such solution is $x_1 = 1, x_2 = -1$.

□

But now we get the following corollary.

COROLLARY 10.1
A set of clauses F is satisfiable if and only if for some matrix M with the same sign pattern as F there is a vector $\vec{x}$ such that $M\vec{x} > \vec{0}$.

It may be esthetically appealing (for some readers) that the characterization of satisfiability given in Corollary 10.1 does not refer to logic at all.

Interesting as it is, Corollary 10.1 refers to an infinite family of matrices. One wonders if we can bound somehow the number of matrices that would have to be checked. The second part of the proof of Proposition 10.7 entails such bound (pretty crude, though). The point here is that it is quite clear that if there is a satisfying valuation then one of the matrices M constructed in the second half of that proof must have a solution for the system of inequalities $M\vec{x} > \vec{0}$. So here are conditions on these matrices M. For all $i, j, 1 \leq i \leq m, 1 \leq j \leq n$:

1. $M \sim M(F)$.
2. For each i, the ith row of M differs from the ith row of $M(F)$ in at most one place.
3. In the place j where these rows differ we must have the equality:
$$M_{i,j} = n \cdot M(F)_{i,j}.$$

Clearly there is only finitely many (to be precise no more than m^n) such matrices M. Thus we get the following corollary.

COROLLARY 10.2
For each set of clauses F, $|F| = m$, $|Var(F)| = n$ there is a set $\mathcal{M}$ of matrices such that

1. $|\mathcal{M}| \leq m^n$.

2. F is satisfiable if and only if for some $M \in \mathcal{M}$, the inequality $M\vec{x} > \vec{0}$ possesses a solution.

Next, we look at Kleene extension theorem (Proposition 2.5(2)). That result implies that whenever $v_3(F) = 1$ and $v \preceq_k w$, then also $w_3(F) = 1$. Now, let us look at the analogous relation for real vectors. Here it is. We write

$$\vec{x}_1 \preceq_k \vec{x}_2$$

if whenever $\vec{x}_1(i) \neq 0$, then $\vec{x}_1(i) = \vec{x}_2(i)$. That is, in transition from $\vec{x}_1$ to $\vec{x}_2$ we can change values but only if that value is 0. For instance $(1, 0, 0.5, -3) \preceq (1, 2, 0.5, -3)$ because we changed the value 0 at the second position to the value 2. Unlike in Kleene logic, relation $\preceq_k$ is not a partial ordering, but it is a preordering. Specifically, it is easy to see that if $\vec{x}_1 \preceq_k \vec{x}_2$ then $v_{\vec{x}_1} \preceq v_{\vec{x}_2}$. Now, we get the following corollary.

COROLLARY 10.3
The set H_F is closed under the preordering $\preceq_k$. That is, if $\vec{x}_1 \in H_F$ and $\vec{x}_1 \preceq_k \vec{x}_2$, then $\vec{x}_2 \in H_F$.

What about autarkies? (It looks like we are fixated by them, we look for them under every rug). Those possess a similar characterization due to Kullmann in terms of linear algebra over ordered rings.

PROPOSITION 10.8
Let F be a CNF formula and let v be a partial valuation. Then v is an autarky for F if and only if for some real matrix M with the same sign pattern as F there exists a vector $\vec{x}$ such that:

1. $M\vec{x} \geq \vec{0}$, and

2. $v_{\vec{x}} = v$.

Proof: Our argument will be similar to that of Proposition 10.7. First, let us assume that v is an autarky for F. Our goal is to construct a matrix M such that $M \sim M(F)$ and a vector $\vec{x}$ such that $M\vec{x} \geq \vec{0}$ and $v_{\vec{x}} = v$. We will construct a desired matrix M by altering entries of matrix $M(F)$. But this time we will only alter those rows that correspond to clauses touched by v. So let us suppose that v touches a clause $C_i \in F$. Then as v is an autarky for F, $v_3(C_i) = 1$. Therefore there is $j(i)$, $1 \leq j(i) \leq n$, such that either $M(F)_{i,j}(i) = 1$ and $p_{j(i)} \in v$ or $M(F)_{i,j(i)} = -1$ and $\neg p_{j(i)} \in v$.

Let us fix such $j(i)$ for each row number i such that v touches C_i, and define a matrix M as follows:

$$M_{i,j} = \begin{cases} M(F)_{i,j} & \text{if } v \text{ does not touch } C_i, \\ M(F)_{i,j} & \text{if } v \text{ touches } C_i \text{ but } j \neq j(i), \\ n \cdot M_{i,j} & \text{if } j = j(i). \end{cases}$$

Then, let us define the vector $\vec{x}$ as follows:

$$\vec{x}(j) = \begin{cases} 1 & \text{if } p_j \in v, \\ -1 & \text{if } \neg p_j \in v, \\ 0 & \text{otherwise.} \end{cases}$$

Clearly, by construction, $M \sim M(F)$ and $v_{\vec{x}} = v$. All we need to show is that $M\vec{x} \geq \vec{0}$.

If v does not touch C_i then the dot product of the ith row of M, M_i, by $\vec{x}$ is 0 because non-zero entries of that row will be zeroed by 0 entries of $\vec{x}$. Indeed, in every place where $M_{i,j}$ is non-zero, $\vec{x}(j)$ is 0, as $v = v_{\vec{x}}$ and v does not touch C_i.

So let us assume that v touches C_i. But v is an autarky for F. Therefore $v_3(C_i) = 1$. Therefore, by construction, like in the proof of Proposition 10.7, we can estimate the dot product of $M_i \cdot \vec{x}$ and we find that it is, actually, strictly greater than 0.

Conversely, let us assume that for some matrix M such that $M \sim M(F)$ there is a vector $\vec{x}$, with $v_{\vec{x}} = v$ and $M\vec{x} \geq \vec{0}$. We want to show that v is an autarky for F. Let C_i be an arbitrary clause in F. If v does not touch C_i then there is nothing to prove. So let us assume that v touches C_i. Then there is some j, $1 \leq j \leq n$ such that $M_{i,j} \neq 0$ and $\vec{x}(j) \neq 0$. Let us fix such j. Two cases are possible.

Case 1: $M_{i,j}$ and $\vec{x}(j)$ have the same sign. If this is the case then, as $M \sim M(F)$ and $v_{\vec{x}} = v$, it must be the case that either p_j occurs in C_i positively and $p_j \in v$, or p_j occurs in C_i negatively and $\neg p_j \in v$. In either case $v_3(C_i) = 1$.

Case 2: $M_{i,j}$ and $\vec{x}(j)$ have opposite signs. Then, since $\sum_{j=1}^{n} M_{i,j} \cdot \vec{x}(j) \geq 0$ and the term $M_{i,j} \cdot \vec{x}(j)$ has negative value, there must be another index j' such that $M_{i,j'} \cdot \vec{x}(j')$ is positive. But then $M_{i,j'}$ and $\vec{x}(j')$ have the same sign and we are back to Case 1. Thus again $v_3(C_i) = 1$.

Since i was arbitrary, we are done, and the proof is now complete. $\square$

There is a corollary to the proof of Proposition 10.8. Let $\vec{e}_i$ be the vector having on the ith coordinate 1, and on the remaining coordinates 0. We then have the following.

COROLLARY 10.4

Let F be a CNF and let v be a partial valuation. Let us fix the orders of clauses in F and of variables. Let C_i be a clause in F. Then v is an autarky for F and v evaluates C_i as 1 if and only if for some matrix M, $M \sim M(F)$, and for some vector $\vec{x}$ such that $v = v_{\vec{x}}$, $M\vec{x} \geq \vec{e}_i$.

10.6 Exercises

1. Write a CNF with at least 5 clauses, each with at least 3 literals. Represent your CNF faithfully in (a) $\{0, 1\}$-integer programming, (b) in $\{-1, 1\}$-integer programming. Conclude that you need to know which of these formalisms you use, as the representations are inequalities, and an inequality has a different *meaning* depending upon which formalism you use.

2. We were a bit cavalier about structural rules; we had only one structural rule (eliminating $l + \bar{l}$). But in fact, if we really want to be *truly formal* we need more structural rules, for instance, rules that allow us to move literals within expressions. Formulate such rule(s).

3. Show the validity of rules for manipulating cardinality constraints. We will discuss these rules in the chapter on knowledge representation, but there is no harm in doing it right now.

4. Show that the representation of clause sets by means of weighted bipartite graphs is, indeed, faithful. That is, different representations of a given CNF as a full, weighted bipartite graph are isomorphic.

5. Show that our discussion of the relationship of matrix operations and permutations of literals (right after we introduced the sign matrix) is valid.

6. Show that when columns of a sign matrix $M(F)$, representing two variables p_i, p_j, are identical, then $\{p_i, \neg p_j\}$ is an autarky for F.

7. Show that when the column of the sign matrix $M(F)$, representing the variable p_i, takes the opposite value to those of p_j (i.e., for all k, $M_{j,k} = -M_{i,k}$) then $\{p_i, p_j\}$ is an autarky for F.

Chapter 11

Coding runs of Turing machines, NP-completeness and related topics

11.1	Turing machines	228
11.2	The language	231
11.3	Coding the runs	232
11.4	Correctness of our coding	233
11.5	Reduction to 3-clauses	237
11.6	Coding formulas as clauses and circuits	239
11.7	Decision problem for autarkies	243
11.8	Search problem for autarkies	245
11.9	Either-or CNFs	247
11.10	Other cases	249
11.11	Exercises	252

This chapter is the first of four that are devoted to issues in knowledge representation. While knowledge representation is an area with practical applications, in this chapter we start with a construction that is theoretical, but carries an important promise. Specifically, in this chapter we encode the problem of Turing machines reaching the final state as a satisfiability problem. To prove that SAT is NP-complete, we will reduce to SAT any language decided by a nondeterministic Turing machine in polynomial time. We need to be very careful so that the reader does not get the impression that SAT solves the halting problem. What we show is that *if* we set the bound on the *length* of the run of the machine then the problem of reaching the final state within such bound can be represented as a SAT problem. In fact, we will find a polynomial $f(\cdot)$ such that the size of the SAT problem solving this limited halting problem is bound by $f(|M| + |I| + k)$, where $|M|$ is the size of the machine, $|I|$ is the size of the input, and k is the bound on the number of steps allowed until we reach the final state.

One consequence of this coding (and proof of correctness of coding) is that every problem in the class NP polynomially reduces to SAT. Thus the message of this chapter is that, at least in principle, SAT can serve as a universal tool for encoding search problems in the class NP.

Next, we also show that we can polynomially reduce satisfiability to satisfiability of sets of 3-clauses. We then show how circuits (and hence arbitrary formulas) can be coded into sets of clauses. We also show that the existence problem for autarkies is NP-complete.

Finally, we show that "mix-and-match" of (most of) polynomial cases (studied in

Chapter 9) results in NP-completeness. Specifically, we show that of 21 possible cases (we refer to Chapter 9 for the definition of 7 "easy" cases) of pairs of "easy" theories, 19 cases lead to classes of CNFs for which the satisfiability problem is NP-complete, whereas two have polynomial complexity.

We refer the reader to the book by M. Sipser [Sip06], Chapter 7, especially Section 7.3, for relevant definitions.

11.1 Nondeterministic Turing machines

A nondeterministic Turing machine is an abstraction of a computing device.

A *Turing machine*, formally, consists of three components. First, a finite set of *states*, S. We will assume that among the elements of S there are two distinguished states: i, the initial state, and f, the final state. Second, the machine M has an *alphabet* Σ. In addition to Σ that is used in describing the input data and transformations on the data that the machine executes, there will be an additional symbol B, a blank symbol, not in Σ. Before we describe the third component of the machine, the set of instructions, let us mention that every machine M has a fixed three-element set of *directions*. Those are: r that is used to denote that machine (to be precise: its read-write head) moves to the right, l, used to denote that the machine moves to the left, and λ, used to denote that the machine stays put (we do not know yet what it means that the machine moves; this will be described shortly). The third part of the machine is the *nondeterministic transition function* δ. The function δ assigns to every pair $\langle s, d \rangle$ where s is a state and d is a datum element (i.e., an element of $\Sigma \cup \{B\}$), a set of triples of the form $\langle s', d', a \rangle$ where $s' \in S$, $d' \in \Sigma \cup \{B\}$, and $a \in \{l, r, \lambda\}$. The set $\delta(s, d)$ may be empty. It should be clear that since S and Σ are finite, each set $\delta(s, d)$ is finite. There are various ways to think about δ, for instance, as a 5-ary relation included in $S \times (\Sigma \cup \{B\}) \times S \times (\Sigma \cup \{B\}) \times \{l, r, \lambda\}$. Now, we define the machine M as a quintuple $\langle S, \Sigma, \delta, i, f \rangle$. By $|M|$, the size of M, we mean the number $|S| + |\Sigma| + |\delta|$. With our conventions, $|M|$ is finite.

A Turing machine M operates on a one-side infinite tape. The tape is divided into cells, enumerated by consecutive non-negative integers. To visualize the operation of the machine, we think about M having a read-write head (very much like older tape devices that stored the data). The cells of the tape contain a symbol of the alphabet Σ or the blank symbol B. The fact that the tape is infinite creates a problem when we want to describe runs of the machine by *finite* propositional theories.

The machine M operates on the tape in a synchronic mode. At any given time moment the read-write head observes a single cell of the tape. We will assume that at the moment 0 the tape observes the cell with index 0. At any moment of time t the machine will be in a specific state s from the set S. So now, let us assume that at the time t the machine M is in the state s and observes a cell n containing the datum (i.e., an alphabet symbol, or a blank) d. Here is what happens now. The machine selects an element $\langle s', d', a \rangle$ belonging to $\delta(s, d)$. This is an instruction to be executed. It is executed as follows: the machine changes its state to s'. It overwrites the datum d in the currently observed cell by d'. Finally, the read-write head may move. The possible move is determined by the third element of the instruction, a. If $a = l$ the read-write head moves one place left, it now observes the cell $n - 1$. If $a = r$, the read-write head moves one place to the right, it now observes the cell $n + 1$. Finally, if $a = \lambda$, the read-write head does not move. The content of no other cell is affected by actions at time t. There are two caveats. First, it may so happen that the set $\delta(s, d)$ is empty. Second, we are at cell with index 0 but the selected instruction tells us to move to the left. In both cases the machine hangs.

In principle the execution may last forever – we can get into an infinite loop, for instance. So now we will introduce an additional limitation. Specifically, we assumed initially that once the machine reached the final state f it halts. Instead we will make a slightly different assumption, namely, that when the machine reaches final state f, then it does not halt but instead it stays in the state f, rewrites the content of the cell with its content (i.e., does not change it), and stays put. This just means that $\delta(f, d) = \{\langle f, d, \lambda \rangle\}$. for all $d \in \Sigma \cup \{B\}$. This assumption allows us to make sure that accepting computations (i.e., ones that reach f) are of the same length.

Let us formalize the notion of the *execution* or *run* of the machine M on an input σ. First, let us look at the tape. At time 0 the first $|\sigma|$ cells of the tape are filled with consecutive elements of σ. The remaining cells contain blanks. So now assume that we have a tape T, that the read-write head observes the cell n, and that the instruction to be executed is $\langle s', d', a \rangle$. Then the new tape T' (result of one-step execution) differs from T in at most one place, n, namely, $T'(n) = d'$. The remaining cells in T' stay the same. So now, we are ready to define one state of the tape and the machine. It is $\langle T, s, n, y \rangle$ where

1. T is a tape (intuitively, the current tape).

2. s is a state (intuitively the current state of the machine).

3. n is a cell number (intuitively the cell currently observed by the read-write head of the machine).

4. y is an instruction $\langle s', d', a \rangle \in \delta(s, T[n])$ (intuitively the instruction to be executed).

The next state of the machine and tape is the tape T', the state s', the index n', and an instruction y' selected nondeterministically from $\delta(s', T[n'])$ where $n' = n - 1$ if $a = l$, $n' = n + 1$ if $a = r$ and $n' = n$ if $a = \lambda$.

Now, let $A = \langle T, s, n, y \rangle$ and let A' be the next state of tape and machine. We call the pair $\langle A, A' \rangle$ a *one-step transition*.

With this definition it is easy to define an execution (run) of the machine M on input σ. It is a sequence $R = \langle A_0, A_1, \ldots, A_i, \ldots \rangle_{i<r}$ subject to the following constraints. A_0 is the state of machine and tape with the tape T_0 with σ in first $|\sigma|$ cells, blanks afterwards, $s = i$, $n = 0$ and y is any instruction in $\delta(i, 0)$. The recursive requirement is that for all $j < r - 1$, $\langle A_j, A_{j+1} \rangle$ is a one-step transition.

To sum up, a run is just a description (pretty redundant!) of what happens to the tape and to the machine. The index r is the *length* of the run.

As we defined it, the run is always an infinite object, because the tape is an infinite object. But we will see that when dealing with runs of finite length we can safely skip this limitation. The reason is that since the read-write head moves only one place along the tape, the run of length r can affect only the first $r-1$ cells of the tape.

So now, we are interested in finite runs. If the run takes r time units, the read-write head, which moves at most one cell per unit, cannot go beyond the $r-1^{th}$ cell, that is, no cell beyond the $r-1^{th}$ cell will ever be observed and thus never changed. This means all cells of the tape will have the unchanged content starting with the rth cell. So now, let us make yet another limiting assumption. Let us fix a polynomial h with positive integer coefficients, with the degree of h at least 1. With this requirement on polynomial h, for all $n \in N$, $n \leq h(n)$. Now, for an input σ of length n we will consider the runs of length $h(n)$.

Given the tape T, let $T[r]$ be the initial segment of the tape T consisting of r cells. Next, when $A = \langle T, s, n, y \rangle$ let $A[r]$ be $\langle T[r], s, n, y \rangle$. Then in any run of length r the initial segment $T[r]$ *uniquely determines* T (just extend $T[r]$ by the cells of T_0 with indices $r, r+1, \ldots$). But now, both $T[r]$ and $A[r]$ are *finite* objects. So the next step is to trim down a run R. Namely, if $R = \langle A_0, \ldots, A_{r-1} \rangle$ define $R[r] = \langle A_0[r], A_1[r], \ldots, A_{r-1}[r] \rangle$. Then again $A[r]$ is finite. But we clearly have the following fact.

PROPOSITION 11.1
If R is a run of length r then $R[r]$ uniquely determines R.

Of course, R uniquely determines $R[r]$ as well. But this means that an infinite object R is uniquely determined by a finite object $R[r]$.

The accepting computation for σ is a run R of length r such that at the end of the run R the machine is in the state f. Here is the question we ask: "Is there an accepting computation for σ of length $h(|\sigma|)$?" Our question, although pertaining to an infinite object (the witnessing run is infinite – the tapes in it are all infinite sequences) reduces to a question on existence of *finite* objects (Proposition 11.1). We will show that we can answer this question with a SAT solver. Moreover, we will show that the CNF to be submitted to the SAT solver will not be "too" big.

This may look pretty awful, but it is quite simple. What it says is that if we were given a SAT solver with enough resources, then for each Turing machine M, and runtime polynomial $h(n)$, we could compute a propositional formula $F_{\sigma,h,M}$ and then check using our SAT solver if $F_{\sigma,h,M}$ is satisfiable. From the satisfying valuation we can then easily recover an accepting computation for M. The key property here is that the size of $F_{\sigma,h,M}$ is tractable (polynomial) in the size of the input, runtime polynomial, and of the machine itself (i.e., in $h(n) + |M|$). This is where we will be careful; we will need to count the number of clauses and assure ourselves that we can compute the number of clauses in $F_{\sigma,p,M}$ as a polynomial in the parameters.

11.2 The language

As should be clear, we will need to have a language with enough propositional variables so we can write appropriate clauses describing operation of the machine. As we did in other places in this book, we will write variables with parameters in them. It looks like we are doing something funny; after all, propositional variables are supposed to *not* have structure. But our parameters within the propositional variables are just subscripts, we just write them in a more intuitive way, that is all.

We will use parameters $0, \ldots, h(n)$ for the indices of the cells. What is important here is that there is $h(n) + 1$, thus a finite number, of them. We will also have parameters s for each state $s \in S$, again a finite number. We will have constants for the symbols of the alphabet (data that can appear in cells). Let us observe that the number of symbols of the alphabet is bounded by the size of the set of instructions (and in particular by the size of the machine M.) The reason for this is that if the symbol d belongs to the alphabet, but does not actually occur in instructions or on tape, it does nothing for us and can be eliminated.

Let us stress (we did this before, but this is the key) the following trivial fact. Once we decide on the available time ($h(n)$), we will not be able to affect more than $h(n)$ cells.

Now, we will be able to list all the predicates. Here is what we will use.

1. The predicate $data(P, Q, T)$. The propositional variable $data(p, q, t)$ will mean "at time t, the cell p contains the symbol q."

2. $position(P, T)$. The propositional variable $position(p, t)$ will mean "at time t the read-write head sees the contents of the cell p."

3. $state(S, T)$. The propositional variable $state(s, t)$ will mean "at time t the read-write head is in the state s."

4. $instr(S, Q, S_1, Q_1, D, T)$. The propositional variable $instr(s, q, s_1, q_1, d, t)$ will mean "at the time t instruction $\langle s_1, q_1, d \rangle$ belonging to $\delta(s, q)$ has been selected for execution."

We are going to use the propositional language, so we need to *ground* our predicates. Here is what will be used to ground it. First, the time moments: $0, \ldots, h(n) - 1$. Then the cell numbers: $0, \ldots, h(n)$. Then the symbols of alphabet and the blank symbol: $\Sigma \cup \{B\}$. We assume that every symbol of the alphabet appears, actually, in the machine M. Finally, the constants for states from the finite set S. As in the case of the alphabet, we will assume that only states in M belong to S. The two limitations (on Σ and on S), bound the sizes of these sets by $|M|$.

So now we have grounded our predicates and gotten our propositional variables. We are ready to specify the encoding of the machine as it runs for the $h(n)$ steps.

11.3 The propositional formula $F_{\sigma,h,M}$

We will now construct the formula $F_{\sigma,h,M}$. This formula will be the conjunction of several clauses. Those clauses will be divided into several groups according to their purpose. Moreover, instead of writing $p_1 \vee \ldots \vee p_k \vee \neg q_1 \vee \ldots \vee \neg q_l$ we will write more intuitively $q_1 \wedge \ldots \wedge q_l \Rightarrow p_1 \vee \ldots \vee p_k$. This is, of course, just "syntactic sugar." When there is no positive atom in the clause, the head of the implication is empty. So, if we write a formula $p \wedge q \Rightarrow$ we really mean $\bar{p} \vee \bar{q}$. Formulas of this sort will occur in the group (3), and also in the group (5.4.4).

(1. Initial conditions) First, we need (unit) clauses to specify the starting point. Recall that i is the initial state.

  (1.1) $data(p, q, 0)$, for $q = \sigma(p)$, $0 \leq p \leq n-1$ (recall that σ is the data, stored in the first n cells of the tape.)

  (1.2) $data(p, B, 0)$, for $n \leq p \leq h(n)$.

  (1.3) $state(i, 0)$.

  (1.4) $position(0, 0)$.

(2. Final condition) We require ending in final state f.

  (2.1) $state(f, h(n))$.

(3. General consistency conditions) The goal of these conditions is to force uniqueness of the execution at any given time. We need three conditions: First, the uniqueness of the position of the read-write head. Second, the uniqueness of the content of each cell at every given time, and finally the uniqueness of instruction selected for execution at every given time.

  (3.1) $state(s_1, t) \wedge state(s_2, t) \Rightarrow$
  for each pair of *different* states s_1, s_2 and for every time moment t, $0 \leq t \leq h(n)$.

  (3.2) $position(p_1, t) \wedge position(p_2, t) \Rightarrow$
  for each pair of *different* cell numbers p_1, p_2, $0 \leq p_1 \leq h(n)$, $0 \leq p_2 \leq h(n)$ and for every time moment t, $0 \leq t \leq h(n) - 1$.

  (3.3) $instr(s_1, q_1, s_2, q_2, d_1, t) \wedge instr(s_3, q_3, s_4, q_4, d_2, t) \Rightarrow$
  for each choice of different tuples $\langle s_1, q_1, s_2, q_2, d_1 \rangle$, and $\langle s_3, q_3, s_4, q_4, d_2 \rangle$ with the obvious conditions on what s_i, q_j, and d_k are.

(4. Selection process for instructions) Once the read-write head is in the state s at time t and points to the position p which holds a datum q, then we have to select one of the instructions for $\langle s, q \rangle$ to be executed. Only instructions in $\delta(s, q)$ can be selected.

(4.1)
$$state(s,t) \wedge position(p,t) \wedge data(p,q,t) \Rightarrow \bigvee_{\langle s_1,q_1,d\rangle \in \delta(s,q)} instr(s,q,s_1,q_1,d,t)$$

(5. Executing instructions) We have to describe faithfully what the machine will do. It has to overwrite the datum q with datum q_1 in the observed cell, but no other cell is touched at this time. It has to change its state. Finally, it has to move the read-write head in the direction specified by the d, unless the read-write head is at the position 0 but the instruction requires that it move to the left. We will have to provide clauses that enforce these conditions.

(5.1) $position(p,t) \wedge instr(s,q,s_1,q_1,d,t) \Rightarrow data(p,q_1,t+1)$
for each cell number p, states s, s_1, symbols q, q_1, and time moment t, with the obvious constraints

(5.2) $data(p,q,t) \Rightarrow position(p,t) \vee data(p,q,t+1)$.
This is the *frame axiom*. It says that the cells other than the one read by the read-write head at the moment t, are not affected at that time.

(5.3) $state(s,t) \wedge instr(s,q,s_1,q_1,d,t) \Rightarrow state(s_1,t+1)$
(with constraints as in (5.1).)

(5.4) We need to describe how the read-write head moves if it can, and how we fail if the head needs to move left of position 0. Four more clauses.

(5.4.1) $position(p,t) \wedge instr(s,q,s_1,q_1,r,t) \Rightarrow position(p+1,t+1)$
with obvious constraints on the position, time and state parameters.

(5.4.2) $position(p,t) \wedge instr(s,q,s_1,q_1,\lambda,t) \Rightarrow position(p,t+1)$
with obvious constraints on the position, time and state parameters.

(5.4.3) $position(p,t) \wedge instr(s,q,s_1,q_1,l,t) \Rightarrow position(p-1,t+1)$
Here the constraint is slightly stronger. p must be non-zero for us to include such clause.

(5.4.4) $position(0,t) \wedge instr(s,q,s_1,q_1,l,t) \Rightarrow$
That is, we fail if the read-write head reads the contents of the cell 0, but the selected instruction forces us to move left.

So now, we have a propositional formula $F_{\sigma,h,M}$. It consists of clauses of groups (1.1)–(5.4.4).

11.4 Correctness of our coding

Now that we have described the coding of a Turing machine M with the initial data σ of size n, and the runtime function h, we have several tasks to do. The first one is to assign to every accepting computation a valuation v that satisfies $F_{\sigma,h,M}$. Second, the other way around, assign to a valuation satisfying $F_{\sigma,h,M}$ an accepting computation of M (running in time $h(n)$). Third, we have to see that we are dealing with bijection between valuations and accepting computations. Fourth, we need to find a polynomial g which bounds the size of $F_{\sigma,g,M}$ in $n + |M|$. While pretty tedious, all these tasks are easy.

First, we will see that we can assign to an accepting computation $\mathcal{Z}$ a valuation v satisfying $F_{\sigma,h,M}$. We need to assign values to propositional variables. Those variables have been defined in Section 11.2.

Let us recall that the computation contains the complete information about the run. We know the initial state of the machine, where the read-write head points at any given time, its state at any given time, and the instruction selected at any given time. Now, let us assume that $\mathcal{Z}$ is an accepting computation. Here is the valuation v that we assign to $\mathcal{Z}$. We have four types of variables (corresponding to four predicates we specified in Section 11.2).

The predicate *data*;

$$v(data(p,q,t)) = \begin{cases} 1 & q \text{ is the content of } p \text{ at moment } t, \\ 0 & \text{otherwise.} \end{cases}$$

The predicate *state*;

$$v(state(s,t)) = \begin{cases} 1 & s \text{ is the state of the read-write head at the moment } t, \\ 0 & \text{otherwise.} \end{cases}$$

The predicate *position*;

$$v(position(p,t)) = \begin{cases} 1 & p \text{ is the position of the read-write head at the moment } t, \\ 0 & \text{otherwise.} \end{cases}$$

The predicate *instr*;

$$v(instr(s,q,s_1,q_1,d,t)) = \begin{cases} 1 & \langle s_1, q_1, d \rangle \text{ is executed on } \langle s, q \rangle \text{ at the moment } t, \\ 0 & \text{otherwise.} \end{cases}$$

So now, all we need to do is to check that the valuation v we defined above satisfies $F_{\sigma,h,M}$. It is quite clear that it does.

Now, what about the other direction? All we need to do is to extract out of a valuation v an accepting computation $\mathcal{Z}_v$ for M. The input is correctly described by (1.1)–(1.4). The final condition (unit) clause states that if what we get is a computation, then it is accepting by (2.1). Constraints of the group (3) tell us that we will never have the head in two different states at once, or pointing to two different cells at once, or having two different instructions for execution. Now what we need is that there is at least one cell that the read-write head points to at any given time, that it is in some state at any given time and that there is an instruction to execute at any given time $t < h(n) - 1$. This actually requires a proof by simultaneous induction, using (4.1) and the clauses (5.1)–(5.3). The clauses (5.4) enforce the correct movements of the read-write head. Thus we hope we have convinced the reader that every valuation v determines an accepting computation of M on an input of size n and of length $h(n)$. The third task is to see that we deal with a bijection between the accepting computations and the satisfying valuations. Here is what we do. We just check that going from v to $\mathcal{Z}$ to $v_{\mathcal{Z}}$ results in the same v, and similarly, that going from $\mathcal{Z}$ to v to $\mathcal{Z}_v$ results in the same $\mathcal{Z}$.

The final step is estimating the size of $F_{\sigma,h,M}$ depending on n, h, and M. For this purpose, we will show that the size of clauses of each group can be bound by a polynomial in n, h, and $|M|$. We need to visit each group separately, then sum up the results. Certainly, the size of clauses of groups (1) and (2) is bound by $h(n) + 3$. Then, there is at most $(h(n))^2 \cdot |S| \leq (h(n) + |M|^3)$ of clauses in (3.1), each of length 2. Similar estimation gives us at most $(h(n))^3$ clauses in (3.2) each of length 2. It is worse in (3.3), but we get the bound $|M|^4 \cdot (h(n))^5 \cdot 3$, again each clause is of size 2. The clauses (4.1) require a bit more effort. Counting the parameters in the premise of the implication we get the bound of $|S|^2 \cdot ((h(n))^2$ clauses. But these clauses have non-constant length. Fortunately, the length of each such clause is bound by $|M| + 3$. Then we have to look up clauses of type (5). But it is quite clear that, reasoning as above, we get a polynomial bound (in n, $h(n)$, and $|M|$) on the size of all of these groups. The final observation is that if we substitute a polynomial into a polynomial, we get another polynomial. So if the function g is a polynomial and we substitute it into another polynomial, the resulting function is a polynomial. Summing up the above we get the following fact.

PROPOSITION 11.2
There is a polynomial $g(x)$ such that $|F_{\sigma,h,M}| \leq g(h(n) + |M|)$.

So, what did we do? We established a polynomial time *reduction* of the problem of existence of accepting computations for the triples consisting of $\langle \sigma, h, M \rangle$ to the satisfiability problem so that the machine M possesses an accepting computation in time $h(n)$ on an input σ of length n if and only if the formula $F_{\sigma,h,M}$ is satisfiable. But we know from [Sip06] that the problem of testing if a triple consisting of Turing machine M has an accepting computation on an input σ, of length $h(n)$ (here h is a polynomial) is NP-complete. To show NP-completeness, we need to show that there is a Turing machine that halts precisely on satisfiable instances of SAT. A moment reflection shows that it is quite simple (although there will be small details that still need some care). Here is what we do. First, we need to develop some scheme for describing clause sets. We will have an alphabet consisting of 0 and 1 and a couple of other symbols that will serve as clause separators and also to handle polarity of variables. We develop a coding scheme where the variable p_n is represented as a string of n 1's. We will use additional alphabet symbols When s represents variable p_n the string as will represent the literal $\neg p_n$. We need to represent clauses. We use a letter b to separate literals within the clause. We use c to be the clause separator. Here is for example the representation of the clause $\neg p_1 \vee p_3$:

$$a1b111c.$$

We will use the letter d to denote the end of the clause set. So, if our clause set $\mathcal{C}$ consists of $\bar{p}_1 \vee p_3, p_2 \vee \bar{p}_3$ we have the following string:

$$a1b111c11ba111cd.$$

Let us observe that we need this or similar coding scheme because we can not represent potentially infinite number of variables directly (the alphabet of the machine must be finite!). We also observe that the size of the representation of $\mathcal{C}$ is linear in the size of $\mathcal{C}$.

Now, the transition relation for the desired machine will consist of two parts. First, the part that generates (nondeterministically) an assignment. This is very simple; at the beginning we find the number k of variables in $\mathcal{C}$ and delimit the segment of length k after the data (i.e. after the representation of $\mathcal{C}$. It is a bit tedious but we can certainly do this. Next, we generate the assignment of variables. We place that assignment in the part of tape that we delineated, that is right after the part of tape that contains the description of the clause set. Once we generated the assignment we need to test if that assignment satisfies all clauses of the input clause set $\mathcal{C}$. Here is the crucial point. If, for some clause C we find that the clause C is not satisfied, the machine hangs. That is we put the machine in a state where there is no instruction to execute. In the alternative situation, when the machine tested that all the clauses of the input are satisfied (i.e. all the clauses are satisfied by the assignment generated by the first part of execution, and there is nothing more to test), then the machine reaches the final state and halts. Clearly, there is an accepting computation of the machine we just described if and only if $\mathcal{C}$ is satisfiable. Indeed, if $\mathcal{C}$ is not satisfiable then no matter which candidate assignment is generated machine will not reach the final state for there will be an unsatisfied clause and so the machine will hang. But if $\mathcal{C}$ is satisfiable with witnessing satisfying assignment v, then the computation that creates v as a candidate assignment and proceed testing satisfaction of all clauses in $\mathcal{C}$ will be accepting.

This slightly informal (but not that much informal) discussion justifies the following fact.

PROPOSITION 11.3 (Cook-Levin theorem)
The problem of testing if a formula F is satisfiable is an NP-complete problem.

If we think a bit about what we did, we actually established a stronger version of this property; not only testing, but also *finding* satisfying valuation is an NP-complete problem, except that it is for *search* problems and not decision problems.

So, here is how the results of our considerations can be interpreted. There is a large class of search problems, called NP-search problems. A classic catalogue of such problems can be found in [GJ79]. Then, the results of this section show that if we have an ideal SAT solver and enough resources, we can solve *every* NP-search problem on such SAT solvers, and nothing else.

11.5 Reduction to 3-clauses

We will now reduce satisfiability of arbitrary sets of clauses to satisfiability of sets of 3-clauses. Before we start, let us observe that we can safely assume that our clause set contains no unit clauses. First, let us recall that the set $\text{BCP}(F)$ consists of the "low-hanging fruit," that is, literals that can be computed from F using unit-resolution. Just for the record, let us observe that one can compute $\text{BCP}(F)$ in linear time in the size of F. Next, let us observe that, by Proposition 8.14, which says that F is satisfiable if and only if $\text{BCP}(F)$ is consistent and the reduct of F by $\text{BCP}(F)$ is satisfiable, we can even assume that $\text{BCP}(F)$ is empty.

We first show how to transform the satisfiability problem for a mix of 2- and 3-clauses to satisfiability of sets of 3-clauses. Let us assume that we deal with a set F of clauses that is a mix of clauses of length 2 and clauses of length 3, but we want to deal (for whatever theoretical reason there could be) only with sets consisting of clauses of length 3. Can we transform F into a set of 3-clauses F' so that F is satisfiable if and only if F' is satisfiable? We can, and here is a construction. Let us get 3 new variables which we call p_F, q_F, and r_F. Let us split F into $F = F_2 \cup F_3$ according to the length of clauses. Now let us form F' as follows.

$$F' = \{l_1 \vee l_2 \vee \neg p_F : l_1 \vee l_2 \in F_2\} \cup F_3 \cup$$
$$\{p_F \vee q_F \vee r_F, p_F \vee \neg q_F \vee r_F, p_F \vee q_F \vee \neg r_F, p_F \vee \neg q_F \vee \neg r_F\}$$

PROPOSITION 11.4
F is satisfiable if and only if F' is satisfiable.

Proof. If v satisfies F define a valuation w as follows.

$$w(x) = \begin{cases} v(x) & \text{if } x \notin \{p_F, q_F, r_F\} \\ 1 & x = p_F \\ 0 & \text{if } x \in \{q_F, r_F\}. \end{cases}$$

Then it is easy to check that $w \models F'$. So F' is satisfiable.
Conversely, assume $w \models F'$. We claim $w \models F$. Let $C \in F$. If $C \in F_3$ then $w \models C$. If $C \in F_2$ then $w \models C \vee \neg p_F$. But w satisfies the last four clauses. But then it must be the case that $w \models p_F$ (this requires checking, we leave this to the reader). But then $w \models l_1 \vee l_2$, as desired. □

An obvious observation is that very few new atoms (actually 3) were needed and the transformation was done in linear time.

Next, let us look at sets of arbitrary clauses, possibly longer than 3. Let F be a finite set of clauses, Let us fix an ordering of propositional variables occurring in F. We can then think about clauses from F as lists of literals (satisfaction is, obviously, preserved when reordering the literals within clauses). Here is what we will do. Given a clause $C \in F$ there are two possible cases: the length of C is at most 3. Then (for a moment) we do nothing. If the length of C is bigger than 3 we get $n - 3$ new propositional variables that we will call $p_{C,1}, \ldots, p_{C,n-3}$. By "new" we mean that for each clause C the atoms $p_{C,i}$ will not be used anywhere else but in redoing C. To simplify notation we will write F as $F_{\leq 3} \cup F_{>3}$, splitting F according to the length of the clauses. The clauses from $F_{\leq 3}$ will not be affected. But for clauses $C \in F_{>3}$ we will now make a construction. Let us assume that $C := l_1 \vee l_2 \vee \ldots \vee l_{n-1} \vee l_n$. We will form a set of *formulas* G_C (but very soon we will get rid of the non-clausal form) inductively: $p_{C,1} \equiv l_1 \vee l_2$, $p_{C,2} \equiv p_{C,1} \vee l_3$, etc. $p_{C,n-3} \equiv p_{C,n-4} \vee l_{n-2}$. Lastly we add the clause $p_{C,n-3} \vee l_{n-1} \vee l_n$. The next thing is to get rid of these equivalences. Here is how we do this. Given a formula

$$m_1 \equiv (m_2 \vee m_3),$$

where m_1, m_2, and m_3 are literals, we assign to it three clauses: $\bar{m}_1 \vee m_2 \vee m_3$, $\bar{m}_2 \vee m_1$, and $\bar{m}_3 \vee m_1$. It is easier to see what it is about if we think in terms of implications; the first clause is $m_1 \Rightarrow m_2 \vee m_3$, the second clause is $m_2 \Rightarrow m_1$, the third one is $m_3 \Rightarrow m_1$. This transformation into clauses preserves semantics. Every valuation satisfying the equivalence satisfies the three clauses and the converse also holds. So now, let us execute the transformations serially: add new variables, write equivalences and a clause (at the end there was an extra clause!), and transform into a set of clauses. Let us call, for a given clause $C \in F_{>3}$, this set of clauses S_C. Now, let us form S_F as

$$S_F = F_{\leq 3} \cup \bigcup_{C \in F_{>3}} S_C.$$

In other words, we just did the transformation for every clause longer than 3, and then took union of all these sets, and threw in $F_{\leq 3}$.

It should be obvious that this is a polynomial reduction: there is a polynomial g such that $|S_F| \leq g(|F|)$ for every set of clauses F. The reason for this is that the number of 3-clauses we construct is bound by $3 \cdot m + 1$ for each clause C of length m and so we certainly bound the size of the whole S_F by $4 \cdot n^2$ where n is the size of F (a very inefficient bound, but we do not care here much).

Now for the serious part. We have the following fact.

PROPOSITION 11.5

The formula F is satisfiable if and only if the formula S_F is satisfiable. In fact, every valuation v of Var_F uniquely extends to a valuation v' of Var_{S_F} satisfying S_F. Conversely, if w satisfies S_F, then $w|_{\mathrm{var}_F}$ satisfies F.

Proof: It will be more convenient to think in terms of equivalences introduced in the theories G_C rather than clauses (the effect of the last transformation outlined above). What happens here is that we have a chain of unique values: the values $v(l_1)$ and $v(l_2)$ force the value $v'(p_{C,1})$, etc. Reasoning by induction (details left to the reader), the values of $v'(p_{C,j})$ are all determined by the values of v on literals $l_1, \ldots, l_{j+1}$, $1 \leq j \leq n-3$. Then, again by induction, we show that $v'(p_{C,j}) = v(l_1 \vee \ldots \vee l_{j+1})$. But then $v'(p_{n-3} \vee l_{n-1} \vee l_n)$ is the same as $v(l_1 \vee \ldots \vee l_n)$. In other words we just waved our hands through the following: given a valuation v, we uniquely extended v to a valuation v' satisfying S_F. The equivalences were satisfied because we extended v to v' to make sure they are satisfied, and the remaining 3-clauses were also satisfied, as we have seen. This completes the argument for the implication $\Rightarrow$.

As concerns the implication $\Leftarrow$, we show by induction, this time going "backward," from $p_{C,n-3}$ to $p_{C,1}$ that

$$w(p_{C,j} \vee l_{j+2} \vee \ldots \vee l_n) = 1.$$

At the end we get
$$w(p_{C,1} \vee l_3 \vee \ldots \vee l_n) = 1.$$

Now we give it the final push. Since $w(p_{C,1} \equiv (l_1 \vee l_2)) = 1$,

$$w(l_1 \vee \ldots \vee l_n) = 1.$$

But now, all the variables of $l_1 \vee \ldots \vee l_n$ are in Var_F, so $w \mid_{var_F} (l_1 \vee \ldots \vee l_n) = 1$, and since we were dealing with an arbitrary clause in F, we are done. □

Thus we polynomially reduced the satisfiability problem for arbitrary sets of clauses to the satisfiability problem for sets of clauses consisting of 2- and 3-clauses only (and by Proposition 11.4 to sets of 3-clauses). This implies the following fact.

PROPOSITION 11.6
The satisfiability problem for sets of 3-clauses is NP-complete, both for decision version and search version of the problem.

11.6 Direct coding of satisfaction of formulas as CNFs with additional variables; combinational circuits

We will now show direct encoding for arbitrary formulas as CNFs. Surprisingly, we will be able to encode a given input formula φ as a CNF G_φ so that the size of G_φ will be polynomial in the size of φ. The price we will need to pay is the introduction of additional variables. The argument given below may serve as an alternative to the construction of the previous section.

Given a finite set of propositional formulas F, its conjunction $\varphi = \bigwedge F$ has the property that for every valuation v, $v \models F$ if and only if $v \models \varphi$. Moreover, $|\varphi| \le 2|F|$. For that reason we will encode single formulas φ as CNFs. Let us recall that the size of formula φ, $|\varphi|$, is the size of its tree.

Of course, there is a CNF formula ψ such that $\psi \equiv \varphi$ is a tautology. It is any of its conjunctive normal forms. But we have seen in Example 3.1 that this may lead to the exponential growth of the CNF. We will now construct a formula ψ which is a CNF, has size linear in the size of φ, and is satisfiable if and only if φ is satisfiable. The only difference with the normal forms of Chapter 3 is that additional variables not present in φ will be present in ψ.

We first need to modify slightly the tree representation of formulas as introduced in Section 2.1. There, we labeled leaves with constants or variables, and internal nodes with functors. Now we modify those labels as follows. The label of every node has two components: the *functor label*, and the *variable label*. For the leaves we just set up the functor labels equal to **nil** and the variable label is their original label. For internal nodes the functor nodes are their original labels, and their variable labels are new variables: one new variable for each internal node. For convenience we will label the root of our tree with the variable label q_{root}. We will denote the tree of the formula φ by T_φ.

Here is an example. Let φ be the formula $(\neg p \wedge q) \vee r$. Its tree has three leaves, labeled with $\langle \textbf{nil}, p \rangle$, $\langle \textbf{nil}, q \rangle$, and $\langle \textbf{nil}, r \rangle$. It has three internal nodes: n_1 with the label $\langle \neg, q_{n_1} \rangle$ (and unique child p), n_2 with the label $\langle \wedge, q_{n_2} \rangle$ (and two children labeled, respectively, n_1 and q), and the node *root*, labeled with $\langle \vee, q_{root} \rangle$, again with two children: n_2 and r.

Next, we talk about subformulas. A node n in the tree T_φ determines a subformula: it is the formula represented by descendants of n. In our case the formula φ has 6 subformulas; three determined by leaves and three determined by nodes n_1, n_2, and *root*. Those last three are: $\neg p$, $\neg p \wedge q$, and φ itself. We will denote by φ_n the subformula of φ determined by n.

Now, by induction of the height of a node n, we will define the set G_n of clauses.

Induction base. For nodes n labeled with variables or constants we define $G_n : \emptyset$.

Inductive step:

(a) When n has a single child m (and thus the functor label of n is the negation symbol $\neg$) we set G_n equal to

$$\{q_n \vee q_m, \neg q_n \vee \neg q_m\}.$$

(b) When the functor label of n is $\wedge$ and n has two children m_1 and m_2 we set G_n equal to:

$$\{\neg q_n \vee q_{m_1}, \neg q_n \vee q_{m_2}, \neg q_{m_1} \vee \neg q_{m_2} \vee q_n\}.$$

(c) When the functor label of n is $\vee$ and n has two children m_1 and m_2 we set G_n equal to:

$$\{\neg q_{m_1} \vee q_n, \neg q_{m_2} \vee q_n, \neg q_n \vee q_{m_1} \vee q_{m_2}\}.$$

(d) When the functor label of n is $\Rightarrow$ and n has two children m_1 and m_2 we set G_n equal to:

$$\{q_{m_1} \vee q_n, \neg q_{m_2} \vee q_n, \neg q_n \vee \neg q_{m_1} \vee q_{m_2}\}.$$

(e) When the functor label of n is $\equiv$ and n has two children m_1 and m_2 we set G_n equal to:

$$\{q_{m_1} \vee q_{m_2} \vee q_n, \neg q_{m_1} \vee \neg q_{m_2} \vee q_n, \neg q_{m_1} \vee q_{m_2} \vee \neg q_n, q_{m_1} \vee \neg q_{m_2} \vee \neg q_n\}.$$

We limited ourselves to four binary functors: $\wedge, \vee, \Rightarrow$, and $\equiv$. Altogether there are 16 binary functors, and we could give an analogous definition for each of these 16 functors.

Now we set $S_\varphi = \bigcup \{G_n : n \text{ is a node in } T_\varphi\}$. In our example, $G_{n_1} = \{q_{n_1} \vee p, \neg q_{n_1} \vee \neg p\}$. We leave to the reader computation of the remaining G_n. It should be clear, though, that the size of S_φ is bound by $c|\varphi|$ for a constant c.

We will now exclude one trivial example, namely, when $\varphi : \bot$. With this exclusion we have the following observation.

PROPOSITION 11.7
The set of clauses S_φ is satisfiable.

Proof: Let us consider any partial valuation v that assigns values to all variables of Var_φ. Then let us extend the partial valuation v to a valuation w_v by setting

$$w_v(x) = \begin{cases} v(x) & \text{if } x \in Var_\varphi, \\ v(\varphi_n) & \text{if } n \text{ is an internal node of } T_\varphi. \end{cases}$$

Now, we check that the clauses of S_φ are satisfied by w_v. Since we constructed S_n inductively, we check that for each n, w_v satisfies S_n. We will do this in just one case of the internal node with functor label $\neg$ and leave the rest to the reader.

In this case, n had one child m and S_n consisted of two clauses $q_n \vee q_m$ and $\neg q_n \vee \neg q_m$. Two cases are possible.

Case 1: $v(\varphi_m) = 0$. Then $v(\varphi_n) = 1$. Thus $v_w(q_m) = 0$ and $v_w(q_n) = 1$. Clearly v_w satisfies both clauses in S_n.

Case 2: $v(\varphi_m) = 1$. Then $v(\varphi_n) = 0$. Thus $v_w(q_m) = 1$ and $v_w(q_n) = 0$. Again v_w satisfies both clauses in S_n. □

Now, we have the following fact that relates the satisfaction of the formula φ with the satisfaction of the set of clauses G_φ. We recall the exclusion of a trivial case of $\varphi : \bot$.

PROPOSITION 11.8

1. Every valuation v defined on Var_φ (but on no other variable) uniquely extends to a valuation w of Var_{G_φ} such that $w \models G_\varphi$.
2. A valuation v of Var_φ satisfies φ if and only if its unique extension w to Var_{G_φ} satisfying G_φ has the property that $w \models q_{root}$.
3. Thus, a formula φ is satisfiable if and only if the CNF $G_\varphi \cup \{q_{root}\}$ is satisfiable.

Proof: We check uniqueness of extension of v to a satisfying valuation w for G_φ by induction of the rank of a node. Then, it follows that this unique extension w is nothing else but v_w constructed in the proof of Proposition 11.7. But we know that the value $w_v(q_{root})$ coincides with that of $v(\varphi)$. Thus $w \models q_{root}$ if and only if $v(\varphi) = 1$. This takes care of (1) and (2). (3) follows from (2). □

Our representation of the formula φ by the set of clauses G_φ is closely related to so-called *combinational circuits* or *circuits*, for short. By a circuit we mean a finite directed acyclic graph where each node is a source, or has just one predecessor, or two predecessors. A node may have many immediate successors. Each node is labeled with a pair. The first element of that pair is the type of the functor, in the parlance of circuits called the *gate*, and it is a unary or binary Boolean operation. The second part of the label is the name of that node. The obvious limitation is that the number of inputs must be the arity of the first part of the label. We assume that different nodes have different names.

Since the graph of such circuit is acyclic, there must be *sources* of that graph, called *inputs*, and sinks of the graph, called *outputs*. Since in an acyclic graph we can assign to each node a rank very much the same way as we did to nodes in trees (after all a tree representation of a formula is a circuit, except that it has just one output, and each node has only one immediate successor). Now, an assignment of Boolean values to the input nodes of a circuit $\mathcal{C}$ uniquely determines the assignment of values to all internal nodes of $\mathcal{C}$ and in particular to output nodes of $\mathcal{C}$. We call an *input-output pair* the pair consisting of the assignment of Boolean values to the inputs, and the values of outputs on these inputs.

By a construction very similar to one used to assign the CNF G_φ to a formula φ we can assign to a circuit $\mathcal{C}$ a CNF $G_\mathcal{C}$. We just compute G_n for each node n and take the union of the resulting sets. Let us observe that $G_\mathcal{C}$ is always satisfiable; it is enough to set the values to the inputs and assign the values to the internal nodes as required by the gates labeling internal nodes. The basic result here is the following.

PROPOSITION 11.9

Satisfying valuations for $G_\mathcal{C}$ are in a bijective correspondence with the input-output pairs for $\mathcal{C}$.

Finally, let us observe that we limited our considerations to circuits with unary and binary gates. But, of course, we can easily generalize to gates with more inputs. Conjunctions or disjunctions of more than two inputs, and especially *ite* gates may be of use in the designs of circuits.

11.7 Decision problem for autarkies

We will now apply the Cook-Levin theorem (Proposition 11.3) to show that the problem of existence of non-empty autarkies (the case of empty autarkies is trivial because every CNF has an empty autarky) is NP-complete. There are, literally, thousands of NP-complete problems. All of them polynomially reduce to SAT, and SAT polynomially reduce to them. The reason why we present this reduction is that having non-empty autarkies is such a desirable property (as we have seen in Section 2.3). So the results of this section may appear negative (autarkies are useful, but it is difficult to find them, except in special cases).

First, we see the following obvious fact.

PROPOSITION 11.10
Let F be a CNF and let v be a consistent set of literals. Then the question if v is an autarky for F can be decided in polynomial time (in the size of F and v).

Thus the problem (language) HA, consisting of those CNFs that *have a non-empty autarky* is in the class NP. Our goal now is to show that, in fact, HA is NP-complete. The argument below, due to M. Truszczyński, is a direct polynomial reduction of SAT to HA.

PROPOSITION 11.11 (Kullmann theorem)
HA is an NP-complete problem.

Proof: We will show a polynomial reduction of the satisfiability problem to HA. To this end let F be a CNF theory and let p_i, $1 \leq i \leq n$, be all propositional variables in F. We introduce n *new* variables q_i, $1 \leq i \leq n$, and define a CNF theory G_F to consist of three groups of clauses:

1. All clauses in F.
2. Clauses $p_i \vee q_i$ and $\neg p_i \vee \neg q_i$, where $1 \leq i \leq n$.
3. Clauses $\neg p_i \vee p_{s(i)} \vee q_{s(i)}$, $p_i \vee p_{s(i)} \vee q_{s(i)}$, $\neg q_i \vee p_{s(i)} \vee q_{s(i)}$, and $q_i \vee p_{s(i)} \vee q_{s(i)}$, where $1 \leq i \leq n$, and the function s is defined by

$$s(i) = \begin{cases} i+1 & \text{if } i+1 \leq n \\ 1 & \text{otherwise.} \end{cases}$$

Clearly, the size of G_F is linear in the size of F.

We will show that F is in SAT (i.e., is satisfiable) if and only if G_F is in HA (i.e., has a non-empty autarky). In this fashion we establish a promised polynomial time reduction of SAT to HA.

($\Rightarrow$) Let v be a set of literals such that for every i, $1 \leq i \leq n$, exactly one of p_i and $\neg p_i$ belongs to v. Moreover, let us assume that v satisfies F. Since F is in SAT, such a set of literals exists – it is just another presentation of a satisfying valuation for F. We define v' as follows:

$$v' = v \cup \{\neg q_i : p_i \in v\} \cup \{q_i : \neg p_i \in v\}.$$

We will show that v' is an autarky for G_F.

First, we note that v' is a complete and consistent set of literals. Therefore, we simply have to show that v' satisfies all the clauses in G_F. Clearly, v (and so also v') satisfies all the clauses in F. By the definition of v', it is also easy to see that all clauses of type (2), that is, the clauses $p_i \vee q_i$ and $\neg p_i \vee \neg q_i$, where $1 \leq i \leq n$, are satisfied by v' as well. Since all clauses of type (3) are subsumed by clauses of type (2), the former ones are also satisfied by v'. It follows that v' is an autarky for G_F.

($\Leftarrow$) Let us assume that v' is a non-empty autarky for G_F. By definition, v' is consistent and v' contains at least one literal. Without loss of generality, we can assume that it is one of $p_1, q_1, \neg p_1$, or $\neg q_1$. The proof in each case will be the same. Let us assume that $p_1 \in v'$. Since the clause

$$\neg p_1 \vee p_2 \vee q_2$$

is touched by v', v' must satisfy $p_2 \vee q_2$. If $p_2 \in v'$ then also $\neg q_2 \in v'$ (by clauses of type (2) in G_F). Likewise, if $q_2 \in v'$ then $\neg p_2$ belongs to v'. Continuing a similar argument we find that $p_3 \vee q_3$ must be satisfied by v', thus v' contains both p_3 and $\neg q_3$ or v' contains $\neg p_3$ and q_3. By an easy induction we find that, in fact, v' must be a complete set of literals. In particular v' touches all clauses of G_F, thus all clauses of F. But then v' is a satisfying valuation for F, and so $v = v' \cap Lit_{\{p_1,\ldots,p_n\}}$ is a satisfying valuation for F, as desired. $\square$

Now, it follows that LEAN, the problem (language) consisting of those finite sets of clauses that are lean (see Chapter 7) is co-NP-complete. Indeed, let us recall Corollary 7.14; a set of clauses F is lean if and only if it has no non-empty autarkies, equivalently, that the largest autark set is empty. In effect, F is in LEAN if and only if it is does not belong to HA. Thus LEAN is the complement of HA, and since the latter is NP-complete, the former is co-NP-complete. Formally we have the following.

COROLLARY 11.1

LEAN is a co-NP-complete problem.

11.8 Search problem for autarkies

We will now see that the search problem for autarkies (find a non-empty autarky if there is one) can be solved using any algorithm that decides *existence* of non-empty autarkies. First, we will need a number of additional properties of autarkies. We stated fundamental properties of autarkies in Section 2.3, Proposition 2.10. Here we give additional, less fundamental properties of autarkies. We use these properties in our algorithm.

Recall that, given a collection of clauses F, F_l is a collection of clauses computed as follows: we eliminate from F all clauses containing l, and in remaining clauses we eliminate $\bar{l}$ (if it occurs there, otherwise the clause is unchanged).

PROPOSITION 11.12
Let F be a CNF, and v a consistent set of literals.

1. *If l is a literal and $l, \bar{l} \notin v$, then v is an autarky for F if and only if v is an autarky for $F \cup \{l, \bar{l}\}$.*

2. *If v is an autarky for F and $l, \bar{l} \notin v$ then*

    (a) v is an autarky for F_l

    (b) v is an autarky for $F_{\bar{l}}$

    (c) v is an autarky for $F_l \cup F_{\bar{l}}$.

3. *If for every literal $l \in Lit$, $F \cup \{l, \bar{l}\}$ has no non-empty autarkies then all non-empty autarkies for F (if any) must be complete sets of literals.*

4. *Let p be a variable. A consistent set of literals v such that v does not contain p nor $\neg p$ is an autarky for F if and only if v is an autarky for the CNF theory $F_p \cup F_{\neg p}$.*

Proof: (1) is obvious.
(2) It is enough to prove the first of three facts. The second is proved very similarly to the first one, and the third one follows from Proposition 2.10 (2). Let $C \in F_l$. Then either $C \in F$, or $C \cup \{\bar{l}\} \in F$. If v touches C then in the first case $v_3(C) = 1$ because v is an autarky for F. In the second case, $v_3(C \cup \{\bar{l}\}) = 1$. But since $v \not\models \bar{l}$ (after all, $\bar{l} \notin v$), it must be the case that $v_3(C) = 1$, as desired.
(3) Let us assume that for every literal $l \in Lit_{Var_F}$, $F \cup \{l, \bar{l}\}$ has no non-empty autarkies. If F has no non-empty autarkies then our conclusion is true. Otherwise, let v be a non-empty autarky for F. If v is not complete, then for some variable p, $p, \neg p \notin v$. But then by (1), v is a non-empty autarky for $F \cup \{p, \neg p\}$, a contradiction.
(4) First, let us see what clauses belong to $F_p \cup F_{\neg p}$. Let us recall that we consider only nontautological clauses. Therefore it cannot be the case that both p and $\neg p$ occur in some clause of F. Next, if both $p, \neg p$ do not occur in a clause C of F then C belongs to F_p thus to $F_p \cup F_{\neg p}$. If $C \in F$ and $p \in C$, then $C \setminus \{p\}$ belongs

to $F_{\neg p}$, thus to $F_p \cup F_{\neg p}$. Likewise, if $C \in F$ and $\neg p$ occurs in C then $C \setminus \{\neg p\}$ belongs to F_p, thus to $F_p \cup F_{\neg p}$. So now we know that $F_p \cup F_{\neg p}$ arises from F by eliminating any occurrence of p (positive or negative) in every clause of F that has such occurrence.

Now we are finally able to prove (4). The implication $\Rightarrow$ has been proved in (2)(c). Conversely, let us assume that v is an autarky for $F_p \cup F_{\neg p}$, $p, \neg p \notin v$. Let C belong to F, v touches C. If $p, \neg p \notin C$ then $C \in F_p \cup F_{\neg p}$ and since v is an autarky for $F_p \cup F_{\neg p}$, v satisfies C. Otherwise, let us assume that $p \in C$ (the case of $\neg p \in C$ is similar). Then $C \setminus \{p\}$ belongs to $F_p \cup F_{\neg p}$ and since v touches C and $p, \neg p \notin v$, v touches $C \setminus \{p\}$. But v is an autarky for $F_p \cup F_{\neg p}$ and so v satisfies $C \setminus \{p\}$. But then v satisfies C, as desired. $\square$

Next, we have a useful fact on autarkies and satisfying valuations.

LEMMA 11.1
Let us assume that all non-empty autarkies for a CNF F are complete. Then for any literal l, autarkies for $F \cup \{l\}$ are precisely those autarkies for F that contain l.

Proof: If v is an autarky for F and $l \in v$, then clearly v is an autarky for $F \cup \{l\}$, because v is an autarky for both F and for $\{l\}$ (Proposition 2.10 (2)).

Conversely, assume that v is an autarky for $F \cup \{l\}$. Then v is an autarky for F (Proposition 2.10 (1)). By our assumption, v is complete. But $\bar{l} \notin v$ because in such case $v_3(l) = 0$. However, our assumption was that v is an autarky for $F \cup \{l\}$, and since v touches l and v is consistent, $v_3(l) = 1$, a contradiction. $\square$

We will now describe how we can compute a non-empty autarky once we have an algorithm HA that *decides* existence of such autarkies.

Proceeding by induction, if F is empty we return the string 'the input formula has no non-empty autarkies' and halt. Also if the algorithm HA returns '*no*' we return the string 'the input formula has no non-empty autarkies' and halt.

Now, for every variable p we run the algorithm HA on the input $F \cup \{p, \neg p\}$. There are two cases to consider.

Case 1: The algorithm HA returns '*no*' on each input $F \cup \{p, \neg p\}$. Then we know that F has only complete autarkies (if any). We can assume that HA returned '*yes*' on the input F (otherwise we would already exit with failure). So now, assuming that the input formula F has variables, we select one such variable p. We run the algorithm HA on a new input $F \cup \{p\}$. If HA returns the value '*yes*,' we add p to the list of literals in the putative autarky and recursively call our algorithm (always choosing a new variable for decision) until an autarky is found. If the algorithm HA returns '*no*' on input $F \cup \{p\}$ it must be the case that $F \cup \{\neg p\}$ has autarkies. Those are all complete, we put $\neg p$ into the putative autarky, and we can select another variable and continue until there are no more variables to decide.

Case 2. The algorithm HA returns '*yes*' on some input $F \cup \{p, \neg p\}$. Then it must be the case that F possesses an autarky, but that autarky v has the property that $p, \neg p \notin v$. We compute $F_p \cup F_{\neg p}$. This new CNF possesses an autarky (Proposition

11.12 (2)). This autarky is an autarky for F. Now we call recursively our algorithm on the input $F_p \cup F_{\neg p}$, by Proposition 11.12(4). This latter set of clauses has less variables (p does not occur in $F_p \cup F_{\neg p}$ and so recursively, our algorithm correctly computes an autarky for $F_p \cup F_{\neg p}$, which is then an autarky for F. □

Thus we found that like in the case of satisfiability, having an algorithm for deciding existence of autarkies allows us to compute autarkies. Moreover, we see that we make a number of calls to HA that is bound by a quadratic polynomial in the number of variables in F.

11.9 Either-or CNFs, first two "easy cases" of SAT together

The results we have seen in various sections of Chapter 9 may have raised the hope that, at least sometimes, a "divide-and-conquer" strategy for testing satisfiability could work. Here is one such putative strategy. Let us assume that a formula F is a conjunction of two formulas, $F_1 \wedge F_2$. Let us assume that each F_i, $i = 1, 2$, is "easy." A good way of thinking is that, for instance, F_1 is a collection of Horn clauses, and F_2 consists of affine formulas (i.e., linear equations). We certainly can solve F_1 easily, and we can also solve F_2 easily. If any of those is unsatisfiable – we are done, F is unsatisfiable. But if each F_i is satisfiable then the idea would be to somehow match a valuation satisfying F_1 with a valuation satisfying F_2. Each of these valuations is just a complete set of literals. If these sets of literals are consistent, then their union is a complete set of literals (in principle the underlying sets of variables could be different!) and so the union, if consistent (or equivalently: a valuation extending both), is a satisfying valuation for F.

Unfortunately, this strategy fails, in a quite spectacular way. It turns out that, essentially (i.e., with exception of trivial cases), "mix-and-match" of easy cases results in an NP-complete case. Thus the 'grand strategy' proposed above cannot work. What we will see is that we will be able to transform a given CNF formula F into another formula F' which can be decomposed into a union $F_1 \cup F_2$ so that F_1 and F_2 separately will be easy to solve. The transformations will always be in polynomial time.

The reader will recall that we considered seven "easy" cases. The classes we considered – positive, negative, Horn, dual Horn, renameable-Horn, Krom, and affine formulas were all "easy." Thus there are $\binom{7}{2} = 21$ cases to consider. It looks like theorems with 21 cases are sort of absurd (unless we are algebraists and we deal with finite groups, that is). We will limit our attention to cases where F is a CNF, or consists of clauses and affine formulas (linear equations).

We first observe that in the case of F_1 consisting of positive clauses and F_2 consisting of dual Horn clauses (and its dual: negative clauses and Horn clauses), the problem remains easy because the union is still dual Horn (or Horn, in the other case). This leaves us with the remaining 19 cases. It turns out that we need a few lemmas that allow us to take care of all cases that need to be handled. The first, prototypical, case is that of the situation when F_1 consists of positive clauses, while F_2 consists of negative clauses. But it turns out that the construction involves, in reality, also the case of Horn clauses, dual Horn clauses, and affine formulas.

Let F be a CNF formula. Recall that in Section 9.6 we introduced a notion of "either-or" (EO) CNFs. Those were CNFs that consisted of clauses that were either positive or negative (*both* could be present in an EO formula).

LEMMA 11.2

For every clause $C : p_1 \vee \ldots \vee p_k \vee \neg q_1 \vee \ldots \vee \neg q_m$ there is a CNF G_C (in the language with possibly additional propositional variables) such that

1. *G_C is an EO-formula.*
2. *There is a bijection between the valuations of propositional variables in Var_C satisfying C and valuations of propositional variables occurring in G_C satisfying G_C.*

Proof: We introduce m new propositional variables $x_1, \ldots, x_m$ and form G_C consisting of three groups of clauses. The first group has just one clause, the other two groups, m each. All these latter clauses will be Krom clauses.

(a) $p_1 \vee \ldots \vee p_k \vee x_1 \vee \ldots \vee x_m$.
(b) $q_1 \vee x_i, 1 \leq i \leq m$.
(c) $\neg q_1 \vee \neg x_i, 1 \leq i \leq m$.

Effectively, the clauses (b) and (c) force that logical value of x_i is opposite to that of q_i. Thus, for any valuation v' satisfying clauses (b) and (c), and defined on all the variables involved below:

$$v'(p_1 \vee \ldots \vee p_k \vee x_1 \vee \ldots \vee x_m) = v'(p_1 \vee \ldots \vee p_k \vee \neg q_1 \vee \ldots \vee \neg q_m).$$

Now, let us assume $v \models C$. We define v' as follows:

$$v'(h) = \begin{cases} v(h) & h \notin \{x_1, \ldots, x_m\} \\ \neg v(q_i) & h = x_i, 1 \leq i \leq m. \end{cases}$$

Now, v' is an extension of v and v' is uniquely determined by v. It is also clear that $v' \models G_C$, by construction. The uniqueness of v' is obvious (remember that we limit ourselves to variables occurring in G_C).
The converse implication is obvious. □

Now, given a CNF F, let us form $F' = \bigcup_{C \in F} G_F$. We then have two facts. First, there is a polynomial p such that for all CNF F, $|F'| \leq p(|F|)$. Second, a valuation v of Var_F satisfies F if and only if F is a restriction of valuation of $Var_{F'}$ that satisfies F'. Thus we exhibited a polynomial reduction of satisfiability to satisfiability restricted to EO. Therefore we have the following corollary.

COROLLARY 11.2
The satisfiability problem restricted to the class EO of CNFs is NP-complete.

Let us look a bit more closely at the formula constructed in Lemma 11.2. Not only is it an EO formula but it is composed of Krom clauses in both (b) and (c). It can also be viewed as composed of positive clauses in (a) and (b) and of Horn clauses (c). We can also treat clauses (b) and (c) jointly and write them as m affine formulas (linear equations): $q_i + x_i = 1$, $1 \leq i \leq m$. All these different representations lead to variations of Lemma 11.2, which are just "layers of sugar." Even more, in the proof of Lemma 11.2 we could tweak the clauses (a), (b), and (c), transforming F to G_C in which the clause (a) is negative, not positive (just having k new atoms $y_1, \ldots, y_k$, and getting the corresponding forms of (b) and (c)).

11.10 Other cases

With all these tweakings and restatements we get two facts which will take care of 18 cases (but one will still stand)!

PROPOSITION 11.13
There is a polynomial p such that for every CNF F there is a formula F' satisfying the following conditions:

1. $size(F') \leq p(size(F))$.
2. $F' = F_1 \cup F_2$.
3. F_1 consists of positive clauses.
4. Either of the following holds:

    (a) F_2 consists of negative clauses; or
    (b) F_2 consists of Krom clauses; or
    (c) F_2 consists of affine formulas (linear equations).

5. There is a bijection between valuations of propositional variables of Var_F satisfying F and valuations of $Var_{F'}$ satisfying F'.

Then we have a similar fact.

PROPOSITION 11.14
There is a polynomial p such that for every CNF F there is a formula F' satisfying the following conditions:

1. $size(F') \leq p(size(F))$.

2. $F' = F_1 \cup F_2$.
3. F_1 consists of negative clauses and
4. Either of the following holds:

   (a) F_2 consists of positive clauses, or
   (b) F_2 consists of Krom clauses, or
   (c) F_2 consists of affine formulas (linear equations).

5. There is a bijection between valuations of propositional variables of Var_F satisfying F and valuations of $Var_{F'}$ satisfying F'.

Propositions 11.13 and 11.14 are sufficient to handle 18 cases out of 19. But one case, "mix-and-match" of Krom clauses and affine formulas, requires a different argument. We will discuss it separately. Right now, we want to show that all 18 cases mentioned above lead to NP-complete problems. Of course, the reader does not expect us to show 18 arguments. To satisfy her curiosity we will handle two cases, in expectation that the remaining 16 can be done by the reader.

Let $\mathcal{K}$ consist of unions of negative CNFs and renameable-Horn CNFs. But it is clear that all positive CNF are renameable-Horn (the permutation Inv is the witness to this fact.). Therefore if F is a union of positive F_1 and negative F_2 (thus an EO CNF) then F is a union of renameable-Horn CNF and negative CNF. As the satisfiability problem for the EO is NP-complete, the satisfiability problem for $\mathcal{K}$ is NP-complete. We randomly select another class out of 18. Now let $\mathcal{K}'$ consist of CNFs F so that $F = F_1 \cup F_2$ where F_1 is Krom and F_2 is negative. We use Proposition 11.14. There we claimed existence of F' consisting of negative clauses and of Krom clauses to which satisfiability of F reduces. Thus again satisfiability for EO CNFs reduces to satisfiability of CNFs in $\mathcal{K}'$, and we are done.

We will now show how a 3-clause can be reduced to one linear equation and two Krom clauses.

LEMMA 11.3
Let $C : l_1 \vee l_2 \vee l_3$ be a 3-clause. Let us form a set F_C consisting of these formulas: $F_C = \{l_1 + y + z, l_2 \vee y, l_3 \vee z\}$. Then every valuation v of variables underlying l_1, l_2, l_3 and satisfying C extends to a valuation v' satisfying F_C and conversely, if $v' \models F_C$ then $v'|_{Var_C} \models C$.

Proof: First, let us assume $v \models C$.
Case 1. $v(l_1) = 1$. Then let us set $v'(y) = v'(z) = 1$.
Case 2. $v(l_1) = 0$. There are 3 subcases.
Case 2.1. $v(l_2) = 0, v(l_3) = 1$. Then let us set $v'(y) = 1, v'(z) = 0$.
Case 2.2. $v(l_2) = 1, v(l_3) = 0$. Then let us set $v'(y) = 0, v'(z) = 1$.
Case 2.3. $v(l_2) = 1, v(l_3) = 1$. We set $v'(y) = 1, v'(z) = 0$. In each of these cases, it is easy to see that the valuation v' defined above satisfies the set of formulas F_C.

Next, suppose for v' defined above that $v' \models F_C$. Again we reason by cases.
Case 1. $v'(l_1) = 1$ then $v'\,|_{Var_C}(l_1) = 1$ and we are done.
Case 2. $v'(l_1) = 0$. Then it must be the case that $v'(y+z) = 1$ because we reduced the equation $l_1 + y + z = 1$. Hence one of $v'(y)$ and $v'(z)$ must be 1 and the other 0. Without loss of generality, we assume $v'(y) = 0$. Then it must be the case that $v'(l_2) = 1$, and so $v'(C) = 1$, thus $v' \models C$. But then $v \models C$, as desired. □

Let us observe that no longer (as for instance in Lemma 11.2) is there a bijection between valuations satisfying C and F_C. Nevertheless, we can always choose different variables y and z for each 3-clause C. This is a linear increase of both the number of variables and of the number of formulas. In this way we have the following fact.

PROPOSITION 11.15

There exists a polynomial p such that for every 3-CNF, F, there is a set of Krom formulas F_1 and set of linear equations F_2 such that $size(F_1 \cup F_2) \leq p(size(F))$ and such that F is satisfiable if and only if $F_1 \cup F_2$ is satisfiable. In fact, a valuation v satisfies F if and only if v extends to a valuation v' such that $v' \models F_1 \cup F_2$.

Proposition 11.15 settles the last case: the 'mix-and-match' of Krom clauses and affine formulas results in an NP-complete satisfiability problem.

The suspicious reader should worry a bit about some cases, though. We stated that mix-and-match of positive and dual Horn formulas was still polynomial But, after all, for instance, positive clauses are renameable-Horn and we still say that mix-and-match is NP-complete! The reason for it is that the collection of renameable-Horn CNFs is *not* closed under unions, so when we take a union of positive clauses and renameable-Horn, the result may be non-renameable-Horn! An example to this effect is easy to devise (in fact it already has been presented earlier).

Now, let us sum up the effects of our arguments in the final part of this chapter as a single proposition.

PROPOSITION 11.16

Let us consider the following classes of formulas $F_1 - F_7$: F_1 : positive formulas, F_2 : negative formulas, F_3 : Horn CNFs, F_4: dual Horn CNFs, F_5: renameable-Horn CNFs, F_6: Krom CNFs, and F_7: affine formulas. Then with the exception of ($i = 1$ and $j = 4$), and ($i = 2$ and $j = 3$), for every pair (i, j) $1 \leq i < j \leq 7$ the satisfiability problem for theories included in the union F_i and F_j is NP-complete.

11.11 Exercises

1. An *ordered decision tree* for a formula $\varphi(x_1, \ldots, x_n)$ is a binary tree where every internal node is labeled with a propositional variable, the leaves are labeled with constants 0 and 1 and on every path from the root the order of labels is the same: $x_1 \prec x_2 \prec \ldots \prec x_n$. Every node has two children: *high* corresponding to the value of the label x_i being 1, and *low* corresponding to the value of the label x_i being 0. Clearly, for every formula φ and listing of variables of that formula we can construct such tree. Do this for φ being $x \vee \neg(y \vee z)$, with the order of variables $z \prec y \prec x$.

2. When the nodes of the ordered decision trees are considered states, and 0 and 1 are symbols of the alphabet, we can treat that tree as a finite automaton. See Sipser [Sip06], p. 31ff. What are the accepting states of this automaton?

3. Continuing the previous problem, we can identify all leaves labeled 0 into one final node, and all leaves labeled 1 into the other final node. Then we no longer deal with a tree, but with an acyclic directed graph. Think about this graph in the context of the previous problem. We still deal with a finite automaton, but now there is just one accepting node. Which one?

4. Isomorphic subgraphs of the graph discussed in the previous problem can be identified. Consider φ: $x \vee y \vee \neg z$. Draw the graph of the diagram as in the previous problem and find isomorphic subgraphs.

5. We may also find ourselves in the situation that once we identify isomorphic subgraphs some tests are *redundant*. The reason is that both decision $x = 0$ and $x = 1$ may result in getting to isomorphic subgraphs which, in the meanwhile, were identified. How would you simplify the graph of φ in such situations?
By the way, in the parlance of decision diagrams, we just go to the *low* and *high* child.

6. Prove Bryant's theorem: application of the three rules (identification of leaves, identification of nodes with isomorphic subgraphs, and elimination of redundant tests until no other reduction possible) results in a unique diagram. This diagram is called *reduced ordered binary decision diagram* (ROBDD) for φ.

7. While ROBDD looks like a finite automaton, representation of such diagram as an automaton is not completely obvious, although certainly possible. Devise a representation of ROBDD as an automaton. The size of the representation will likely be bigger than the size of the diagram.

8. Of 21 possible combinations of "mix-and-match" the author discussed only 6, leaving the remaining 15 for you. Of course, Propositions 11.13 and 11.14 contain (at least the author claims so) all that is needed to settle these remaining 15 cases. Choose randomly three mix-and-match cases not handled by the author and prove NP-completeness.

Chapter 12

Computational knowledge representation with SAT – getting started

12.1 Encoding into SAT, DIMACS format 254
12.2 Knowledge representation over finite domains 261
12.3 Cardinality constraints, the language L^{cc} 267
12.4 Weight constraints 273
12.5 Monotone constraints 276
12.6 Exercises 283

In this chapter we will discuss the issues associated with the use of SAT (i.e., satisfiability) for actual computation. The idea is to use the propositional logic (and its close relative, predicate logic) as knowledge representation language. We will assume that we have a piece of software, called *solver*, which does the following: it accepts CNFs (say, F) and returns a string 'Input theory unsatisfiable' or a satisfying valuation for F. We will assume that *solver* is correct and complete. That is, if there is a satisfying valuation v for F, it returns one. If there is none, it returns the string 'Input theory unsatisfiable.' This is an idealized situation for a variety of reasons. First, F itself may be too big to be handled by the program *solver*. Second, since our resources are finite, it may be that the program *solver* will need more time for processing F than we have (after all, we all sooner or later expire, and DPLL searches the entire tree of partial valuations of Var which is exponential in the size of Var). Third, *solver*, and this is a minor, but very real, issue, like every humanly written program, may have bugs. But we will assume that we live in an ideal world. Arbitrary large finite CNFs are accepted, the *solver* returns the solution (if one exists) immediately, and *solver* is a correct implementation of DPLL or any other complete algorithm for SAT.

We will also discuss the use of *predicate logic* for knowledge representation. The logic we will study is predicate calculus over finite domains, that is, the universes of structures will be finite. There will be no function symbols (equivalently we may have functions, but they need to be explicitly represented by tables of values). The semantics is limited to Herbrand models as introduced in Section 12.2. In such an approach formulas of predicate calculus are just shorthand for propositional formulas (i.e., quantifiers are shorthand for conjunctions and disjunctions ranging over the universe of the model). While grounded in the studies of logicians of the past, this approach treats formulas of predicate calculus as *propositional schemata* [ET01].

Next, we extend the language of propositional logic by means of cardinality atoms (also known as cardinality constraints). This is a generalization of propositional variables in the spirit of integer inequalities with integer coefficients. We study several generalizations. First, cardinality atoms, correspond to inequalities with coefficients from $\{-1, 0, 1\}$ and solutions in $\{0, 1\}$. Second, we relax the requirement to arbitrary integer coefficients, but still restricting the solutions to values in $\{0, 1\}$. Then we discuss abstract versions of these constraints, called monotone, anti-monotone, and convex constraints and characterize those in terms of cardinality constraints.

12.1 Encodings of search problems into SAT, and DIMACS format

If we are going to use *solver* or any other SAT-solving software to compute satisfying valuations, we need to agree how are we going to represent the input, i.e., the CNFs which will be solved. There must be a format for describing the CNFs, and the SAT community devised such a format. This format is commonly called the *DIMACS format* after the Center for Discrete Mathematics and Computer Science, currently at Rutgers University. In this format, the propositional variables are represented as positive integers. Integer 0 is used as the end-of-clause symbol. This, in particular, allows for multi-line clauses. The negation symbol in this formalism is the familiar 'minus' sign: $-$.

To make things easier on the solver, the first line of the encoding has the format

$$p \quad cnf \quad V \quad C.$$

Here p is a reserved symbol (the beginning of the CNF), *cnf* is again a reserved word, expressing the fact that the rest of the data is to be interpreted as the set of clauses. V is an integer, the number of propositional variables occurring in the data so encoded. Finally, C is the number of clauses in the formula.

No programming can be done without the comments. For that reason, the lines starting with a reserved letter c are treated as comment lines. It should be observed that the choice of the header, and in particular the string *cnf*, was also fortunate. If a different class of formulas is used (we will see at least one example in subsequent sections) a different reserved string could be used.

We illustrate the concept of coding CNFs with the following example.

Example 12.1

We will express the CNF $\{p \vee q, \neg p \vee q, \neg p \vee \neg q\}$ in *DIMACS* format. We encode the variables p and q with the integers 1 and 2, respectively. The input formula now looks as follows:

c This is an encoding for a sample CNF
p cnf 2 3

```
1 2 0
-1 2 0
-1 -2 0
```

Of course, the parser will disregard lines starting with c. Let us look at line (2). We see that the data start there. There are two variables and three clauses. Then the clauses follow in subsequent lines. Let us observe that as long as we know that 1 denotes p and 2 denotes q (we could use a table for that purpose) the DIMACS file determines the CNF. □

We will now encode in DIMACS format a more complex problem (there will be 64 propositional variables), but still quite simple. The problem we want to solve is to find a "completed grid" for a "mini sudoku," a simpler version of a currently popular numerical puzzle. This is a first step in designing such puzzles.

Example 12.2

Our goal is to fill a 4×4 grid with numbers 1, 2, 3, and 4 so that every row, every column, and every quadrant has occurrences of those numbers exactly once. Mathematically speaking, we are looking for a Latin square with additional constraints. Here is the intuition. We will have propositional variables which will have the following meaning: "cell with coordinates (i, j) contains the number k." There are 16 cells, and each, potentially, contains a number in the range (1..4). Thus we will need 64 variables $p_{i,j,k}$. The idea is that $p_{i,j,k}$ expresses the fact that the cell $c_{i,j}$ contains k. These statements are either true or false. Once we have a grid, it obviously tells us which statements are true, and which are false. For instance, the grid

1	1	2	3
2	1	3	4
2	1	3	4
1	3	4	2

is a 4×4 grid. Here, for instance, $c_{1,3} = 2$. Thus in this grid $p_{1,3,2}$ is true. The reader will notice, though, that this particular grid does not satisfy our constraints. One reason is that it is not a Latin square.

Now, we will enumerate the propositional variables expressing the statements $c_{i,j} = k$. There are 64 of those statements and we will use variables $p_1, \ldots, p_{64}$ to enumerate them. Here is how we do this. We assign to the statement $c_{i,j} = k$ the variable p_l where

$$l = ((i - 1) + (j - 1) \cdot 4 + (k - 1) \cdot 16) + 1.$$

Clearly, what we do is to use a quaternary number system. We shift (i, j, k) one digit back, assign to it the number for which it is a quaternary expansion, and finally shift it forward by 1. The reason is that first we have digits (1..4) but the digits in the quaternary expansion are (0..3), and we also need the number 0 for the end-of-clause. So, for instance, the statement "the cell with the index $(2, 3)$ holds the number 2," i.e., $c_{2,3} = 2$ is the Boolean variable p_{26} because $(2-1)+(3-1)\cdot 4+(2-1)\cdot 16+1$ is 26. It is a simple exercise in arithmetic to see that all values from the range $(1..64)$ are taken.

Now to clauses. There will be many of those clauses, and we will have to wave our hands through the parts of computation. Needless to say, nobody would write them by hand; a simple script would do, though.

So here are our constraints. There will be four groups:

Group 1: For each row i with i in $(1..4)$ and each value k, k in $(1..4)$, there is a column j in $(1..4)$ so that the cell $c_{i,j}$ contains k. Let us fix i and k, say $i = 2$ and $k = 3$. Then our constraint is that:

$$c_{2,1} = 3 \lor c_{2,2} = 3 \lor c_{2,3} = 3 \lor c_{2,4} = 3.$$

But now, the statement $c_{2,1} = 3$ is the propositional variable p_{34} (please check!). Likewise, the statement $c_{2,2} = 3$ is p_{38}, then $c_{2,3} = 3$ is p_{42} and $c_{2,4} = 3$ is p_{46}. Thus our constraint becomes:

$$p_{34} \lor p_{38} \lor p_{42} \lor p_{46}.$$

In DIMACS format we have this clause:

$$34 \quad 38 \quad 42 \quad 46 \quad 0.$$

There will be 16 clauses of this sort (4 rows, 4 values).

Group 2: For each column j with j in $(1..4)$ and each value k, k in $(1..4)$ there is an index i in $(1..4)$ so that the cell $c_{i,j}$ contains k. Let us fix j and k, say $j = 1$ and $k = 4$. Then our constraint is that:

$$c_{1,1} = 4 \lor c_{2,1} = 4 \lor c_{3,1} = 4 \lor c_{4,1} = 4.$$

This is the disjunction (check the calculations, please!):

$$p_{49} \lor p_{50} \lor p_{51} \lor p_{52}.$$

In DIMACS format we have this clause:

$$49 \quad 50 \quad 51 \quad 52 \quad 0.$$

There will be 16 clauses of this sort (4 columns, 4 values).

Group 3: Each quadrant contains each of the numbers $(1..4)$. You will recall that we did not define what quadrant is. The first quadrant consists of cells $c_{1,1}, c_{1,2}, c_{2,1}$, and $c_{2,2}$. The second quadrant consists of cells $c_{1,3}, c_{1,4}, c_{2,3}$, and $c_{2,4}$. We leave to the reader to identify the remaining quadrants. So, for each quadrant, and for each value k, the quadrant must contain that value k. For instance, the third quadrant must contain the value 1. This is the following constraint:

$$c_{3,1} = 1 \lor c_{3,2} = 1 \lor c_{4,1} = 1 \lor c_{4,2} = 1.$$

When we compute the indices of the propositional variables occurring in the formula above, we get:

$$p_3 \vee p_7 \vee p_4 \vee p_8.$$

In DIMACS format we have this clause:

$$3\ \ 4\ \ 7\ \ 8\ \ 0.$$

There will be altogether 16 clauses (4 quadrants, 4 values).

Group 4: Our final group of constraints ensures that no cell contains two different values. What it means is that for each cell $c_{i,j}$ and for each pair k_1, k_2 of different values in $(1..4)$ the statement $c_{i,j} = k_1 \wedge c_{i,j} = k_2$ is false. That is, the clause

$$\neg(c_{i,j} = k_1) \vee \neg(c_{i,j} = k_2)$$

is true. We can assume $k_1 < k_2$. Then there are exactly 6 such choices of k_1 and k_2. Let us look at one example of such clause. Let us select $i = 2$, $j = 3$, $k_1 = 1$, and $k_2 = 4$. This determines the clause:

$$\neg(c_{2,3} = 1) \vee \neg(c_{2,3} = 4).$$

Moving to the propositional variables encoding the constraints we get the clause:

$$\neg p_{10} \vee \neg p_{58}.$$

In DIMACS format:

$$-10\ \ -58\ \ 0.$$

Altogether there are 16×6 of those constraints (16 cells, 6 clauses per each cell), i.e., 96 clauses of group (4).

An inquisitive reader should ask what constraint guarantees that each cell will, actually, hold a value. The reason for this is that (for instance) each row has four cells. There are four values. Since no cell holds two values (constraint (4)), it must be the case that each cell must contain a value. This is the familiar principle that holds for finite sets (but not for infinite sets): If two finite sets X and Y have the same size then any injection of X into Y is automatically a bijection.

Now, let us count our clauses. There are 16 clauses in each of groups (1) through (3). Then there are 96 clauses of group (4). Altogether we have 144 clauses. We are not claiming that our encoding is optimal. Anyway, the first line of the resulting file expressing our problem in DIMACS format will be:

$$p\ \ \ cnf\ \ \ 64\ \ \ 144.$$

It will be followed for 144 clauses, each ending in 0. We assume this file is called 44grid.dim (any reasonable naming convention would do).

What happens now? We call our program *solver* on the input 44grid.dim. The *solver* returns a satisfying valuation v. There are many such valuations, we got back one.

It evaluates all 64 variables ($p_1..p_{64}$). We display only those variables that were returned as *true*.

$$\{p_3, p_5, p_{12}, p_{14}, p_{18}, p_{24}, p_{25}, p_{31}, p_{36}, p_{38}, p_{43}, p_{45}, p_{49}, p_{55}, p_{58}, p_{64}\}.$$

The remaining 48 variables were returned as *false*.
Now, what does it mean that (say) the variable p_{55} is *true*? We need to see what p_{55} meant. To this end let us expand the number 55 and get:

$$55 = 2 + 4 + 48 + 1 = (3-2) + (2-1) \cdot 4 + (4-1) \cdot 16 + 1.$$

Now we can read off this representation that p_{55} represents the following constraint on the grid: $c_{3,2} = 4$. This means that the cell with coordinates $(3,2)$ (third row, second column) holds number 4. Similarly, for the variable p_{18} we compute:

$$18 = (2-1) + (1-1) \cdot 4 + (2-1) \cdot 16 + 1,$$

that is, $c_{2,1} = 2$ (assuming we did not make a mistake). Proceeding similarly with the remaining 14 variables, we find the grid:

4	1	2	3
2	3	4	1
1	4	3	2
3	2	1	4

which, indeed, satisfies our constraints. □

At first glance, our problem of finding the grid satisfying various constraints did not involve logic at all. But by the choice of a suitable knowledge representation scheme we reduced our problem to a satisfiability problem. How did the process work? The first observation is that it would be convenient to have a one-to-one correspondence between solutions to the problem and the satisfying valuations for the encoding. We could relax this requirement a bit: every solution should be encoded by a satisfying assignment, and every satisfying assignment should encode some solution. Otherwise, either some solutions may be missed or the *solver* may return a valuation which does not decode to a solution. There are two tacit assumptions here. The first of these two is that we have an encoding function that computes a CNF out of the problem. The second assumption is that we are able to decode out of satisfying valuation the solutions to the original problem.
To sum up, in the scheme we used, there are three distinct phases:

1. Encoding
2. Solving
3. Decoding

This is a general scheme for using the SAT solver as a constraint solver. We will see below that the first point, *Encoding*, creates an opportunity for the simplification of the process. In the spirit of this book, with everything treated, if possible, in a formal

Computational knowledge representation with SAT – getting started 259

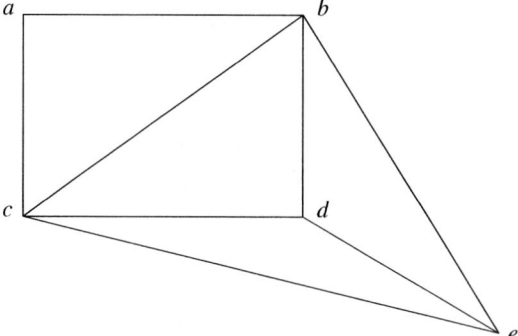

FIGURE 12.1: Graph G

fashion, we describe this entire scheme as follows. *A search problem*, Π ([GJ79]), consists of *instances*. Every instance $I \in \Pi$ has an associated set of *solutions*, Sol_I (it may be empty). We require existence of two *feasible* (i.e., computable in polynomial time) functions e and d. The function e is defined on Π and assigns to $I \in \Pi$ its *encoding* $F_{I,\Pi}$. We assume that $F_{I,\Pi}$ is a CNF (although, in general, natural encodings are not necessarily CNFs). The decoding function d has two arguments: an instance and a valuation. We assume that whenever I is an instance and if a valuation v satisfies $e(I)$, that is, $F_{I,\Pi}$, then $d(I,v)$ is a solution for the instance I, i.e., $d(I,v) \in Sol_I$. In other words, $d(I,v)$ decodes a satisfying assignment (provided one is returned) into a solution. If there is no satisfying assignment for $F_{I,\Pi}$ then d returns a string 'no solutions.'

Let us look at what happens in our *mini sudoku grid* Example 12.2. Our search problem has just one instance. We encoded this single instance as a SAT problem (a CNF to be solved), and then solved it with *solver*, and decoded one of (there were many) the solutions to our instance.

To sum up, we require that we have two feasible maps:

1. $e : \Pi \to Form$ (usually the values will be CNFs), and
2. $d : \Pi \times Val \xrightarrow[onto]{1-1} Sol,$

where $Sol = \bigcup_{I \in \Pi} Sol_I$. The additional requirement is that the decoding function actually decodes solutions. That is, $d(e(I), v) \in Sol_I$ if $v \models e(I)$, and $d(e(I))$ is the string 'no solution' if there are no solutions for the instance I.

But before we study the subtleties of this scheme (in subsequent sections), let us look at another example of coding, solving, and decoding.

Example 12.3

We will now use our solver *solver* to color a graph. Our graph G has 5 vertices, 8 edges, and is presented in Figure 12.1.

Our problem is to find the coloring of vertices of the graph G with three colors, *red*, *blue*, and *green*, so that every pair of vertices connected with an edge has different colors. Such coloring is called *valid 3-coloring*, or simply *3-coloring* of G. We will need only 15 propositional variables. Variables p_1, p_2, p_3 will encode the fact that the vertex a is colored, respectively, with *red*, *blue*, and *green*. Similarly, p_4, p_5, p_6 will denote that the vertex b is colored with the colors, respectively, *red*, *blue*, and *green*, etc. Once we know the intended meaning of the propositional variables, we are ready to write clauses expressing constraints of the problem.

Group 1: We write a constraint requiring that each vertex possesses a color. There will be five clauses; there are five vertices. Here is one such clause which expresses the constraint that the vertex b possesses a color:

$$p_4 \lor p_5 \lor p_6.$$

The corresponding clause in DIMACS format will be:

$$4 \quad 5 \quad 6 \quad 0.$$

Group 2: We express the constraint that no vertex has two colors. Thus, for each vertex x and each pair of different colors c_1 and c_2 (three such pairs) we need to assert that that it is not true that x both has the color c_1 and the color c_2. Let us do this for the vertex c. There will be three such clauses; we list all of them:

$$\neg p_7 \lor \neg p_8, \quad \neg p_7 \lor \neg p_9, \quad \neg p_8 \lor \neg p_9.$$

This means that various clauses will be add to the DIMACS format file, for instance:

$$-7 \quad -8 \quad 0.$$

Altogether we will produce 15 clauses in this group.

Group 3: Here we need to take care of the constraint that the endpoints of each edge have different colors. We have eight edges here. For each such edge we need to put three clauses that assert that the colors of endpoints are different. Here is one such example. In our graph vertices b and c are connected by an edge. Therefore we need three constraints. We list one of those; we hope the reader writes the remaining two.

It is not the case that both b and c are red.

These three constraints are expressed by clauses (now we list all three):

$$\neg p_4 \lor \neg p_7, \quad \neg p_5 \lor \neg p_8, \quad \neg p_6 \lor \neg p_9.$$

In DIMACS format the first one of these is:

$$-4 \quad -7 \quad 0.$$

Altogether (8 edges, 3 colors) we will write 24 clauses in group (3).

Summing up the numbers, we see that in our encoding (there are different encodings, too!) there are 44 clauses. Thus the first line of our file mycolor.dim containing the clauses describing the constraints of the problem will be:

$$c \quad cnf \quad 15 \quad 44.$$

What will we do now? We submit the resulting file to the *solver*. Not surprisingly the *solver* will return no satisfying valuation (in fact our graph G has no valid 3-coloring). Since there is a one-to-one correspondence between the valid 3-colorings of G and satisfying valuations for our encoding we conclude that there is no solution for the 3-coloring problem for our graph. □

12.2 Knowledge representation with predicate logic over finite domains, and reduction to SAT

We will now study predicate logic as a knowledge representation formalism. Predicate logic is an extension of propositional logic. But in addition to connectives ¬, ∧, ∨, and others we discussed in Chapter 2, there are two constructs that are very useful in modeling. Those are object variables and quantifiers. Those who need information about syntax and semantics of predicate logic will find it in [NS93] or the classic [Sh67]. Before we start talking about the minutiae of the representation with predicate logic, let us realize that there are several fundamental obstacles to the use of predicate logic in knowledge representation. The first of those is the issue of infinite models. If we admit infinite models then, certainly, we will not be able to list them. We can (sometimes) query them (this is the basis of PROLOG programming language), but listing them is impossible. The second fundamental problem is the Church undecidability theorem; the set of tautologies of predicate logic admitting non-unary predicates is not computable. Because of the completeness theorem for predicate calculus, if we allow for all possible models then there cannot be a decision procedure for sentences true in all those models. The third one is the Gödel incompleteness theorem; every consistent axiomatizable theory containing enough of arithmetic is incomplete. Thus there is no computable axiomatization of all sentences true in the standard model of arithmetic. Since one still needs to reason in predicate logic, techniques of automated theorem proving are used. These are sound but not complete.

Hence, if we want to use the language of predicate logic, we need to introduce limitations. Here are the limitations we will impose both on syntax and on semantics of predicate logic. First, we will limit ourselves to *finite* theories. Second, we will not allow function symbols. Third, we will allow only models with finite universes. Without limiting our approach we will assume that the *only* relational structures that we will consider are structures where the universe is composed of all constant terms occurring in the theory. Those are so-called Herbrand models. In our case, since there are no function symbols, the universe of the Herbrand models consists of the constants of the language.

Let L_σ be a language of predicate logic based on the signature σ. We will omit the subscript σ. This signature has predicate symbols $P_1, \ldots, P_k$ and constant symbols $a_1, \ldots, a_m$. The set $\mathcal{C}$ consists of constants of L. The arity of P_j is denoted l_j. Let T be a set of formulas of L. We will assume that all constants of L occur in formulas of T. The Herbrand universe $\mathcal{H}_L$ consists of the constant terms of L. Since there are no function symbols, $\mathcal{H}_L = \mathcal{C}$.

A Herbrand relational structure (Herbrand structure, for short) for L is a structure

$$\mathcal{M} = \langle \mathcal{H}_L, r_1, \ldots r_k, a_1, \ldots, a_m \rangle.$$

Here r_j, $1 \leq j \leq k$ are relations (of arity l_j) on set $\mathcal{C}$. They are interpretations of predicate symbols $P_1, \ldots, P_k$. This means that r_j is l_j-ary relation.

Thus Herbrand structures (in our case) have the universe consisting of constants, nothing else. Since we consider finite theories T, this universe is finite.

In addition to the Herbrand universe, we have the *Herbrand base*. The Herbrand base of the language L consists of atomic sentences of L, that is, expressions of the form

$$P_j(d_1, \ldots, d_{l_j}),$$

where P_j is a predicate symbol and d_i's are (not necessarily different) elements of $\mathcal{C}$. Here is an example. Let us assume our language L has two predicate symbols: P (binary) and Q (unary) and three constants: a, b, and c. Then the Herbrand base H_L of L consists of 12 objects. Nine of those are generated from P: $P(a,a), P(a,b), P(a,c), P(b,a)$, etc., and three generated from Q: $Q(a), Q(b)$, and $Q(c)$.

PROPOSITION 12.1
There is a bijective correspondence between Herbrand structures for L and the subsets of the Herbrand base H_L of L.

Proof: Here is the desired correspondence. When $\mathcal{M}$ is:

$$\langle \mathcal{H}_L, r_1, \ldots r_k, a_1, \ldots, a_m \rangle,$$

$S_\mathcal{M}$ is:

$$\bigcup_{j=1}^{k} \{P_j(d_1, \ldots, d_{l_j}) : \langle d_1, \ldots, d_{l_j} \rangle \in r_j\}.$$

We need to show that the correspondence $\mathcal{M} \mapsto S_\mathcal{M}$ is one to one, and "onto."

First, we show that the mapping is one to one. If two Herbrand structures for L, $\mathcal{M}_1$ and $\mathcal{M}_2$ are different, then, since the universes are the same, there must be a predicate symbol P_j, $1 \leq j \leq k$ such that its interpretations in $\mathcal{M}_1$, $r_{j,1}$ and in $\mathcal{M}_2$, $r_{j,2}$ are different. But then there is a tuple $\langle d_1, \ldots, d_{l_j} \rangle$ such that $\langle d_1, \ldots, d_{l_j} \rangle$ belongs to $r_{j,1} \setminus r_{j,2}$ or $\langle d_1, \ldots, d_{l_j} \rangle$ belongs to $r_{j,2} \setminus r_{j,1}$. But then $P_j(d_1, \ldots, d_{l_j})$

belongs to $S_{\mathcal{M}_1} \setminus S_{\mathcal{M}_2}$ or $P_j(d_1, \ldots, d_{l_j})$ belongs to $S_{\mathcal{M}_2} \setminus S_{\mathcal{M}_1}$.
Next, we need to show that the mapping is "onto." Given S consisting of atomic sentences of L, let us define, for each j, $1 \leq j \leq k$,

$$r_j = \{\langle d_1, \ldots d_{l_j}\rangle : P_j(d_1, \ldots d_{l_j}) \in S\}.$$

Then, setting $\mathcal{M} = \langle \mathcal{C}, r_1, \ldots, r_k, a_1, \ldots, a_m \rangle$, we see that $S_{\mathcal{M}} = S$. □

Let us identify the Herbrand base with a set of propositional variables. This is what we really did in our Example 12.2. If we do this, subsets of the Herbrand base H_L are nothing more than the valuations of H_L. We can, then, investigate propositional satisfaction of those valuations. Recall that there is a one-to-one correspondence between subsets of H_L and valuations of H_L. In Chapter 2 we gave a formal definition of satisfaction of *propositional* formulas by subsets of the set of variables.

There is nothing deep in Proposition 12.1, but it turns out that there is a strong connection between the satisfaction in Herbrand structure $\mathcal{M}$ and the propositional satisfaction by set $S_{\mathcal{M}}$.

PROPOSITION 12.2

Let L be a language of predicate logic with a finite number of relational symbols, a finite number of constant symbols, and no function symbols. For every sentence Φ of the language L, there is a propositional formula F_Φ such that for every Herbrand structure $\mathcal{M}$ for L,

$$\mathcal{M} \models \Phi \text{ if and only if } S_{\mathcal{M}} \models F_\Phi.$$

Proof: Before we prove our proposition let us observe that the symbol $\models$ is used in *different* meanings on both sides of the equivalence. On the left-hand side we talk about satisfaction for *predicate logic* structures, on the right-hand side we talk about *propositional* satisfaction.

We will construct the sentence F_Φ by induction on the complexity of sentence Φ.

First, let the sentence Φ be atomic. There are two cases to consider: first, when Φ is of the form $P_j(d_1, \ldots, d_{l_j})$, and the second, when Φ is the equality $c = d$ where c and d are constants.

(1) When Φ is of the form $P_j(d_1, \ldots, d_{l_j})$, then $\mathcal{M} \models \Phi$ if and only if $\langle d_1, \ldots, d_{l_j}\rangle \in r_j$, which is equivalent to $P_j(d_1, \ldots, d_{l_j}) \in S_{\mathcal{M}}$, which is equivalent to $S_{\mathcal{M}} \models P_j(d_1, \ldots, d_{l_j})$. Thus we define $F_\Phi = \Phi$.

(2) When Φ is $c = d$, then there are two possibilities: either c is identical to d or not. In the first case we define F_Φ as $\top$, otherwise we set F_Φ equal to $\bot$. Now, $\mathcal{M} \models \Phi$ precisely when c is identical to d, i.e., precisely if $S_{\mathcal{M}} \models F_\Phi$.

Now, the definition of satisfaction both in predicate and propositional case is such that the inductive step for the propositional connectives $\neg, \wedge, \vee, \Rightarrow$, and $\equiv$ is obvious. Specifically, we set: $F_{\neg \Phi}$ is: $\neg F_\Phi$, $F_{\Phi_1 \vee \Phi_2}$ is $F_{\Phi_1} \vee F_{\Phi_2}$, etc.

So, all we need to do is to define $F_{\forall_x \psi(x)}$ and $F_{\exists_x \psi(x)}$, and prove the property postulated in the assertion.

When Φ is $\forall_x \psi(x)$, then we form sentences $\psi(c)$ for each constant c of L. There are only finitely many such sentences because there are finitely many constants. By inductive assumption (sentence $\psi(c)$ is simpler), we have propositional formulas $F_{\psi(c)}$. We now define:

$$F_{\forall_x \psi(x)} = \bigwedge_{c \in \mathcal{C}} F_{\psi(c)}.$$

We now need to prove that:

$$\mathcal{M} \models \forall_x \psi(x) \text{ if and only if } S_\mathcal{M} \models \bigwedge_{c \in \mathcal{C}} F_{\psi(c)}.$$

But according to the definition of satisfaction, $\mathcal{M} \models \forall_x \psi(x)$ if and only if for every constant c, $\mathcal{M} \models \psi(c)$. By inductive assumption $\mathcal{M} \models \psi(c)$ if and only if $S_\mathcal{M} \models F_{\psi(c)}$. But $\mathcal{C}$ is finite, so we can form $\bigwedge F_{\psi(c)}$. Clearly, then, $S_\mathcal{M} \models \bigwedge F_{\psi(c)}$. The converse reasoning is similar.

When Φ is $\exists \psi(x)$, we set:

$$F_{\exists_x \psi(x)} = \bigvee_{c \in \mathcal{C}} F_{\psi(c)}.$$

The argument is analogous. $\square$

So now let us define a theory T in predicate calculus to be *Herbrand-consistent*, if it possesses a Herbrand model. We then have the following corollary.

COROLLARY 12.1

Let L be a language of predicate calculus without function symbols, and with a finite number of predicate and constant symbols. Let T be a set of sentences of L. Then T is Herbrand-consistent if and only if the propositional theory $\{F_\Phi : \Phi \in T\}$ is satisfiable. In fact, there is a one-to-one correspondence between the Herbrand models of T and the propositional assignments of the Herbrand base evaluating $\{F_\Phi : \Phi \in T\}$ as 1.

Now, it should be quite clear what Corollary 12.1 says. As long as we have in mind a Herbrand model of a finite theory with no function symbols, all we need to do is to test the *propositional* theory $\{F_\Phi : \Phi \in T\}$ for satisfiability. We can read off such satisfying valuations from Herbrand models of T, that is, desired structures.

But the propositional theory $\{F_\Phi : \Phi \in T\}$ does not need to be clausal. We will see below under what conditions on Φ, and more generally T, we can read off directly clausal theory out of T. But in the general case we can always reduce (possibly using additional propositional variables) any propositional theory F to a *clausal* theory G_F so that there is a bijective correspondence between valuations satisfying F and valuations satisfying G_F (Proposition 11.8(2)). So, to use *solver* to find if a theory T (let us have in mind the restrictions on the language) has a Herbrand model (and if so to find one), we proceed in two steps. First we translate T to a finite collection of propositional formulas $\{F_\Phi : \Phi \in T\}$. If the resulting theory is clausal, we use *solver* directly. If the resulting theory is not clausal, we transform it to the clausal theory as described in Chapter 11. Assuming the resulting clausal theory is satisfiable we can read off the Herbrand model of input theory T either in one step (if there were no additional variables) or in two steps (first eliminating the additional variables and then computing the model).

Example 12.4 (Example 12.2 revisited)
We now code the problem of finding the 4×4 grid in predicate logic. There will be, as before, four conditions. One of those (on quadrants) will require four formulas to code. Others will be much simpler. Here is the desired theory T.

1. (every value taken in each row) $\forall_{i,k} \exists_j p(i,j,k)$
2. (every value taken in each column) $\forall_{j,k} \exists_i p(i,j,k)$
3. (every value taken in each quadrant, four clauses)

    (a) (every value taken in the first quadrant)
    $\forall_k (p(1,1,k) \lor p(1,2,k) \lor p(2,1,k) \lor p(2,2,k)$
    (b) (every value taken in the second quadrant)
    $\forall_k (p(1,3,k) \lor p(1,4,k) \lor p(2,3,k) \lor p(2,4,k)$
    (c) (every value taken in the third quadrant)
    $\forall_k (p(3,1,k) \lor p(3,2,k) \lor p(4,1,k) \lor p(4,2,k)$
    (d) (every value taken in the fourth quadrant)
    $\forall_k (p(3,3,k) \lor p(3,4,k) \lor p(4,3,k) \lor p(4,4,k)$

4. (single value in each cell) $\forall_{i,j,k_1,k_2}(p(i,j,k_1) \land p(i,j,k_2) \Rightarrow k_1 = k_2)$

Herbrand models of this theory describe the desired grids. □

The reader will notice that there are two discrepancies between our encoding of Examples 12.2 and 12.4. The first difference comes from the fact that the universal quantifier will be transformed into conjunction. Altogether we get 7 formulas after we translate from predicate calculus, not 128. The second difference is with the formula (4). In Example 12.2 we translated this constraint directly in clausal form. Not so now: it will be conjunction of implications. We will now address both issues raised by this translation.

To address the first one (conjunction of formulas, when translating universally quantified formula) we observe that for every valuation v and for every set $\{\varphi_j : j \in J\}$ of propositional formulas

$$v \models \bigwedge_{j \in J} \varphi_j \text{ if and only if for all } j \in J, v \models \varphi_j.$$

This means that we can transform conjunctions of formulas into *sets* of formulas and do not change the status of satisfiability. This is precisely why in Example 12.2 we expressed constraints (1) – (3) of Example 12.4 by 16 clauses, each. Then, referring to the issue with the translation of the fourth constraint, let us look at the formulas we get. In principle there should be 4^4 (i.e., 256) propositional formulas generated from the formula (4). First, let us look at the formulas where k_1 and k_2 are equal. Each such formula is translated (recall the way we treated translation of equality!) as

$$p(i,j,k_1) \land p(i,j,k_2) \Rightarrow \top.$$

This formula, regardless of constants i, j, k_1, k_2 (but with $k_1 = k_2$), is a tautology. Thus it can be safely eliminated. There are 4×16 (4 common values of k_1 and k_2, 16 values of the pair (i, j)) of such true constraints. This leaves us with 192 remaining formulas. But it is easy to observe that half of those, in effect, will be repeated, due to the commutativity of conjunction and symmetry of equality relation. This leaves us still with 96 constraints. Those constraints look like this:

$$p(i,j,k_1) \land p(i,j,k_2) \Rightarrow \bot.$$

But these last formulas are equivalent to the formulas

$$\neg p(i,j,k_1) \lor \neg p(i,j,k_2),$$

which we computed in intuitive manner. Thus, in effect, an intelligent transformation software (one that accepts predicate logic theories as inputs, and produces propositional theories as outputs) should be able to compute propositional theory such as in Example 12.2 from predicate theories such as in Example 12.4. Such software is called a *grounder*. There are several issues associated with grounders. First, the *grounder* must produce the output in the format understood by *solver*. Second, once the *grounder* produces the input formula (for simplicity let us assume in DIMACS format), the meaning of variables is lost. Thus the *grounder* must make aware to the *solver* the meaning of propositional variables. It is worth mentioning that grounders have been implemented.[1] To sum up, as of today we are not aware of full grounders (those that would force us to use a two-step process described above), but grounders accepting important classes of formulas exist.

Our discussion of translation of predicate logic to propositional logic makes clear that if the input formula has the universal quantifier as a top connective, then the translation convenient from the point of view of the *solver* is not the one that translates the input to a single propositional formula, but rather to a set of formulas (a formula for each constant). This is what we did in Example 12.2. If the input formula starts with a string of universal quantifiers, then we can repeat the elimination. In effect what we use is the following principle. Let $\Phi = \forall_{x_1} \forall_{x_2} \ldots \forall_{x_n} \psi(x_1, \ldots, x_n)$. Then for every Herbrand structure $\mathcal{M}$

$$\mathcal{M} \models \Phi \text{ if and only if } \bigwedge_{c_1 \in \mathcal{C}, \ldots, c_n \in \mathcal{C}} S_{\mathcal{M}} \models F_\psi(c_1, \ldots, c_n).$$

It is then quite natural to introduce the notion of *clause-like* formulas of predicate logic. Those are of the form $\forall_{x_1} \forall_{x_2} \ldots \forall_{x_n} \psi(x_1, \ldots, x_n)$ where $\psi(x_1, \ldots, x_n)$ is a disjunction of literals.

---

[1] In a non-SAT context, Niemelä and collaborators (the *lparse* software, a component of *smodels* ASP solver package) implemented a grounder. Then, the grounder *psgrnd* was implemented by Truszczyński and collaborators [EIMT06]. Truszczyński's grounder handles only a proper subset of predicate logic.

The *grounder* translates clause-like formulas directly to sets of clauses, and those can serve as an input to the *solver*. We will see below that there is a larger class of formulas which can be translated easily to the sets of ground clauses.

12.3 Cardinality constraints and the language L^{cc}

We observed in Section 10.1 that the clause $l_1 \vee \ldots \vee l_k$, where $l_1, \ldots, l_k$ are literals, can be understood under a suitable interpretation as a *pseudo-Boolean* inequality (also known as a *pseudo-Boolean constraint*):

$$l_1 + \ldots + l_k \geq 1.$$

The limitations were: first, the negated variables $\neg x_j$ were written as integer terms $1 - x_j$. Second, the integer solution had to be "pseudo-Boolean," that is, taking the values 0 and 1 only. Of course, collections of clauses corresponded to pseudo-Boolean systems of inequalities.

So, in this section we will generalize this interpretation by allowing the right-hand side to be bigger than 1. Then we will generalize it even further, allowing the coefficients to be integers other than -1 and 1.

We will identify a clause $l_1 \vee \ldots \vee l_k$ with the integer inequality $l_1 + \ldots + l_k \geq 1$, with the convention that $\neg x$ is $1 - x$. The meaning of both forms is the same: at least one of the literals of the clause is true, i.e., takes the value 1. With this interpretation it is very natural to generalize; we just vary the right-hand side. Instead of asserting that at least one of the literals among $l_1, \ldots, l_k$ is true, we assert that at least m of them are true. Thus, formally, in the integer-inequalities world, we have an inequality

$$l_1 + \ldots + l_k \geq m.$$

On the other side, in logic, we have a new kind of expression (a generalized propositional variable) $m\{l_1, \ldots, l_k\}$ asserting that from the literals $l_1, \ldots, l_k$ at least m are true. We call such an expression a *cardinality constraint*.

Clearly it is possible to eliminate expressions $m\{l_1, \ldots, l_k\}$. But the cost is significant. It is easy to write a DNF ψ consisting of $\binom{k}{m}$ elementary conjunctions each of length m such that $m\{l_1, \ldots, l_k\}$ is equivalent to ψ. It is also quite easy to eliminate such an expression using additional variables in linear time. We will not do this now; we leave the task to the reader (the second task requires some ingenuity, but not much).

First, let us observe that as long as we care about pseudo-Boolean solutions, *every* inequality of the form

$$a_1 x_1 + \ldots + a_k x_k \geq r$$

with all the coefficients a_is in the set $\{-1, 1\}$ is a cardinality constraint. To see this, consider the number h

$$h = |\{j : a_j = -1\}|$$

and rewrite the inequality as above into the form

$$h + a_1x_1 + \ldots + a_kx_k \geq r + h.$$

Then the number h on the left-hand side is split into h 1s which are added to those x_is for which $a_i = -1$. Now we get the form of our inequality,

$$l_1 + \ldots + l_k \geq r + h,$$

where each l_i is x_i (if $a_i = 1$), or $1 - x_i$ (if $a_i = -1$).
Let us look at the example. Our inequality is

$$x_1 - x_2 - x_3 - x_4 \geq -1.$$

Adding 3 to both sides we get:

$$x_1 + (1 - x_2) + (1 - x_3) + (1 - x_4) \geq 3 - 1,$$

that is, the cardinality constraint

$$2\{x_1, \neg x_2, \neg x_3, \neg x_4\}.$$

So now, we will officially admit into our language *two* new kinds of expressions: $m\{l_1, \ldots, l_k\}$ and $\{l_1, \ldots, l_k\}n$. Here is the semantics of these expressions. We define:

$$v \models m\{l_1, \ldots, l_k\} \quad \text{if} \quad |\{l_j : v \models l_j\}| \geq m.$$

Here m is the *lower bound* of our cardinality constraint. We will call such constraints *lower-bound cardinality constraints*. But once we do this, it is only natural to introduce another kind of cardinality constraint: $\{l_1, \ldots, l_k\}n$ with the semantics:

$$v \models \{l_1, \ldots, l_k\}n \quad \text{if} \quad |\{l_j : v \models l_j\}| \leq n.$$

We will call those *upper-bound cardinality constraints*. It is quite clear what this new cardinality constraint $\{l_1, \ldots, l_k\}n$ is. This is just a pseudo-Boolean inequality:

$$l_1 + \ldots + l_k \leq n.$$

As before, every pseudo-Boolean inequality

$$a_1x_1 + \ldots + a_kx_k \leq r$$

with all the coefficients a_is in the set $\{-1, 1\}$ is a cardinality constraint of this second kind. We observe that the propositional variables themselves are cardinality constraints: the cardinality constraint $1\{x\}$ has precisely the same semantics as the propositional variable x.

At this moment, the reader will ask if we really need two kinds of cardinality constraints. Indeed, after all, the constraint $\{l_1, \ldots, l_k\}n$ means "at most n of literals in $\{l_1, \ldots, l_k\}$ are true." But this latter expression is, clearly, equivalent to "at least

$k - n$ of literals in $\{\bar{l}_1, \ldots, \bar{l}_k\}$ are true," i.e., $(k-n)\{\bar{l}_1, \ldots, \bar{l}_k\}$. The answer to this objection is similar to the answer we offer to the question "Is the implication functor needed?" After all we can eliminate implication using negation and disjunction. We just say that we do not *need* implication, but it is *convenient* in knowledge representation. The situation is analogous here: we do not need the upper-bound cardinality constraints, but they are convenient as we will see in Proposition 12.3.

Now, let L^{cc}_{Var} be the language such that the atomic expressions are of the form $m\{l_1, \ldots, l_k\}$ and $\{l_1, \ldots, l_k\}n$. Due to the fact that x is equivalent to $1\{x\}$, we can assume that all we have as atomic expressions are $m\{l_1, \ldots, l_k\}$ and $\{l_1, \ldots, l_k\}n$. Thus the language L^{cc}_{Var} contains L; its formulas are just convenient expressions allowing us to write formulas of L_{Var} more compactly. We will not write the subscript *Var*, of course.

The connectives (functors) of L^{cc} are the usual connectives of L. The semantics is provided by means of valuations of the set *Var*. The conditions on satisfaction relation are the two conditions listed above (for the cardinality constraints $m\{l_1, \ldots, l_k\}$ and $\{l_1, \ldots, l_k\}n$), and the standard inductive conditions (as specified in Chapter 2) for more complex formulas.

By identifying a valuation v with a set M of variables on which v takes value 1, we get a definition of satisfaction for a set M of variables. We will later see that cardinality constraints (of special kinds) offer a natural way to define families of sets. Right now, let us look at the example of satisfaction of a formula of L^{cc} by a set of variables. For instance, the set $\{p\}$ satisfies the cardinality constraint $2\{p, q, \neg r\}$. Why? Because two literals satisfied by our set $\{p, q, \neg r\}$, namely p and $\neg r$, are satisfied by $\{p\}$.

We also need to be a bit careful. There is nothing that precludes the presence of both x and $\neg x$ in the set $\{l_1, \ldots, l_k\}$. It should be clear that if this happens, we can erase both x and $\neg x$ and decrement the bound (either the lower or the upper) by 1. This does not change the semantics.

The language L^{cc} admits both conjunctive and disjunctive normal forms. In particular we can talk about *cardinality clauses* i.e. clauses where the atoms are cardinality constraints. It turns out that the language L^{cc} has a stronger property: we can eliminate negation from the language. In the statement of proposition below we will assume that the length of lower and upper bounds in cardinality constraints counts as 1. With these assumptions we have the following fact.

PROPOSITION 12.3
For every formula φ of L^{cc}_{Var} there is a formula ψ in L^{cc}_{Var} such that:

1. *The length of ψ is at most twice the length of φ.*
2. *ψ is semantically equivalent to φ.*
3. *ψ has no occurrence of negation symbol.*

Proof: Let us observe that $\neg(m\{l_1, \ldots, l_k\})$ is equivalent to $\{l_1, \ldots, l_k\}(m-1)$, and $\neg\{l_1, \ldots, l_k\}n$ is equivalent to $(n+1)\{l_1, \ldots, l_k\}$. Since L^{cc} satisfies DeMorgan laws, every formula φ of L^{cc} is equivalent to a formula ϑ of L^{cc} where negation occurs only in front of cardinality constraints. Moreover, the length of ϑ is at most twice that of φ. Then, we can eliminate negation in front of cardinality atoms of ϑ as indicated above. □

Let us observe that the fact that $\neg p$ is equivalent to $\{p\}0$ implies by itself a weaker form of Proposition 12.3; we can find a formula with no negation symbol just from the CNF for L, but the size of this formula will be exponential in that of φ.

Next, we focus on the issue of the existence of DPLL-like algorithms for testing satisfiability of sets of clauses of L^{cc}. In an abstract setting there are two ingredients that allow for designing such an algorithm. The first of those is the availability of some form of the Kleene theorem on preservation of Boolean values when we extend partial valuations. Second, we need some form of BCP. It turns out that both are available for L^{cc}.

To establish the first component, the necessary version of the Kleene theorem on extensions (Proposition 2.5, point (2)), let us define the three-valued evaluation function for partial valuations and cardinality constraints as follows:

$$(v)_3(m\{l_1,\ldots,l_k\}) = \begin{cases} 1 & \text{if } |\{j : (v)_3(l_j) = 1\}| \geq m \\ 0 & \text{if } |\{j : (v)_3(l_j) \neq 0\}| \leq m - 1 \\ u & \text{otherwise} \end{cases}$$

and

$$(v)_3(\{l_1,\ldots,l_k\}n) = \begin{cases} 1 & \text{if } |\{j : (v)_3(l_j) = 0\}| \geq k - n \\ 0 & \text{if } |\{j : (v)_3(l_j) = 1\}| \geq n + 1 \\ u & \text{otherwise.} \end{cases}$$

What is the intuition? We commit to $m\{l_1,\ldots,l_k\}$ if the current *partial* valuation v carries enough information to make sure that however we extend v in the future to a valuation w, w will always evaluate $m\{l_1,\ldots,l_k\}$ as 1. Likewise, we commit to negation of $m\{l_1,\ldots,l_k\}$ if we already know that regardless of how we extend v in the future we will not be able to make it 1. Intuitions of the upper bound cardinality constraints are similar.

We observe that if v is a *complete* valuation and $(v)_3(m\{l_1,\ldots,l_k\}) = 1$ then $v \models m\{l_1,\ldots,l_k\}$, and likewise, if $(v)_3(\{l_1,\ldots,l_k\}n) = 1$ then $v \models (\{l_1,\ldots,l_k\}n$.
Here is the Kleene theorem for the language L^{cc}.

PROPOSITION 12.4
Let v_1, v_2 be two partial valuations of the set Var, and let φ be a formula of L^{cc}_{Var}. If $v_1 \preceq_k v_2$ then $(v_1)_3(\varphi) \preceq_k (v_2)_3(\varphi)$.

Proof: Let us recall that $v' \preceq_k v''$ means that:

1. $Dom(v') \subseteq Dom(v'')$ and
2. $v''|_{Dom(v')} = v'$.

It should be clear that the only expressions that need to be considered are those of the form $m\{l_1,\ldots,l_k\}$ and $\{l_1,\ldots,l_k\}n$. The rest of the argument goes exactly as in the proof of Theorem 2.5, part (2).

So, let us consider the expression $E = m\{l_1,\ldots,l_k\}$. If $v' \preceq_k v''$ and $(v')_3(E) = 1$, then v' evaluates at least m of literals $l_1,\ldots,l_k$ as 1. But by the Kleene theorem (Proposition 2.5, part (2)) v'' also evaluates each of those literals (evaluated by v' as 1) as 1. Therefore $(v'')_3(E) = 1$. If $(v')_3(E) = 0$, then the number of literals evaluated by v' as non-zero (1 or u) is at most $m - 1$. But then the same happens with respect to v'', so $(v'')_3(E) = 0$. Finally, if $(v')_3(E) = u$, then regardless of the value of $(v'')_3(E)$, $(v')_3(E) \preceq_k (v'')_3(E)$.

Now, let us consider the case of the expression $E = \{l_1,\ldots,l_k\}n$. The argument is similar. If $(v')_3(E) = 1$, then already at least $k - n$ of literals of $\{l_1,\ldots,l_k\}$ are evaluated as 0 by v'. Then the same happens in the case of v'' and so $(v'')_3(E) = 1$. If $(v')_3(E) = 0$ then already $n + 1$ of literals of $\{l_1,\ldots,l_k\}$ are evaluated as 1 by v'. Then the same happens for v''. The case of $(v')_3(E) = u$ is obvious.

As mentioned above, the rest of the argument follows exactly as in the case of the original Kleene theorem. □

COROLLARY 12.2

Let φ be a formula of L_{Var}^{cc}, let v be a partial valuation of Var, and let w be a valuation of Var so that $v \preceq_k w$. If $(v)_3(\varphi) = 1$, then $w \models \varphi$ and if $(v)_3(\varphi) = 0$, then $w \models \neg\varphi$.

The reader may get the impression that we could define other constraints and the Kleene theorem would still hold regardless of their semantics. This is not the case in general (although in this chapter we will see another class of constraints, extending that of the cardinality constraints, where the Kleene theorem holds). The example is that of *parity constraints*, constraints that hold if the number of literals from a given set and evaluated as 1 is (say) even. Here, we may have a situation where one partial valuation evaluates an even number of literals from $\{l_1,\ldots,l_k\}$ as 1, but its extension fails to do so.

Proposition 12.4 and Corollary 12.2 form a basis for the correctness of the backtracking search algorithm for searching for a valuation that satisfies formulas of L^{cc}. Indeed, let us assume we perform a backtracking search for a satisfying valuation for a theory T. If at some point we established that a formula $\varphi \in T$ is evaluated (in three-valued logic) by a partial valuation v as 1 then *all* partial valuations in the subtree of extensions of v will definitely also evaluate φ as 1, and so we will not need to test those extensions for satisfaction of φ. If, on the other hand, v already evaluates φ as 0, then *all* extensions of v evaluate φ as 0, and so we can safely backtrack, for there is no satisfying valuation below v. This is not the case, of course, for parity constraints. Until we find ourselves at a point where all variables in the set $\{l_1,\ldots,l_k\}$ have assigned logical values, we cannot be sure if such a constraint is satisfied.

Next, we look at the other ingredient of DPLL, the Boolean constraint propagation

(BCP). It turns out that there are *two* types of BCP present in the case of cardinality constraints.

The first one is the familiar BCP-like technique: if in a clause E, all but one cardinality constraints (recall that clauses have no negated cardinality constraints) are already assigned the value 0 (that is, are false), then the last one must be assigned the value 1.

The second BCP-like mechanism for cardinality constraints consists of two forms. The first one is a positive BCP and does the following: if $m\{l_1,\ldots,l_k\}$ is a cardinality constraint (either from the original input theory, or derived) and precisely $k - m$ literals in $\{l_1,\ldots,l_k\}$ are already evaluated as 0 then the *remaining* literals in $\{l_1,\ldots,l_k\}$ must all be assigned the value 1. Then, there is a negative form of BCP, when dealing with the constraints of the form $\{l_1,\ldots,l_k\}n$. Namely, if $k - n$ of literals in $\{l_1,\ldots,l_k\}$ have been assigned the value 1, then *all* the remaining literals must be assigned the value 0. Moreover, it should be observed that cardinality constraints admit two *simplification rules* that generalize the unit resolution rule. The first one is:

$$\frac{m\{l_1,\ldots,l_k\} \quad l_1}{(m-1)\{l_2,\ldots,l_k\}}.$$

This rule simplifies the constraint $m\{l_1,\ldots,l_k\}$ as follows: if we need to make m out of literals in $\{l_1,\ldots,l_k\}$ true, but we already made one of them true, then we need to make $(m-1)$ of the remaining ones true. The second principle directly generalizes modus ponens:

$$\frac{m\{l_1,\ldots,l_k\} \quad \bar{l}_1}{m\{l_2,\ldots,l_k\}}.$$

This second rule says that if we want to make m of literals out of $\{l_1,\ldots,l_k\}$ true, but one of literals $l_1,\ldots,l_k$ is already made false, then out of the remaining ones we still have to make m literals true.

There are two analogous rules for the upper-bound cardinality constraints, as well. Here they are:

$$\frac{\{l_1,\ldots,l_k\}n \quad l_1}{\{l_2,\ldots,l_k\}(n-1)},$$

and

$$\frac{\{l_1,\ldots,l_k\}n \quad \bar{l}_1}{\{l_2,\ldots,l_k\}n}.$$

The simplification rules provided above indicate the presence of several strategies for DPLL-like algorithms for satisfiability of L^{cc} formulas.

One may think, at first glance, that cardinality constraints are exotic. But it is not the case. In fact every student of mathematics sees in her studies expressions that may look like this: $\exists!_x \varphi(x)$. Those may be written in this, or some other fashion, but they express the property that there is exactly one x such that $\varphi(x)$. But a moment's reflection indicates that this is, in fact, a cardinality constraint. Indeed, if all available constants are $a_1,\ldots,a_k$, then $\exists!_x \varphi(x)$ is just the conjunction

$1\{\varphi(a_1),\ldots,\varphi(a_k)\} \wedge \{\varphi(a_1),\ldots,\varphi(a_k)\}1$. It is natural to write such a conjunction as $1\{\varphi(a_1),\ldots,\varphi(a_k)\}1$.

We will show that the use of cardinality constraints allows us to write constraints more concisely. To this end we come back to our mini sudoku grid Example 12.2. Our goal is to describe the constraints of that example using cardinality constraints.

Example 12.5

Here is how we express the constraint that the cell (i, j) contains exactly one number from the range $(1..4)$:

$$1\{p_{i,j,1}, p_{i,j,2}, p_{i,j,3}, p_{i,j,4}\}1.$$

We have 16 such constraints (one for each cell).
Then we need to impose a constraint that i^{th} row contains number k (in exactly one place):

$$1\{p_{i,1,k}, p_{i,2,k}, p_{i,3,k}, p_{i,4,k}\}1.$$

Again, we have 16 such constraints (one for each row i and value k).
Then we need to impose a constraint that j^{th} column contains number k:

$$1\{p_{1,j,k}, p_{2,j,k}, p_{3,j,k}, p_{4,j,k}\}1.$$

We have 16 such constraints (one for each column j and value k).
Then all we need to do is to take care of quadrant constraints. There are four quadrants and four values. There will be 16 quadrant constraints altogether. We list only one of these, leaving the rest of them to the reader. We will impose the constraint that the number 3 appears in exactly one place in the fourth quadrant:

$$1\{p_{3,3,3}, p_{3,4,3}, p_{4,3,3}, p_{4,4,3}\}1.$$

Then, all we need to do is to make sure that the *solver* is capable of handling cardinality constraints, and then, using the coding scheme of Example 12.2, transform our constraints into the form acceptable to *solver*. Actually, East and Truszczyński *aspps* handles cardinality constraints (and variables) so the whole idea is not vacuous. There exist versions of DIMACS format for cardinality constraints, and surely there will be more of such formats in the future. □

12.4 Weight constraints, the language L^{wtc}

We will now generalize the cardinality constraints to a wider class of constraints, called *weight constraints*. Like in the case of cardinality constraints there will be *lower-bound weight constraints* and *upper-bound weight constraints*. It turns out that most properties of cardinality constraints can be easily lifted to the case of weight constraints.

A lower-bound weight constraint is an expression of the form

$$a_1 x_1 + \ldots + a_k x_k \geq m,$$

where $a_1, \ldots a_k, m$ are integers. An upper-bound weight constraint is an expression of the form

$$a_1 x_1 + \ldots + a_k x_k \leq n,$$

where $a_1, \ldots a_k, n$ are integers. We will be interested in pseudo-Boolean solutions to such inequalities; that is, we accept only those solutions that take values 0 and 1. First, like in the case of cardinality constraints, and in a very similar fashion, we move from variables to literals and in this process we make all coefficients $a_1, \ldots, a_k$ positive. Indeed, let us look at an example which is pretty general. Let our constraint be

$$-x_1 - 2x_2 + 3x_3 \geq 0.$$

We add to both sides 3 (because 3 is the sum of absolute values of negative coefficients in our inequality) and after simple transformation we get:

$$(1 - x_1) + 2(1 - x_2) + 3x_3 \geq 3.$$

So now, we can assume that the coefficients are positive; however, we combine not variables, but literals (with $\neg x$, i.e., $\bar{x}$ interpreted as $1 - x$). To make this technique general, let us define

$$h = \sum \{|a_j| : a_j < 0\}.$$

As in the example above, we add h to both sides of the inequality and get a new inequality

$$b_1 l_1 + \ldots b_k l_k \geq m'.$$

Here

$$l_i = \begin{cases} x_i & \text{if } a_i > 0 \\ 1 - x_i & \text{if } a_i < 0. \end{cases}$$

But now all the coefficients $b_j, 1 \leq j \leq k$ are positive. The number m' does not have to be positive, but the restriction requiring solutions to be pseudo-Boolean eliminates such inequalities for they are always true.

The lower-bound weight constraints, after the transformation described above, are uniquely determined by three parameters: first, the set of literals $\{l_1, \ldots, l_k\}$; second, the weight function that assigns to each l_i its weight a_i (in our setting only weight functions taking positive integer values are admitted), and finally the lower bound m. We write such constraint E as:

$$m\{l_1 : a_1, \ldots, l_k : a_k\}.$$

Now it is easy to see how the weight constraints generalize the cardinality constraints. Namely, cardinality constraints are weight constraints with the weight identically equal to 1. Since cardinality constraints generalize propositional variables, the same happens for weight constraints; the propositional variable x is the lower-bound weight constraint $1\{x:1\}$.

Similarly to lower-bound weight constraints we can introduce *upper-bound weight constraints*. Now, treating weight constraints as a new kind of variable we get a language L^{wtc}_{Var}, by closing weight constraints (lower-bound and upper bound) under propositional connectives: $\neg, \wedge, \vee$, etc. We will omit the subscript *Var* if it is clear from the context.

First, we need to define semantics for the language L^{wtc}. As before the semantics is provided by valuations of the set *Var* into $\{0, 1\}$. All we need to do is to define when a valuation v satisfies lower-bound and upper-bound weight constraints. Here is how we define that.

$$v \models m\{l_1 : a_1, \ldots, l_k : a_k\} \quad \text{if} \quad \sum\{a_i : v(l_i) = 1\} \geq m.$$

Similarly for upper-bound weight constraints we set:

$$v \models \{l_1 : a_1, \ldots, l_k : a_k\}n \quad \text{if} \quad \sum\{a_i : v(l_i) = 1\} \leq n.$$

This definition of satisfaction is coherent with our interpretation of variables in L^{wtc} propositional variable x:

$$v \models x \quad \text{if and only if} \quad v \models 1\{x:1\}.$$

The properties of cardinality constraints and of the language L^{cc} can be mostly lifted to the context of L^{wtc} *verbatim*. For instance, it is quite clear that the negation-elimination theorem (Proposition 12.3) holds in the context of L^{wtc}.

To see that we can design a DPLL-like algorithm for finding satisfying valuations for sets of clauses of L^{wtc}, we need to check that the Kleene theorem holds for theories in L^{wtc}. To do so, we need to define the three-valued evaluation function $(v)_3$ for L^{wtc} such that:

1. For complete valuations such a function coincides with the satisfaction by valuations.
2. The evaluation function is monotone in Kleene ordering.

Here is such an evaluation function satisfying both conditions.

$$(v)_3(m\{l_1 : a_1, \ldots, l_k : a_k\}) = \begin{cases} 1 & \text{if } \sum\{a_j : (v)_3(l_j) = 1\} \geq m \\ 0 & \text{if } \sum\{a_j : (v)_3(l_j) \neq 0\} \leq m - 1 \\ u & \text{otherwise.} \end{cases}$$

For upper-bound weight constraints we define:

$$(v)_3(\{l_1 : a_1, \ldots, l_k : a_k\}n) = \begin{cases} 1 & \text{if } \sum\{a_j : (v)_3(l_j) \neq 0\} \leq n \\ 0 & \text{if } \sum\{a_j : (v)_3(l_j) = 1\} \geq n + 1 \\ u & \text{otherwise.} \end{cases}$$

A simple proof of the preservation (Kleene) theorem can be given for both properties (1) and (2). We formulate it formally as follows.

PROPOSITION 12.5
If v', v'' are partial valuations of Var, $v' \preceq_k v''$, and $\varphi \in L_{Var}^{wtc}$ then

$$(v')_3(\varphi) \preceq_k (v'')_3(\varphi).$$

In particular, if $(v')_3(\varphi) = 1$ then $(v'')_3(\varphi) = 1$, and if $(v')_3(\varphi) = 0$ then $(v'')_3(\varphi) = 0$.

Both techniques for simplifications of cardinality constraints lift to the context of weight constraints. There is a slight difference in the simplification rules, and now we list their versions for weight constraints.

$$\frac{m\{l_1 : a_1, \ldots, l_k : a_k\} \quad l_1}{(m - a_1)\{l_2 : a_2, \ldots, l_k : a_k\}}$$

and

$$\frac{m\{l_1 : a_1, \ldots, l_k : a_k\} \quad \bar{l}_1}{m\{l_2 : a_2, \ldots, l_k : a_k\}}.$$

The rules for upper-bound cardinality constraints generalize in a similar fashion. Let us observe that the solver *smodels* [SNS02] admits (in a slightly different context, namely of ASP (see Chapter 14) weight constraints.

12.5 Monotone and antimonotone constraints, characterization by means of cardinality constraints

The reader will observe that in Section 12.3 we changed the terminology; instead of talking about *propositional variables* we started to talk about *constraints*. In Section 12.3 we defined a new language, L^{cc}. But the atomic expressions of that language were not propositional variables (even though propositional variables could be expressed as cardinality constraints, i.e., atomic expressions of L^{cc}), but the expressions mX and Xn where X was a set of literals. We called those atomic building blocks cardinality constraints. Why did we called them constraints, and what is a constraint? The point of view we adopt for this section is that a constraint is *any* set of complete valuations. This concept certainly generalizes propositional variables (because we can identify the propositional variable x with the set of valuations that satisfy x) and more generally, a formula. Indeed we proved, in Chapter 3, Proposition 3.15, that as long as the set of variables is finite, every set of valuations is of the form $Mod(\varphi)$ for some propositional formula φ. Thus every constraint is (or at least can be identified with) a formula. But we can choose different formulas to *represent* a constraint, and it makes a difference in performance if, for instance, our *solver* handles only certain types of formulas. To sum up, every constraint has a representation as a formula, but there are different representations and we can certainly investigate how can we represent constraints efficiently.

It will be convenient in this section to look at representations of valuations as subsets of Var. Thus sets of valuations will be identified with the families of subsets of Var, i.e., subsets of $\mathcal{P}(Var)$.

With the perspective on constraints discussed above, let us look at some examples. The constraint determined by the formula $x \land \neg y$ is the collection $\{M \subseteq Var : x \in M \land y \notin M\}$. The cardinality constraint $2\{x, \neg y, \neg z\}$ is the set of those subsets of Var that contain x but do not contain at least one of y, z or do not contain any of x, y, z (but please check!).

So now, we will classify constraints. A constraint $\mathcal{C}$ is *monotone* if whenever $M \in \mathcal{C}$, and $M \subseteq N \subseteq Var$, then also $N \in \mathcal{C}$. Likewise, we define constraint $\mathcal{C}$ to be *antimonotone* if whenever $M \in \mathcal{C}$, and $N \subseteq M$, then $N \in \mathcal{C}$. Finally, we call a constraint $\mathcal{C}$ *convex* ([LT06]) if whenever M_1, M_2 belong to $\mathcal{C}$, and $M_1 \subseteq N \subseteq M_2$ then $N \in \mathcal{C}$. We will see that both monotone and antimonotone constraints are convex.

Recall that a formula φ of the propositional language L is *positive* if its negation-normal form does not contain negation or $\bot$.

For instance, the formula $(x \lor y) \land (x \lor z)$ is positive, but also the formula $\neg z \Rightarrow x$ is positive (even though it contains negation).

We then have the following observation.

PROPOSITION 12.6
The collection of monotone constraints is closed under unions and under intersections.

Proposition 12.6 implies the following fact.

PROPOSITION 12.7
If φ is a positive formula, then $Mod(\varphi)$ is a monotone constraint.

Proof: By induction on the complexity of the formula $NNF(\varphi)$. Our assertion is certainly true if $\varphi = x$, for a propositional variable x or if $\varphi = \top$. This is the induction basis. The inductive step follows from Proposition 12.6. □

Next, we say that a formula φ is *negative* if $\varphi = \neg \psi$ where ψ is positive. We then immediately have the following.

PROPOSITION 12.8
If φ is a negative formula, then $Mod(\varphi)$ is an antimonotone constraint.

All monotone constraints are represented by positive formulas. Here is how this can be done (the representation can be enormous). Let C be a monotone constraint. Since Var is finite, there is only a finite number of inclusion-minimal sets in C, and every set in C contains an inclusion-minimal set in C. Let $\langle M_1, \ldots, M_p \rangle$ be a listing of all inclusion-minimal sets in C. Let us consider the following formula φ_C (which is, clearly, positive): $\bigvee_{i=1}^{p} \bigwedge M_i$. We then have the following.

PROPOSITION 12.9
If C is a monotone constraint, then $C = Mod(\varphi_C)$.

Using the distributivity laws of propositional logic we can transform φ_C (in principle, the size may be cosmic) to its CNF form. The resulting formula will consist of positive clauses because distributivity laws do not introduce negation.

But what is the formula φ_C for a monotone constraint C? It is in fact, a disjunction of certain lower-bound cardinality constraints. Indeed, let us look at an example. Let C be defined by the positive formula $x \wedge (y \vee z)$. Then there are two equivalent formulas from L^{cc} that represent C. One of those is $1\{x\} \wedge 1\{y, z\}$. The other one is $2\{x, y\} \vee 2\{x, z\}$.

Generalizing from this example, we introduce two concepts: positive lower-bound cardinality constraint, and positive upper-bound cardinality constraint.[2] A *positive lower-bound cardinality constraint* is a cardinality constraint of the form $m\{x_1, \ldots, x_k\}$ where all $x_1, \ldots, x_k$ are propositional variables. Likewise *positive upper-bound cardinality constraint* is a cardinality constraint of the form $\{x_1, \ldots, x_k\}n$, where all $x_1, \ldots, x_k$ are propositional variables.

PROPOSITION 12.10
Let C be a subset of $\mathcal{P}(Var)$. The following are equivalent:

1. *C is a monotone constraint*
2. *There exist positive lower-bound cardinality constraints $E_1, \ldots, E_p$ such that*
$$C = Mod(E_1 \vee \ldots \vee E_p).$$

Proof: To see (1) $\Rightarrow$ (2), we select the inclusion-minimal subsets $X_1, \ldots, X_p$ in C, we compute the numbers $r_i = |X_i|$, $i = 1, \ldots, p$ and then set $E_i = r_i X_i$. It is then easy to see that
$$C = Mod(E_1 \vee \ldots \vee E_p).$$

For the implication (2) $\Rightarrow$ (1), let us observe that each positive lower-bound cardinality constraint is a monotone constraint, so their disjunction is also a monotone constraint. □

---

[2] We are certainly aware of the fact that concepts requiring a term consisting of more than three words are, usually, meaningless.

We will now generalize the second form of representation of monotone constraints via conjunction of positive lower-bound cardinality constraints. The proof we give below has an advantage of actually showing what these cardinality constraints are (in our example we got them from the distributivity laws, without any indication what those constraints are. The following result is due to Z. Lonc. The benefit of the proof is that we actually construct the cardinality constraints representing a monotone constraint. First, we need to define the notion of a hitset for a family of sets. Given a family $\mathcal{F}$ of subsets of Var, a *hitset* for $\mathcal{F}$ (also known as a transversal for $\mathcal{F}$) is any set $X \subseteq Var$ such that for all $Y \in \mathcal{F}$, $X \cap Y \neq \emptyset$. Given a family $\mathcal{F}$, the set of all hitsets for $\mathcal{F}$ is, obviously, a monotone constraint. The reason is that if X has a nonempty intersection with all sets in $\mathcal{F}$, then every bigger set Y also has a nonempty intersection with all sets in $\mathcal{F}$.

PROPOSITION 12.11
Let $\mathcal{C}$ be a subset of $\mathcal{P}(Var)$. The following are equivalent:

1. $\mathcal{C}$ is a monotone constraint

2. There exist positive lower-bound cardinality constraints $E'_1, \ldots, E'_r$ such that
$$\mathcal{C} = Mod(E'_1 \wedge \ldots \wedge E'_r).$$

Proof: The implication (2) $\Rightarrow$ (1) is similar to the one in our proof of Proposition 12.10 because the collection of monotone constraints is closed under intersections. For the implication (1) $\Rightarrow$ (2), let us assume the $\mathcal{C}$ is a monotone constraint. Let $\mathcal{D} = \mathcal{P}(Var) \setminus \mathcal{C}$. Then, because $\mathcal{C}$ is a monotone constraint, $\mathcal{D}$ is an antimonotone constraint. Moreover $\mathcal{C} \cap \mathcal{D} = \emptyset$. Since Var is finite, there are only finitely many sets in $\mathcal{D}$, and in particular finitely many inclusion-maximal sets in D. Let $\mathcal{B}$ be the collection of all inclusion-maximal sets in $\mathcal{D}$, and let $\mathcal{A}$ be the family of complements of sets in $\mathcal{B}$. That is:
$$\mathcal{A} = \{ Var \setminus X : X \in \mathcal{B}\}.$$
We list the sets in $\mathcal{B}$, $\langle Y_1, \ldots, Y_p \rangle$, and all the elements of $\mathcal{A}$, $\langle X_1, \ldots, X_p \rangle$.
We claim that $\mathcal{C}$ *is precisely the set of hitsets for the family* $\mathcal{A}$.
Proof of the claim: Recall that $X_i = Var \setminus Y_i$,
(a) If $M \in \mathcal{C}$, and $X_i \in \mathcal{A}$, then if $M \cap X = \emptyset$, then $M \subseteq Y_i$. But $Y_i \in \mathcal{D}$ by construction. Since $M \subseteq Y_i$, $M \in \mathcal{D}$. But $\mathcal{C}$ and $\mathcal{D}$ are disjoint, a contradiction.
(b) Conversely, let N be a hitset for $\mathcal{A}$. If $N \notin \mathcal{C}$, then $N \in \mathcal{D}$. But then for some i, $1 \leq i \leq p$, $N \subset Y_i$. But then $N \cap X_i = \emptyset$, which contradicts the fact that N is a hitset for $\mathcal{A}$. $\square$Claim
Claim proven, let us prove the implication (1) $\Rightarrow$ (2). Recall that $\mathcal{A} = \langle X_1, \ldots, X_p \rangle$. The fact that a set M is a hitset for $\mathcal{A}$ means that M has a nonempty intersection with each X_i, that is, satisfies the cardinality constraint $1X_i$, for each i, $1 \leq i \leq p$. Let us consider the formula of L^{cc}
$$\varphi = 1X_1 \wedge \ldots \wedge 1X_p.$$

Clearly, $M \models \varphi$ if and only if M is a hitset for $\mathcal{A}$. By our claim, $M \models \varphi$ if and only if $M \in \mathcal{C}$. Thus we constructed the desired conjunction of positive lower-bound cardinality constraints that defines $\mathcal{C}$. □

Putting together the previous two results we get the following.

PROPOSITION 12.12
Let $\mathcal{C}$ be a constraint in the set Var. The following are equivalent:

1. *$\mathcal{C}$ is a monotone constraint.*

2. *There exist positive lower-bound cardinality constraints $E_1, \ldots, E_p$ such that*
$$\mathcal{C} = Mod(E_1 \vee \ldots \vee E_p).$$

3. *There exist positive lower-bound cardinality constraints $E'_1, \ldots, E'_r$ such that*
$$\mathcal{C} = Mod(E'_1 \wedge \ldots \wedge E'_r).$$

Let us observe that the property of cardinality constraints described in Proposition 12.12 does not hold for propositional variables. Specifically, the constraint $\mathcal{C} = Mod(x \vee y)$ cannot be expressed as the $Mod(\varphi)$ where φ is a conjunction of literals. (This actually requires a proof!)

For the antimonotone constraints we will prove a property analogous to one described in Proposition 12.11. But this time, instead of positive lower-bound cardinality constraints we will have positive upper-bound cardinality constraints. We can give direct proof, following the proof of Proposition 12.11. The argument is pretty much the same, but the crucial claim that we proved inside of the proof of Proposition 12.11 does not concerns hitsets, but rather so-called *antitransversals*. An antitransversal for a family $\mathcal{A}$ is a set M such that for all $X \in \mathcal{A}$, $M \not\subseteq X$. Here is a claim that is used in the direct argument characterizing antimonotone constraints as conjunctions of positive upper-bound cardinality constraints. We leave the proof to the reader; the argument is very similar to one made in the proof of the claim in the proof of Proposition 12.11.

LEMMA 12.1
A family $\mathcal{C}$ of subsets of Var is an antimonotone constraint if and only if for some family $\mathcal{A}$, $\mathcal{C}$ is the family of all antitransversals of $\mathcal{A}$.

With Lemma 12.1, we proceed as follows. We first find the family $\mathcal{A}$ for which the desired $\mathcal{C}$ is the family of antitransversals. This is the family $\mathcal{A}$ of complements of the minimal sets in the positive constraint that is the complement of $\mathcal{C}$. Let $\langle X_1, \ldots, X_p \rangle$ be the listing of $\mathcal{A}$, $k_i = |X_i|$, $1 \leq i \leq p$. Then the fact that M is an antitransversal for $\mathcal{A}$ is equivalent to

$$\neg(X_1 \subseteq M) \wedge \ldots \wedge \neg(X_p \subseteq M).$$

This, in turn, is equivalent to

$$M \models \neg(k_1 X_1) \wedge \ldots \wedge \neg(k_p X_p).$$

But now,
$$M \models \neg(k_i X_i) \text{ if and only if } M \models X_i(k_i - 1).$$

Thus, M is an antitransversal for $\mathcal{A}$ if and only if

$$M \models X_1(k_1 - 1) \wedge \ldots \wedge X_p(k_p - 1).$$

In this fashion we get the following result.

PROPOSITION 12.13
Let $\mathcal{C}$ be a constraint in the set Var. The following are equivalent:

1. *$\mathcal{C}$ is an antimonotone constraint.*
2. *There exist positive upper-bound cardinality constraints $E_1, \ldots, E_p$ such that*
$$\mathcal{C} = Mod(E_1 \vee \ldots \vee E_p).$$
3. *There exist positive upper-bound cardinality constraints $E'_1, \ldots, E'_r$ such that*
$$\mathcal{C} = Mod(E'_1 \wedge \ldots \wedge E'_r).$$

We observe that there is another technique for proving Proposition 12.13. This is based on the properties of the permutations of literals. First of all, the permutation lemma (Proposition 2.23) lifts to the language of cardinality constraints L^{cc} verbatim (this requires a proof which is not too difficult). Then we consider the permutation $\pi : x \mapsto \bar{x}$ for all variables x (no renaming, changing sign of every variable). This permutation, when lifted to subsets of *Var* (one needs to be a bit careful at this point) maps every set of variables to its complement, and so it maps the Boolean algebra $\langle \mathcal{P}(Var), \subseteq \rangle$ to the Boolean algebra $\langle \mathcal{P}(Var), \supseteq \rangle$. With the mapping π, the cardinality constraint $m\{x_1, \ldots, x_k\}$ becomes the constraint $m\{\neg x_1, \ldots, \neg x_k\}$, that is (this is a crucial observation!), the constraint $\{x_1, \ldots, x_k\}(k - m)$. So this is what really happens here. The positive lower-bound cardinality constraint is transformed by our permutation π to a positive upper-bound cardinality constraint (and the bound changes). Let us observe that all these considerations based on permutations require additional groundwork that may be of independent interest and may be useful in other contexts.

Now, let us discuss convex constraints. Given a constraint $\mathcal{C}$, we can assign to $\mathcal{C}$ two constraints, one monotone, another antimonotone, as follows. We define $\underline{\mathcal{C}} = \{M : \exists_{Y \in \mathcal{C}} M \subseteq Y\}$, and $\overline{\mathcal{C}} = \{M : \exists_{Y \in \mathcal{C}} Y \subseteq M\}$. It is easy to see that $\underline{\mathcal{C}}$ is an antimonotone constraint, while $\overline{\mathcal{C}}$ is a monotone constraint. We then have the following fact.

PROPOSITION 12.14

If C is a convex constraint then $C = \underline{C} \cap \overline{C}$. In other words, the sets in C are precisely those that belong to the antimonotone constraint $\underline{C}$ and to the monotone constraint $\overline{C}$.

Proof: Clearly, if $M \in C$ then $M \in \overline{C}$ and $M \in \underline{C}$. Thus $C \subseteq \underline{C} \cap \overline{C}$.
Conversely, if $M \in \underline{C} \cap \overline{C}$, then there are $M' \in C$ and $M'' \in C$ such that

$$M'' \subseteq M \subseteq M'.$$

But C is convex, so $M \in C$. □

Next, we have the following lemma.

LEMMA 12.2

1. Every monotone constraint is convex and every antimonotone constraint is convex.

2. The intersection of two convex constraints is convex, and so intersection of a monotone constraint and of an antimonotone constraint is convex.

But now, Propositions 12.11 and 12.13 imply the following property.

PROPOSITION 12.15

Let C be a constraint over a finite set of propositional variables Var. The following are equivalent:

1. C is a convex constraint.

2. There is a collection of positive cardinality constraints (some may be lower bound and some upper bound) $\{E_1, \ldots, E_p\}$ such that

$$C = Mod(E_1 \wedge \ldots \wedge E_p).$$

Proof: First, let us assume that C is convex. Then $C = \underline{C} \cap \overline{C}$. But for the monotone constraint $\overline{C}$, there is a set of positive lower-bound constraints $E'_1, \ldots, E'_r$ so that $\overline{C} = Mod(E'_1 \wedge \ldots \wedge E'_r)$, and for the antimonotone constraint $\underline{C}$ there is a set of positive upper-bound constraints $E''_1, \ldots, E''_s$ so that $\underline{C} = Mod(E''_1 \wedge \ldots \wedge E''_s)$. Thus

$$C = Mod(E'_1 \wedge \ldots \wedge E'_r \wedge E''_1 \wedge \ldots \wedge E''_s).$$

Conversely, given a set $\{E_1, \ldots, E_p\}$, of positive cardinality constraints, we can separate constraints in $\{E_1, \ldots, E_p\}$ into $\{E'_1, \ldots, E'_r\}$ consisting of lower-bound constraints and $\{E''_1, \ldots, E''_s\}$ consisting of upper-bound constraints. We then get $C_1 = Mod(E'_1 \wedge \ldots \wedge E'_r)$ which is monotone and $C_2 = Mod(E''_1 \wedge \ldots \wedge E''_s)$ which is antimonotone. Since $C = C_1 \cap C_2$, the desired implication follows from Lemma 12.2. □

12.6 Exercises

1. Assume we want to represent as a SAT problem not the 4×4 sudoku grids, but the standard 9×9 sudoku grids. How could this be done? Also, how many propositional variables would you need?

2. Write a script (in Perl, Python, or some other language) that produces encoding of sudoku. What will you do to make sure that every run returns a different solution?

3. Continuing the sudoku theme. Now that your favorite SAT solver returned a *sudoku solution*, design a script that will compute out of this solution a *sudoku problem*. The idea is that when you add to the theory designed in problem (1) the solution obtained in (2), the resulting theory has just one satisfying assignment. Remove atoms of the solution one by one until you get more than one satisfying assignment (of course you return the last partial assignment that maintains the invariant).

4. Write to Professor Raphael Finkel (raphael@cs.uky.edu) if you find Problem (3) cryptic.

5. The SEND-MORE-MONEY puzzle is to assign to letters $\{D, E, M, N, O, R, S, Y\}$ digits in $(0..9)$ so that different letters are assigned different digits, and

$$\begin{array}{r} S\ E\ N\ D \\ +\ M\ O\ R\ E \\ \hline M\ 0\ N\ E\ Y \end{array}$$

No 0s at the beginning of a number. Describe this problem as a SAT problem (using a similar technique to one we used for sudoku-like puzzles). Specifically create variables of the form $x_{X,i}$ where $X \in \{D, E, M, N, O, R, S, Y\}$, $i \in (0..9)$ and write suitable clauses.

6. Devise a technique for elimination of cardinality constraints. Specifically, given a finite set of formulas F in the language of cardinality constraints $\mathcal{C}$, extend your language by additional propositional variables (which will interpret cardinality constraints occurring in F and possibly additional cardinality constraints). The following should hold for your transformation $Tr(F)$: an assignment v satisfies the set of formulas F if and only if for some interpretation w of new variables the assignment $v \cup w$ satisfies $Tr(F)$. Moreover, the size of $Tr(F)$ should be bounded by a linear polynomial in the size of F. Four simplification rules of Section 12.3 and the technique used in the proof of reducing SAT to 3-SAT could be used.

7. Design a DIMACS-like representation for clauses built of cardinality constraints.

8. Continuing the previous problem, do the same thing for weight constraints.

9. Generalize the argument for the Kleene theorem for cardinality constraints (Proposition 12.4) to the case of weight constraints.
10. Prove that union of two monotone constraints is monotone, and intersection of two monotone constraints is monotone.
11. Give an example of two convex constraints such that their union is not convex.
12. What about the difference of two convex constraints?

Chapter 13

Computational knowledge representation with SAT – handling constraint satisfaction

13.1 Extensional and intentional relations, *CWA*	285
13.2 Constraint satisfaction and SAT	292
13.3 Satisfiability as constraint satisfaction	297
13.4 Polynomial cases of Boolean CSP	300
13.5 Schaefer dichotomy theorem	305
13.6 Exercises	317

In this chapter we show a generic use of SAT as a vehicle for solving constraint satisfaction problems over finite domains. First, we discuss extensional and intentional relations. Roughly, extensional relations are those that are stored (they are our data). Intentional relations are those relations that we search for (and the *solver* computes). The difference between the two is expressed by so-called closed world assumption (CWA). After illustrating what this is all about (and showing how CWA is related to the existence of the least model) we discuss constraint satisfaction problems (CSP). We show how the propositional logic (and the version of predicate logic considered above) can be used to reformulate and solve CSPs. Then we show that, in fact, the search for satisfying valuations is a form of CSP. We also discuss the Schaefer theorem on classification of Boolean constraint satisfaction problems.

13.1 Extensional and intentional relations, closed world assumption

The motivation for this section comes from the following example. Let us assume that we have a very simple graph G, with three vertices, $\{a, b, c\}$, and just two edges, (a, b) and (a, c). Now, our goal is to select one of those two edges. We will use for this our *solver*. We represent the edges of our graph by means of a predicate *edge*, asserting two propositional variables: $edge(a, b)$ and another $edge(a, c)$ true. Now, we consider a new predicate symbol *sel* and we write a theory in predicate logic (but no function symbols). Here is the theory:

285

1. $\exists_{x,y} sel(x,y)$.
2. $\forall_{x,y}(sel(x,y) \Rightarrow edge(x,y))$.
3. $\forall_{x,x_1,y,y_1}(sel(x,y) \wedge sel(x_1,y_1) \Rightarrow (x=x_1) \wedge (y=y_1))$.

The meaning of these formulas is clear: (1) tells us that something is selected. When we translate (1) into a propositional formula, we get one disjunction of length 9. The formula (2) requires that whatever we select will be an edge. This formula generates nine implications. Finally, the formula (3) requires that just one edge has been selected. We leave to the reader to check that after grounding there will be 81 ground instances of formulas for type (3). We also leave to the reader transformation of our formulas to the conjunctive normal form (a bit, but really only a bit, of work is required). The theory S consists of formulas (1)–(3). But the theory S does not take into account our data on the graph G. So we need to add to our grounding of the theory the *data D* on our graph, namely the unit clauses $edge(a,b)$ and $edge(a,c)$. It appears this is all we need to do. Let us call the resulting theory T.

We then submit the CNF T to the *solver*. Many satisfying valuations will be generated. One of those will be the valuation evaluating the following five variables as true, the rest as false:

$$sel(b,c), \; edge(b,c), \; edge(a,b), \; edge(a,c), \; edge(c,c).$$

This is strange, and certainly not what we expected. We selected an edge that is not in the graph at all! Moreover, we added two edges to the graph (the edge from b to c and a loop on c). What is the reason? What happens is that the theory T does not prevent us from *adding* new edges. Once we do this, we could *select* a new edge. On the other hand, our theory prevents us from selecting two different edges, so only one edge is the content of the predicate *sel*.

So we need to do something to prevent this phenomenon (making true *phantom*, i.e., unintended variables) happen. This technique is called *closed world assumption* (*CWA*). The reader will recall that we mentioned *CWA* before.

Let R be a k-ary relation on the set $\mathcal{C}$, and let p be a k-ary predicate letter used to interpret R. The *diagram* of R, $diag(R)$ is the following set of (propositional) literals:

$$\{p(a_1,\ldots,a_k) : (a_1,\ldots,a_k) \in R\} \cup \{\neg p(a_1,\ldots,a_k) : (a_1,\ldots,a_k) \notin R\}.$$

Let us see what happens if, instead of adding to the theory S the units D describing edges of G, we add to S the diagram of *edge*. Let us call this resulting theory T'.

This diagram consists of nine, not two literals, namely, two positive units considered above, and seven negative literals. We list two of these seven: $\neg edge(b,c)$ and $\neg edge(c,c)$. Now, because we added these seven additional literals, *every* satisfying valuation will not include variables describing "phantom" edges such as the ones that appeared in the valuation exhibited above. Thus, the abnormal edges disappear, and we can see that T' has the desired property, that is, it possesses only two satisfying valuations. One evaluates as *true* variables

$$sel(a,b), \; edge(a,b), \; edge(a,c),$$

and another evaluates as *true*

$$sel(a,c), \ edge(a,b), \ edge(a,c).$$

Let us observe that CWA is used by humans every day. For instance, if I call my travel agent and ask about a direct flight from Lexington, KY to San Jose, CA, he will tell me there is none. How does he know? He looks at the table of the flights originating from Lexington. He does not see one going directly to San Jose. He performs the closed world assumption, and deduces that there is none. In other words, he derives the literal $\neg connect(LEX, SJC)$ but from the $diag(connect)$ and *not* from the set of tuples containing the positive information about relation *connect*. We will now make this a bit more general. This approach comes from the following observation: the *diagram* of R consists of the positive part (just facts about R) and negative part. This latter part consists of negations of those facts which cannot be derived from the positive part. Now, let T be an arbitrary propositional theory. We define

$$CWA(T) = T \cup \{\neg p : p \in Var \wedge T \not\models p\}.$$

In our example (when T was the positive part of the diagram of a relation) $CWA(T)$ was consistent. But this is not always the case. Let T be $\{p \vee q\}$. Clearly $T \not\models p$ and likewise $T \not\models q$. We get

$$CWA(T) = \{p \vee q, \neg p, \neg q\}.$$

We observe that this latter theory is inconsistent.

Motivated by this example we say that a theory T is *CWA-consistent*, if the theory $CWA(T)$ is consistent. The question arises which theories are CWA-consistent. Here is a characterization. It generalizes quite a simple argument for the case of finite theories. But we will give a full argument without a finiteness assumption. We will have to use the Zorn lemma. The reader who does not like such arguments will recall that for *finite* posets the Zorn lemma is obvious. So if she does not like arguments using transfinite induction she will have to limit herself to a finite case. We will prove that CWA-consistency characterizes precisely those propositional theories that possess a least model. We first have a lemma which is best formulated in terms of sets of variables rather than valuations.

LEMMA 13.1
If a propositional theory T is satisfiable, then T possesses a minimal model. In fact, for every model M of T, there is a minimal model N of T such that $N \subseteq M$.

Proof: Let us observe that in the case of finite theories T the lemma is obvious. But in the general case, we need quite strong means in the argument, namely the Zorn lemma. Anyway, all we need to show is that given a set of variables M the family $\mathcal{M}_{M,T} = \{N : N \in Mod(T) \wedge N \subseteq M\}$ is closed under intersections of well-ordered, descending chains. So let us look at such an inclusion-descending chain

$\langle N_\xi \rangle_{\xi < \beta}$. If β is a successor, i.e., $\beta = \gamma + 1$, then our chain has its last element. That element, as the sequence $\langle N_\xi \rangle_{\xi < \beta}$ is descending, is N_γ, which belongs to $\mathcal{M}_{M,T}$. If β is a limit ordinal, then for each formula $\varphi \in T$, there is $\alpha < \beta$ such that for all $p \in Var$, $p \in \bigcap_{\xi < \beta} N_\xi$ if and only if $p \in N_\alpha$. But then

$$\bigcap_{\xi < \beta} N_\xi \models \varphi \quad \text{if and only if} \quad N_\alpha \models \varphi$$

by Proposition 2.3. Since $N_\alpha \models T$, and in particular $N_\alpha \models \varphi$, $\bigcap_{\xi < \beta} N_\xi \models \varphi$. Since φ was an arbitrary formula in T, we are done. □

We will now prove the characterization of *CWA*-consistent theories.

PROPOSITION 13.1
Let T be a propositional theory. Then T is CWA-consistent if and only if T possesses a least model.

Proof. First, let us assume that T possesses a least model M_0. We claim that M_0 is a model of $CWA(T)$. All we need to show in this part of our proposition is that for an arbitrary propositional variable p whenever $T \not\models p$, then $M_0 \models \neg p$. But if $T \not\models p$ then there is a model N of T such that $N \models \neg p$, that is, $p \notin N$. But M_0 is a least model of T, thus $M_0 \subseteq N$. But then $p \notin M_0$, i.e., $M_0 \models \neg p$, as desired.

So now, let us assume that T is satisfiable, but has no least model. We claim that T possesses two different minimal models. First, by Lemma 13.1, T possesses at least one minimal model, say M_0. Since M_0 is not a least model, there is a model M_1 of T such that $M_0 \not\subseteq M_1$ and $M_1 \neq M_0$. But it cannot be the case that $M_1 \subseteq M_0$ because M_0 is minimal and $M_0 \neq M_1$. Now, again using Lemma 13.1, let us consider a minimal model M_2 of T included in M_1. We claim that M_2 is different from M_0. Indeed, M_2 cannot be included in M_0 because in such case, as M_0 is minimal, $M_0 = M_2$ and so $M_0 \subseteq M_1$ contradicting the choice of M_1. And it is also impossible that M_0 is included in M_2, because in such case M_0 is included in M_1, again a contradiction.

Thus, assuming that T has no least model, it has at least two distinct minimal models, N_0 and N_1. Now, for each $p \notin N_0$, $T \not\models p$ because $N_0 \models \neg p$. Likewise, for each $p \notin N_1$, $T \not\models p$ because $N_1 \models \neg p$. But we claim that

$$T \cup \{\neg p : p \notin N_0\} \cup \{\neg p : p \notin N_1\}$$

is unsatisfiable. Indeed, if $M \models \{\neg p : p \notin N_0\} \cup \{\neg p : p \notin N_1\}$, then $M \subseteq N_0 \cap N_1$. But then M cannot be a model of T because both N_0 and N_1 are different minimal models of T, so their intersection does *not* contain a model of T. But

$$(T \cup \{\neg p : p \notin N_0\} \cup \{\neg p : p \notin N_1\}) \subseteq CWA(T)$$

so $CWA(T)$ is unsatisfiable. □

In Section 9.2 we proved that every consistent Horn theory possesses a least model (Corollary 9.4). Thus we get the following fact, due to R. Reiter.

COROLLARY 13.1 (Reiter Theorem)
If H is a consistent Horn theory, then H is CWA-consistent.

Every consistent Horn theory H can be viewed as a specification of the least model of its program part, H_1. This least model is computed as a least fixpoint of a monotone operator. Thus we can view consistent Horn theories as specifications of inductively defined sets. This direction has been pursued by recursion theorists and a large body of knowledge is available on this topic.

Going back to closing *some* relations under CWA, we have seen that we do this if we need to make sure that unexpected tuples do not show in the solutions. When we need to make sure that those relations are fixed, we have two options: either to define (and explicitly introduce into the theory) the diagram of such relation R, or to declare such relation to be *extensional* and expect that the *grounder* will make sure that the diagram of those relations is computed and added. We will illustrate this second approach with the following example.

Example 13.1

Our goal now is to solve a simple logical puzzle of the sort used to teach logic to lawyers. We will assume that we have at our disposal a *grounder* that understands the declarations of the form

$$\text{ext}(rel\_symb),$$

where rel_symb is a relational symbol. This declaration tells the grounder that the relation rel_symb is declared *extensional*, that is, that only those propositional variables of the form $rel\_symb(a_1, \ldots, a_k)$ are true that are explicitly included in the theory. In other words, the CWA is supposed to be applied to the relations declared *extensional*. The *grounder* must compute their diagram (in fact its negative part) and add it to the input theory. This is what we did manually in our edge-selection example. Relations that are not extensional are *intentional*.

So here is the promised puzzle. While planning to attend the Beaux Arts Ball, three ladies, *Ann, Barb*, and *Cindy* (abbreviated a, b, c, resp.), painted their hair *red, green*, and *blue* (abbreviated r, g, b, resp.), so that every lady has hair of a different color, and each of them wears a dress: *sari, Chanel costume*, and *kimono* (abbreviated s, cc, and k). Each lady wears a single dress, and each dress is worn.

Here is what we are told: *First, Ann wears the kimono. Next, Barb has a green hair, but she does not wear the sari. Finally, the lady with blue hair does not wear the kimono.* Our goal is to establish the color of hair and the dress of each lady.[1]

Here is how we formalize it. We will be using six extensional predicates. First, three of those are binary and we give them the names *name_hair, name_dress*, and *hair_dress*. There will be three more predicates, all unary: $dom1$, $dom2$, and $dom3$.

---
[1] Certainly not a very *complex* puzzle!

TABLE 13.1: Listing for binary extensional relations of the puzzle

name_color		name_dress		color_dress	
a	r	a	k	r	s
a	b	b	s	r	k
a	g	b	k	r	cc
b	g	c	s	g	s
c	r	c	k	g	cc
c	b	c	cc	g	k
c	g			b	g
				b	cc

TABLE 13.2: Listing for unary extensional relations of the puzzle

dom1	dom2	dom3
a	g	s
b	r	cc
c	b	k

We list first three relations in Table 13.1. The tables of our relations express the possible values our relations can take. For instance, we know that Barb has green hair. Thus the values (b, r), and (b, b) are *absent* in the relation *name_color* since we know that every lady has a unique color of hair. Next, we define unary predicates (describing the domains of binary predicates. Those are shown in Table 13.2.

We now declare all six of relations listed in Tables 13.1 and 13.2 extensional. This has the effect of adding (fortunately, we do not have to do this by hand, *grounder* will do this for us) all sorts of *negative* literals. For instance, we will have a negative literal $\neg name\_color(b, r)$. Now, we will have an intentional ternary predicate *sel*. Here are four groups of clauses involving both extensional and intentional predicates.

Group 1 – Existence of values.
$\forall_x \exists_{y,z}\, sel(x, y, z).$
$\forall_y \exists_{x,z}\, sel(x, y, z).$
$\forall_z \exists_{x,y}\, sel(x, y, z).$

Group 2 – Uniqueness.
$\forall_{x,y_1,y_2,z_1,z_2} (sel(x, y_1, z_1) \wedge sel(x, y_2, z_2) \Rightarrow y_1 = y_2 \wedge z_1 = z_2).$
$\forall_{y,x_1,x_2,z_1,z_2} (sel(x_1, y, z_1) \wedge sel(x_2, y, z_2) \Rightarrow x_1 = x_2 \wedge z_1 = z_2).$
$\forall_{z,x_1,x_2,y_1,y_2} (sel(x_1, y_1, z) \wedge sel(x_2, y_2, z) \Rightarrow x_1 = x_2 \wedge y_1 = y_2).$

Group 3 – Fixed domains of predicates.
$\forall_{x,y,z} (sel(x, y, z) \Rightarrow dom1(x) \wedge dom2(y) \wedge dom3(z)).$

TABLE 13.3: Solution to the name-color-dress puzzle

Name	Color	Dress
Ann	Red	Kimono
Barb	Green	Chanel
Cindy	Blue	Sari

Group 4 – Correct choices.
$\forall_{x,y,z}(sel(x,y,z) \Rightarrow name\_hair(x,y))$.
$\forall_{x,y,z}(sel(x,y,z) \Rightarrow name\_dress(x,z))$.
$\forall_{x,y,z}(sel(x,y,z) \Rightarrow hair\_dress(y,z))$.

Here is what those formulas do for us. First, Group 1. It tells us that each value of each attribute (name, hair color, dress) is actually used. Thus Ann has to have some dress, and the sari will be owned by a person with some hair color. Then, Group 2. This tells us that each person has a unique hair color and a dress, that a dress has a unique owner, etc. Next, Group 3. All it does for us is to make sure that when we select a color of hair, we will not get back a dress, or name. Similar conditions are enforced for names and dresses. Finally, Group 4. This is the most interesting of all; let us see what did we do by providing extensional relations *name_hair*, *name_dress* and *hair_dress*. We specified the *local constraints*. Now, we require that the tuples conform to all these local constraints, globally. For instance, we required Barb to have green hair. Thus we enforced a local constraint: the tuples (b,r) and (b,b) were absent from our constraint. By means of extensionality of the relation *name_hair*, we enforced the presence of negative literals $\neg name\_hair(b,r)$ and $\neg name\_hair(b,b)$. This is how we made sure that the local properties of the relation *name_hair* are made global. Let us observe that there is no local constraint on the color of Cindy's hair. This constraint was discovered during the search process.

Once we listed our theory T (groups of formulas (1)–(4)), and the relations describing the constraints of our problem (there were six such relations: *name_hair*, *name_dress* and *hair_dress* and three unary relations describing the domains of attributes), we submit T to the *grounder* (yes, the formulas are not CNFs, but this is easy to fix, and also we must declare the six relations extensional, to avoid spurious solutions). The *grounder* produces a *propositional* CNF $ground(T)$ which is then submitted to the *solver*. The *solver* will return a (unique) model of our grounded theory $ground(T)$. This model will consist of positive literals in extensions of relations *name_hair*, *name_dress*, and *hair_dress* and three domains, and just three atoms: $sel(a,r,k)$, $sel(b,g,cc)$, and $sel(c,b,s)$. The solution is presented in Table 13.3. □

Of course, many other logic puzzles (especially of the sort published in puzzle journals and using so-called hatch-grids) can be solved by a technique very similar to the one discussed in our Example 13.1.

Group 3 (consisting of a single axiom which generates three clauses) amounts to strong typing of object variables that occur within the predicate *sel*. That is, the type of the first variable is the user-defined type *dom*1, the type of the second variable is *dom*2, etc. The mechanism for enforcing strong types of object variables has been, actually, implemented in the solver *aspps* [ET01] (along with the grounder *psgrnd* which allows for separation of data and the program, and in this process recognizes extensional relations). Generally, one can define a type system suitable for problem-solving with SAT, but the only types that can be so defined must be finite.

13.2 Constraint satisfaction problems and satisfiability

In this section we investigate the relationship between constraint satisfaction problems and satisfiability. We will show how the constraint satisfaction problems can be coded as satisfiability problems. Later we will also show how satisfiability can be treated as constraint satisfaction. Our exposition of constraint satisfaction will have a database-theoretic flavor. The technique of extensional relations will play a major role in our presentation.

Like in databases, we will deal with relations which will be subsets of Cartesian products of *domains*. Formally, a constraint satisfaction problem $\mathcal{P}$ ([Apt03]) consists of:

1. A finite number of *variables*, $X = \{x_1, \ldots, x_k\}$[2]. A fixed listing of X, $\langle x_1, \ldots, x_k \rangle$ is called a *scheme* of $\mathcal{P}$.

2. A finite number of sets, indexed with elements of X, $D_{x_1}, \ldots, D_{x_k}$. Those sets are called *domains* of variables (resp. $x_1, \ldots, x_k$). Domains of variables do not have to be different – the situation is similar to that in databases; different attributes can have the same or related domains.

3. A finite number of relations $R_1, \ldots, R_l$. Each relation R_j, $1 \leq j \leq l$, is assigned its *local scheme* $R_j \mapsto \{x_{j,1}, \ldots, x_{j,i_j}\}$ so that

$$R_j \subseteq D_{x_{j,1}} \times \ldots \times D_{x_{j,i_j}},$$

for $1 \leq j \leq l$.

We define the notion of a solution to a problem $\mathcal{P}$. A *solution* to the problem $\mathcal{P}$ is a sequence $\langle a_1, \ldots, a_k \rangle$ such that

(a) For each j, $1 \leq j \leq k$, $a_j \in D_j$
(b) For each j, $1 \leq j \leq l$, $\langle a_{j,1}, \ldots, a_{j,r_j} \rangle \in R_j$.

This is a very general framework, and it has several natural interpretations. Here is one that uses databases. The problem of existence of a solution to the constraint satisfaction problem $\mathcal{P}$ is precisely the question if the natural join of relations $R_1, \ldots, R_l$

---

[2]The same construct occurs in database theory; there the variables are called *attribute names*.

is non-empty. There is a corresponding search problem. In this version not only do we want to find if the problem $\mathcal{P}$ possesses a solution, but we also want to find one (if a solution exists). It may look like all we need to do is to compute the join and then see if the result is non-empty. Unfortunately, the amount of work needed to do so may be prohibitive. We will see, however, that a satisfiability solver may be of use.

Let us look at an example of a constraint satisfaction problem.

Example 13.2

The scheme of our problem has four variables: x, y, z, and t. We have four domains, each consisting of integers (but not all), with the domains $D_x = D_z = \{n : n \in N \land n \text{ is odd}\}$ and the domains $D_y = D_t = \{n : n \in N \land n \text{ is even}\}$. Then we have two relations: R_1 with a scheme $\{x, y, z\}$, $R_1 \subseteq D_x \times D_y \times D_z$, $R_1 = \{(x, y, z) : x + y = z\}$, and R_2 with a scheme $\{x, y, t\}$, $R_2 \subseteq D_x \times D_y \times D_t$, $R_2 = \{(x, y, t) : x \cdot y = t\}$. We specified a constraint satisfaction problem $\mathcal{P}$. What about solutions? Our problem $\mathcal{P}$ possesses solutions, in fact, infinitely many solutions. One solution is $(1, 2, 3, 2)$. Another is $(5, 4, 9, 20)$. Indeed, in this latter case, $5 + 4 = 9$, and $5 \cdot 4 = 20$. □

In Example 13.2 the domains were infinite. We cannot handle infinite domains in the context of SAT (but look below for special cases). So, for a moment we will assume that all domains are finite. Let $\mathcal{P}$ be a constraint satisfaction problem where all domains $D_i, 1 \leq i \leq k$, are finite. We will now construct a CNF formula $F_\mathcal{P}$ such that there is a bijective correspondence between the solutions to $\mathcal{P}$ and satisfying valuations for $F_\mathcal{P}$. Assuming that m is a bound on the arity of relations in $\mathcal{P}$ our encoding will be polynomial of degree m in the size of $\mathcal{P}$. Before we do this, it should be clearly stated that the CSP community developed a large number of techniques for solving CSPs. Those techniques, usually, amount to transforming CSPs to simpler CSPs, for instance, by eliminating some values from the domains. We will use one of these simplification rules below. Some of those techniques translate to simplifications similar to BCP.

So we will translate a CSP problem to SAT. There are various possible translations; we will present one. What we do amounts to associating with a CSP problem $\mathcal{P}$ a Herbrand structure, then adding to it a number of intentional predicates so that the Herbrand models of the resulting theory will be in a one-to-one correspondence with solutions to the original CSP. We will *not* compute the entire join of relations constituting $\mathcal{P}$; all we want is one tuple in the join.

First, we need to specify the language. There will be extensional predicates and then intentional predicates. The set of constants will consist of the union of the domains, $\bigcup_{i=1}^{k} D_i$. There will be $k + l$ extensional predicates (recall that k is the number of variables, and l is the number of relations). Of these, first k will be used to describe domains of variables, $D_i, 1 \leq i \leq k$, and then l predicates will be used to describe relations $R_j, 1 \leq j \leq l$. The first k predicates will be unary. For lack of better notation we will call them $dom_i, 1 \leq i \leq k$. For the remaining l relations we will have predicates of various arities. Assuming R_j is r_j-ary relation, p_j will be the r_j-ary predicate symbol. Then we will have additional k unary predicates sel_i. These will be intentional predicates. The propositional formula $F_{\mathcal{P}}$ will be the result of grounding a formula of predicate calculus, $G_{\mathcal{P}}$. The formula $G_{\mathcal{P}}$ will consist of four groups (i.e., is a conjunction of formulas that we will list in four groups) and of a number of literals. This latter part describes extensional relations.

Group 1: $\exists_x sel_i(x)$, for each i, $1 \leq i \leq k$.
Group 2: $\forall_x (sel_i(x) \Rightarrow dom_i(x))$, for all i, $1 \leq i \leq k$.
Group 3: $\forall_x \forall_y (sel_i(x) \wedge sel_i(y) \Rightarrow x = y)$, for all i, $1 \leq i \leq k$.
Group 4: $\forall_{y_{j,1}} \ldots \forall_{y_{j,r_j}} (sel_{j,1}(y_{j,1}) \wedge \ldots \wedge sel_{j,r_j}(y_{j,r_j}) \Rightarrow p_j(y_{j,1}, \ldots, y_{j,r_j}))$.

The theory $G_{\mathcal{P}}$ consists of:

(a) Formulas of groups (1)–(4).
(b) Diagrams of unary relations dom_i, $1 \leq i \leq k$.
(c) Diagrams of relations R_j, $1 \leq j \leq l$.

Intuitively, formulas of groups (1) and (2) tell us that for each variable x_i some element of the domain D_i will be selected. This we do for *each* variable. Formulas of group (3) tell us that only one such element will be selected from the domain of each variable x_i (i.e., D_i) by means of the predicate sel_i. The reader should have in mind that the intuition is that we assign values to variables, not domains. Finally, formulas of group (4) enforce the constraints that the projections (according to schemes of relations) of the selected sequence $\langle y_1, \ldots, y_k \rangle$ belong to R_j, $1 \leq j \leq l$. Then, in (b) and (c) we enforce that the domains D_i and relations R_j are extensional so as to avoid spurious solutions.

Now, we are ready to see what $F_{\mathcal{P}}$ is; it is the grounding of $G_{\mathcal{P}}$.

Let us look again at Example 13.2, but now let us limit our domains as follows. $D'_x = D'_z = \{1, 3, 5\}$, $D'_y = D'_t = \{0, 2, 4, 6\}$. The relations (constraints) are the restrictions of the original relations to the corresponding domains. Now, $(1, 2, 3, 2)$ is a solution, but $(3, 4, 7, 12)$ is not, because (for instance) 7 does not belong to D'_z.

Before we prove the promised bijection result, let us observe that the *grounder* can significantly simplify the groundings of the formulas of groups (1)–(4). For instance, in (2), whenever $x \notin D_i$, our ground theory will contain $\neg dom_i(x)$. Therefore, by contraposition, we derive $\neg sel_i(x)$. In the dual case, when $x \in D_i$, the formula $dom_i(x)$ is in $F_{\mathcal{P}}$, so the formula $sel_i(x) \Rightarrow dom_i(x)$ will be satisfied by any valuation making $F_{\mathcal{P}}$ true, and so can be safely eliminated. The net effect is that the formulas in (1) can be reduced (without changing semantics) to a formula $\exists_x (sel_i(x) \wedge dom_i(x))$ which generates a shorter disjunction. We also observe that the equality relation is also extensional and thus it immediately reduces in grounding the formulas of group (3) to clauses of length 2. Finally, in the formulas of group (4) we can bind (like we did in the case of formulas of group (1)) quantifiers to corresponding domains. We now state and prove the fundamental result on the use of SAT solvers for finite domains constraint satisfaction.

PROPOSITION 13.2
Let $\mathcal{P} = \langle X, D_{x_1}, \ldots, D_{x_k}, R_1, \ldots, R_l \rangle$ be a constraint satisfaction problem with all domains finite.

1. If $\vec{y} = (y_1, \ldots, y_k)$ is a solution for $\mathcal{P}$ then the set of propositional variables $M_{\vec{y}}$ consisting of

    (a) $dom_i(x)$, for $x \in D_{x_i}$, $1 \leq i \leq k$
    (b) $p_j(y_{j,1}, \ldots, y_{j,r_k})$, for $(y_{j,1}, \ldots, y_{j,r_j}) \in R_j$, $1 \leq j \leq l$
    (c) $sel_i(y_i)$, $1 \leq i \leq k$

    satisfies the propositional formula $F_{\mathcal{P}}$.

2. Conversely, if M is a model of $F_{\mathcal{P}}$ then

    (a) For all i, $1 \leq i \leq k$, $D_i = \{x : dom_i(x) \in M\}$
    (b) For all j, $1 \leq j \leq l$, $R_j = \{(y_{j,1}, \ldots, y_{j,r_j}) : p_j(y_{j,1}, \ldots, y_{j,r_j}) \in M\}$
    (c) For each i, $1 \leq i \leq k$ there is a unique x such that $sel_i(x) \in M$
    (d) If y_i is the unique x such that $sel_i(y) \in M$, $1 \leq i \leq k$, then the sequence $\vec{y} = (y_1, \ldots, y_k)$ is a solution to $\mathcal{P}$.

3. The mapping $M_{\vec{y}} \mapsto \vec{y}$ is one to one, and hence a bijection.

Proof: (1) By construction, $M_{\vec{y}}$ satisfies the sets of formulas $diag(D_i)$, $1 \leq i \leq k$, and $diag(R_j)$, $1 \leq j \leq l$. This implies that the ground formulas that we added to groups (1)–(4) are satisfied by $M_{\vec{y}}$. Moreover, since $\vec{y}$ is a sequence, for each i, $1 \leq i \leq k$, there is only one propositional variable of the form $sel_i(x)$ in $M_{\vec{y}}$. Since $\vec{y}$ is a solution, that x must belong to D_i. These facts imply that the groundings of formulas in groups (1), (2), and (3) are satisfied by $M_{\vec{y}}$. Finally, because $\vec{y}$ is a solution the groundings of formulas of group (4) are also satisfied.
(2) Since M is a model of $F_{\mathcal{P}}$, and $F_{\mathcal{P}}$ contains diagrams of all D_i, $1 \leq i \leq k$, and R_j, $1 \leq j \leq l$, (a) and (b) are true. The condition (c) follows from the presence of groundings of formulas of groups (1) and (3), and the condition (d) from the presence of groundings of formulas of groups (2) and (4).
(3) If M_1, M_2 are two different models of $F_{\mathcal{P}}$, then they cannot differ on variables of the form $dom_i(x)$, $1 \leq i \leq k$, or on variables of the form $p_j(a, \ldots, b)$, $1 \leq j \leq l$, because both M_1, M_2 satisfy diagrams of unary relations D_i, $1 \leq i \leq k$ and also diagrams of relations R_j, $1 \leq j \leq l$. Thus if $M_1 \neq M_2$ then there must be i, $1 \leq i \leq k$ so that M_1, M_2 differ on a variable of the form $sel_i(x)$ for some x (in fact some $x \in D_i$). But then the corresponding sequences are different. □

There is nothing deep in Proposition 13.2. It just tells us that the construction we proposed is correct.

Now, let us discuss possible strengthening of Proposition 13.2. First, the question arises if we could do something with infinite domains. We need a definition. Given a constraint satisfaction problem $\mathcal{P} = \langle X, D_1, \ldots, D_k, R_1, \ldots, R_l \rangle$, the variable x in the scheme of $\mathcal{P}$, and a set $Y \subseteq D_x$, the restriction of $\mathcal{P}$ to Y, $\mathcal{P}|_{x,Y}$ is the constraint satisfaction problem with the same scheme, but with two changes:

1. The domain of the variable x is now Y (no other domains change).

2. Whenever x is in the scheme of relation R_j, $x = a_{j,m}$, then instead of R_j we have

$$R'_j = R_j \cap (D_{j,1} \times \ldots \times D_{j,m-1} \times Y \times D_{j,m+1} \times \ldots \times D_{j,r_j}).$$

We had to be careful in our definition because different variables may have same domains, and when we restrict, we restrict on one domain, namely that of the variable x. Next, we define, for a relation R with the name $x = x_{j,m}$ in its scheme, $\pi_x(R)$ is the set of those elements of $D_{x_m^j}$ which actually occur in tuples of R_j on the mth position. We then have the following fact which we will leave without the proof.

LEMMA 13.2

If $\mathcal{P}$ is a constraint satisfaction problem, R_j is one of its relations, $x = x_m^j$ is one of the variables in the scheme of R_j, and $Y = \pi_x(R_j)$, then the constraint satisfaction problems $\mathcal{P}$ and $\mathcal{P}|_{x,Y}$ have exactly the same solutions.

Lemma 13.2 tells us that elimination of unused values from the domain does not change the set of solutions. But this is a great news, for it tells us that if a relation R_j is finite then we can limit its domains, one after the other and limit all these domains to finite sets! Moreover, if every variable x_i has the property that *some* relation R_j with x_i in its scheme has a finite projection on the x_i coordinate then we can find a finite-domain CSP $\mathcal{P}'$ so that $\mathcal{P}$ and $\mathcal{P}'$ have the same solutions! But the latter problem can, in principle, be solved by a SAT solver. To formalize it, we call a CSP $\mathcal{P}$ locally finite if every variable x_i has the property that *some* relation R_j with x_i in the scheme of R_j has a finite projection on the x_i coordinate. We then have the following property.

PROPOSITION 13.3

If a CSP is locally finite then there is a finite CSP $\mathcal{P}'$ so that $\mathcal{P}$ and $\mathcal{P}'$ have the same solutions.

We get the following corollary.

COROLLARY 13.2

If $\mathcal{P}$ is a locally finite constraint satisfaction problem then there is a finite CNF $F_\mathcal{P}$ such that there is a bijection between solutions to $\mathcal{P}$ and satisfying valuations of $F_\mathcal{P}$.

TABLE 13.4: Relation R_φ for the formula $\varphi := (x \wedge y) \vee \neg z$

0	0	0
0	1	0
1	0	0
1	1	0
1	1	1

13.3 Satisfiability as constraint satisfaction

In Section 13.2 we saw that the finite-domain constraint satisfaction problems (and even some CSP problems with infinite domains) can be encoded as propositional theories so that there is a bijective correspondence between the solutions to a finite-domain CSP $\mathcal{P}$ and the satisfying valuations for the propositional encoding, the theory $G_\mathcal{P}$. In the process we introduced an intermediate encoding in the predicate calculus language. This last encoding, called $F_\mathcal{P}$ in Section 13.2, represented solutions to $\mathcal{P}$ as Herbrand models of $F_\mathcal{P}$.

In this section we look at the representation of propositional satisfiability as constraint satisfaction. The departure point is that if φ is a propositional formula in variables $x_{i_1}, \ldots, x_{i_k}$, then we can represent the truth table for φ as a certain k-ary relation. Here is how. Let us recall that the truth table $\mathcal{T}_\varphi$ is a table with 2^k rows and $k+1$ columns. The first k columns of $\mathcal{T}_\varphi$ represent all valuations. The last column lists the values of φ under these valuations. Now, let us prune from $\mathcal{T}_\varphi$ those rows where φ takes the value 0. That is, we leave only those rows where the last column takes the value 1. Then there is no reason to have that last column at all – it is constant 1 – so we can prune it too. We call the resulting table R_φ. Hence R_φ has precisely k columns and at most 2^k rows. Let us look at an example. Let φ be the formula $(x \wedge y) \vee \neg z$. The truth table for φ has 8 rows, 4 columns. But the table R_φ (shown in Table 13.4) has 5 rows and 3 columns. It should now be clear that the table R_φ lists as its rows all valuations that satisfy φ.

Now, given a finite theory T, we define a constraint satisfaction problem $\mathcal{C}_T$ as follows (the assumption that T is finite is immaterial, except that our constraint satisfaction problems had a finite number of tables, so we limit ourselves to finite T to avoid the problems with the definition). The variable names (i.e., the scheme) of $\mathcal{C}_T$ are the propositional variables of T, i.e., it is Var_T. For each propositional variable x, the domain D_x is the two-element set *Bool* (that is, $\{0,1\}$). The relations of $\mathcal{C}_T$ are R_φ with φ ranging over T. The scheme of each R_φ is Var_φ. Then we have the following fact, which is pretty obvious.

PROPOSITION 13.4
Satisfying valuations for T are precisely the solutions for $\mathcal{C}_T$.

Proof: Let the propositional variables of T be $\langle x_1, \ldots, x_k \rangle$. Let us list in every relation R_φ the variables in the order inherited from $\langle x_1, \ldots, x_k \rangle$ (this may require permuting columns in R_φ, but does not affect the fact that R_φ lists all valuations satisfying φ). The solutions of $\mathcal{C}_T$ are assignments to variables $x_1, \ldots, x_k$, that is, the valuations of Var_T. Now, if $\langle a_1, \ldots, a_k \rangle$ is a solution to $\mathcal{C}_T$, then let v be a valuation defined by $v(x_i) = a_i$, $1 \leq i \leq k$. Since $\langle a_1, \ldots, a_k \rangle$ is a solution for $\mathcal{C}_T$, for each $\varphi \in T$, with variables $\langle x_{i_1}, \ldots, x_{i_r} \rangle$, $\langle a_{i_1}, \ldots, a_{i_r} \rangle$ is a row in R_φ so $v|_{\{x_{i_1}, \ldots, x_{i_r}\}} \models \varphi$ and so $v \models \varphi$. Thus $v \models T$.

Conversely, if $v \models T$ then for every $\varphi \in T$, $v \models \varphi$. Assuming the variables of φ are $\langle x_{i_1}, \ldots, x_{i_r} \rangle$, $v|_{\{x_{i_1}, \ldots, x_{i_r}\}} \models \varphi$ (Localization Lemma, Proposition 2.3). Thus $\langle a_{i_1}, \ldots, a_{i_r} \rangle$ is a row in R_φ, that is, $\langle a_1, \ldots, a_k \rangle$ is a solution to $\mathcal{C}_T$. □

We will now use Proposition 13.4 to prove an NP-completeness result.

PROPOSITION 13.5

Existence of a solution for constraint satisfaction problems is an NP-complete problem.

Proof: Let C be the language consisting of those constraint satisfaction problems that possess a solution. We need to show C is NP-complete. First, we need to see that C is in the class NP. To this end we need to be able to check that the given assignment is a solution to a CSP problem. So, let $\mathcal{P} = \langle X, D_{x_1}, \ldots, D_{x_k}, R_1, \ldots, R_l \rangle$ be a constraint satisfaction problem and let $\langle a_1, \ldots, a_k \rangle$ be a solution to $\mathcal{P}$. We need to estimate the cost of checking that $\langle a_1, \ldots, a_k \rangle$ is a solution. But two things need to be tested. First, that for all i, $1 \leq i \leq k$, $a_i \in D_{x_i}$ and second, that for all j, $1 \leq j \leq l$, $\langle a_1^j, \ldots, a_{i_j}^j \rangle \in R_j$. Both tasks can be done in linear time in the size of $\mathcal{P}$.

For completeness, we need to reduce some known NP-complete problem to C. The known NP-complete problem that we reduce to C is the problem of satisfiability of sets of 3-clauses. We proved in Proposition 11.6 that the language consisting of sets of 3-clauses is NP-complete. Let F be a set of 3-clauses. The corresponding CSP problem $\mathcal{P}_F$ has variables that are propositional variables of F, all domains are *Bool*, relations are R_C for $C \in F$, and the schemes of relations consist of variables occurring in those clauses.

Each relation R_C has exactly seven rows (because a 3-clause is evaluated as *false* by exactly one row in its table). Therefore, we find that the size of the problem $\mathcal{P}_F$ is bound by $7 \cdot |F|$ where $|F|$ is the size of F. By Proposition 13.4 any algorithm solving $\mathcal{P}_F$ produces satisfying valuations for F. □

Proposition 13.4 tells us that, in principle, if we have a software program P that solves finite-domain constraint satisfaction problems then, given a finite propositional theory T, we can solve the satisfiability problem for T by using P. All we have to do is to compute, for each $\varphi \in T$ the table R_φ, construct (using these tables) the problem $\mathcal{C}_T$ and then use $\mathcal{C}_T$ as an input to P. But we should not put too much hope in this approach. The reason for this is that the size of the representation may grow exponentially in the size of T. Let us look at an example. Let T consist of a single clause $C = l_1 \vee \ldots \vee l_m$, where $l_1, \ldots, l_m$ are literals in variables $x_1, \ldots, x_m$, respectively. Then the problem $\mathcal{C}_T$ has just one table R_C, but this table has $2^m - 1$ rows. The only row missing in R_C is $\langle \bar{l}_1, \ldots, \bar{l}_m \rangle$. Generally, when the formulas of T have at most k variables, the size of the tables in $\mathcal{C}_T$ is at most 2^k. For instance, if all formulas in T have at most three variables, the tables of $\mathcal{C}_T$ have at most eight rows. Let us note that our observation on the size of the tables for clauses can be reversed; if a relation R_φ has exactly $2^k - 1$ rows, then φ is equivalent to a clause $\bar{l}_1 \vee \ldots \vee \bar{l}_k$ where $\langle l_1, \ldots, l_k \rangle$ is the only row missing in R_φ. But every relation with binary domains (i.e., with *Bool* as the domain of all variables) is missing a certain number of rows (in the extreme case it may be missing none, or all rows). The following proposition restates, in effect, the existence and uniqueness of the complete disjunctive normal form theorem (Proposition 3.17). We write $\bar{R}$ for $Bool^n \setminus R$.

PROPOSITION 13.6
Let R be a relation, $R \subseteq Bool^n$. Then $R = R_\varphi$ where $\varphi = \bigwedge_{\langle l_1,\ldots,l_n\rangle \in \bar{R}} \bar{l}_1 \vee \ldots \vee \bar{l}_n$.

Proof: Let us first observe that the following equality holds for all formulas φ and ψ that have the same propositional variables:

$$R_{\varphi \wedge \psi} = R_\varphi \cap R_\psi.$$

This is just a restatement of one of the clauses of the definition of satisfaction relation. Iterating this equality for conjunctions of more formulas we get

$$R_{\bigwedge_{\langle l_1,\ldots,l_n\rangle \in \bar{R}} \bar{l}_1 \vee \ldots \vee \bar{l}_n} = \bigcap_{\langle l_1,\ldots,l_n\rangle \in \bar{R}} R_{\bar{l}_1 \vee \ldots \vee \bar{l}_n} = \bigcap_{\langle l_1,\ldots,l_m\rangle \notin R} R_{\bar{l}_1 \vee \ldots \vee \bar{l}_n}$$

$$= \bigcap_{\langle l_1,\ldots,l_n\rangle \notin R} (Bool^n \setminus \{\langle l_1, \ldots, l_n\rangle\}).$$

Thus our assertion reduces to

$$R = \bigcap_{\langle l_1,\ldots,l_n\rangle \notin R} (Bool^n \setminus \{\langle l_1, \ldots, l_n\rangle\}).$$

But this is quite simple. If $\langle h_1, \ldots, h_n \rangle \in R$, then for all $\langle l_1, \ldots, l_n \rangle \in \bar{R}$, we have $\langle h_1, \ldots, h_n \rangle \neq \langle l_1, \ldots, l_n \rangle$. Thus whenever $\langle h_1, \ldots, h_n \rangle \in R$, $\langle h_1, \ldots, h_n \rangle \in Bool^n \setminus \{\langle l_1, \ldots, l_n\rangle\}$ for $\langle l_1, \ldots, l_n \rangle \in \bar{R}$. Therefore for $\langle h_1, \ldots, h_n \rangle \in R$, $\langle h_1, \ldots, h_n \rangle \in \bigcap_{\langle l_1,\ldots,l_n\rangle \notin R}(Bool^n \setminus \{\langle l_1, \ldots, l_n\rangle\})$.
For converse inclusion, if $\langle h_1, \ldots, h_n \rangle \in R$, $\langle h_1, \ldots, h_n \rangle \in \bigcap_{\langle l_1,\ldots,l_n\rangle \notin R}(Bool^n \setminus \{\langle l_1, \ldots, l_n\rangle\})$, then for each $\langle l_1, \ldots, l_n \rangle \in \bar{R}$, $\langle h_1, \ldots, h_n \rangle \neq \langle l_1, \ldots, l_n \rangle$. Thus $\langle h_1, \ldots, h_n \rangle \in R$. □

We also observe that the solutions of the constraint satisfaction problem $\mathcal{P}$ where $\mathcal{P} = \langle X, D_1, \ldots, D_m, R_1, \ldots, R_n, R \cap S \rangle$ are the same as those of $\mathcal{P}'$ where

$\mathcal{P}' = \langle X, D_1, \ldots, D_m, R_1, \ldots, R_n, R, S \rangle$ because when tables have the same scheme then their join coincides with their intersection. Therefore Proposition 13.6 has the following consequence (which amounts to transforming a constraint satisfaction problem to yet another satisfiability problem; this is totally obvious in the case of binary domains, but can be used for yet another encoding of CSP as SAT in the general case, too).

PROPOSITION 13.7
For every finite-domain constraint satisfaction problem $\mathcal{P}$ (and in particular for every constraint satisfaction problem with domains Bool) there is a finite-domain constraint satisfaction problem $\mathcal{P}'$ such that $\mathcal{P}$ and $\mathcal{P}'$ have precisely the same solutions, and every table in $\mathcal{P}'$ has exactly one row missing.

Proposition 13.7 is a normal form result, and does not contribute directly to techniques of solving CSP or SAT.

13.4 Polynomial cases of Boolean constraint satisfaction

Propositions 13.4 and 13.7 tell us that, in effect, logic provides the syntax for constraint satisfaction problems with binary (Boolean) domains. Those are usually called Boolean CSPs. To some extent, the clausal representation and relational representation are two extremes. Clauses provide a concise representation of tables with one missing row. In general, we need a large number of clauses to represent a CSP. At the other end, tabular (i.e., relational) representation may be also very large. Somewhat in the middle is the representation of Boolean CSPs by means of formulas of propositional logic (we know, by Proposition 13.6, that every Boolean table possesses such representation, but one table may have many equivalent representations). We have techniques for processing representation by means of formulas, for instance, the tableaux method, studied in Section 8.3. Other techniques are also available.

When the tables are given explicitly, there are six classes of Boolean constraint satisfaction problems which allow for feasible processing. Those are the so-called Schaefer classes. We will discuss them now.

Let $\mathcal{F}$ be a class of formulas. A relation $R \subseteq Bool^n$ is called $\mathcal{F}$-*relation* if there exist formulas $\varphi_1, \ldots, \varphi_n$ all in $\mathcal{F}$ such that $R = R_{\varphi_1 \wedge \ldots \wedge \varphi_n}$. Given the truth table of the formula φ, its *positive part* consists of the listing of valuations that satisfy φ. Thus the $\mathcal{F}$-relations are positive parts of truth tables of formulas that are conjunctions of formulas from $\mathcal{F}$. So, for instance, every relation R is an $\mathcal{F}$-relation where $\mathcal{F}$ is a set of clauses. Let us look at another example, the table shown in the Table 13.5. This table is a Krom-relation, i.e., definable by a formula which is a conjunction of

TABLE 13.5: Relation R_φ for the formula $\varphi := (x \vee y) \wedge (\bar{x} \vee z) \wedge (\bar{y} \vee z)$

0	1	1
1	0	1
1	1	1

2-clauses.

Below, we investigate six classes of relations. We call them *Schaefer classes* of Boolean relations.

1. **1**-satisfiable, that is, satisfiable by means of a valuation that is constant, with the value 1 at every variable
2. **0**-satisfiable, that is, satisfiable by means of a valuation that is constant, with the value 0 at every variable
3. Krom relations, i.e., positive parts of truth tables of 2-clauses
4. affine relations, i.e., positive parts of truth tables of affine equations
5. Horn relations, i.e., positive parts of truth tables of Horn formulas
6. dual Horn relations, i.e., positive parts of truth tables of dual Horn formulas.

Before we prove the next result, so-called "easy part of Schaefer theorem" [Sch78], we will focus on the Horn constraint satisfaction problems. Let us assume that all the domains of variables are *Bool* (so we do not need to list them.) Now, let $\mathcal{P} = \langle Var, R_1, \ldots, R_m \rangle$ be a Boolean constraint satisfaction problem where all the relations R_i, $1 \leq i \leq m$ are Horn, that is, for each R_i, $1 \leq i \leq m$, there is a formula H_i, which is a conjunction of Horn clauses such that R_i is the positive part of the truth table for H_i. In other words, the rows of R_i are precisely valuations satisfying H_i. We can think about R_i as a collection of subsets of X_i where X_i is the scheme of R_i. Then that family is closed under intersections of non-empty families (Theorem 9.1).

We have the following lemma.

LEMMA 13.3

Let $\mathcal{P} = \langle R_1, \ldots, R_m \rangle$ be a Horn CSP. If v_1, v_2 are solutions to $\mathcal{P}$, then so is their bitwise conjunction, $v_1 \wedge v_2$.

Proof: Let R_i be any of the relations of $\mathcal{P}$ and let X_i be its scheme. For any assignment $v : Var \to Bool$ and $Y \subseteq Var$, $v|_Y$ is the restriction of v to Y. Then, clearly,

$$(v_1 \wedge v_2)|_{X_i} = v_1|_{X_i} \wedge v_2|_{X_i}.$$

Now, $v_1|_{X_i} \wedge v_2|_{X_i}$ belongs to R_i as R_i is closed under bitwise $\wedge$. As R_i is arbitrary, $v_1 \wedge v_2$ is a solution to $\mathcal{P}$. □

TABLE 13.6: Algorithm computing a solution to Horn CSP $\mathcal{P}$, if one exists

---

Algorithm HornCSP
Input: a Horn Boolean constraint satisfaction problem $\mathcal{P} = \langle R_1, \ldots, R_m \rangle$. The scheme of $\mathcal{P}$ is X, the schemes of R_i are X_i, $1 \leq i \leq m$. $\mathcal{M}_i$ is the family of subsets of X_i representing R_i, that is, R_i lists as rows characteristic functions of sets in $\mathcal{M}_i$.
$M := \emptyset$
$\mathcal{T} :=$ The set of those R_i, $1 \leq i \leq m$, that $M \cap X_i \notin \mathcal{M}_i$
    while ($\mathcal{T} \neq \emptyset$)
    {
        select (one constraint R_j from $\mathcal{T}$);
        if (there exists $U \in \mathcal{M}_j$ such that $M \cap X_j \subseteq U$)
        then
        {
        $S :=$ the least $V \in \mathcal{M}_j$ such that $M \cap X_j \subseteq V$;
            $M := M \cup S$
            $\mathcal{T} :=$ the set of R_i so that $M \cap X_i \notin \mathcal{M}_i$;
        }
        else
        {
            return('No solution exists')
        }
return (M)

As we are dealing with finite sets, we have the following corollary which we state in a set-form.

COROLLARY 13.3

Let $\mathcal{P} = \langle R_1, \ldots, R_m \rangle$ be a Horn CSP. Then the collection $\mathcal{M}$ of solutions to $\mathcal{P}$ is closed under non-empty intersections.

We get a following corollary.

COROLLARY 13.4

If a Horn Boolean CSP possesses a solution, then it possesses a least solution.

We will now give an algorithm to compute a solution of a Horn Boolean CSP, and then will prove its correctness. It will be convenient to represent a table R by a set of subsets of propositional variables. Because we think in terms of sets, set-theoretic notation will be used.

PROPOSITION 13.8

The algorithm HornCSP returns a solution to a Horn CSP $\mathcal{P}$ if and only if such solution exists.

Proof. First, let us assume that the contents of the variable M has been returned. This will occur when the contents of the variable $\mathcal{T}$ is empty, i.e., if M satisfies all constraints $R_1, \ldots, R_m$. Thus, if M is returned, it is a solution.

Now, we need to show that if a solution exists then one will be returned. First, let us observe that the set of unsatisfied constraints (relations R_i which are not satisfied) does not change monotonically as we iterate the **while** loop. In particular $\mathcal{T}$ may incorporate constraints that previously were satisfied but because M changed, no longer are satisfied. However, the variable M grows at every step. The reason is that in each iteration of the loop the set $S \setminus M \neq \emptyset$ and so M, becoming $M \cup S$, is strictly bigger. This means that our algorithm will terminate after at most r iterations of the **while** loop, where r is the size of X.

Let us assume that a solution to $\mathcal{P}$ exists. Then, by Corollary 13.4, there exists a least solution for $\mathcal{P}$. Let us call that solution M_0. We claim that for each iteration of the **while** loop, the contents of the variable M is included in M_0. We prove our claim by induction on the number of iterations of the loop. The claim is certainly true initially, since then the content of M is the empty set (this is how M is initialized, see Table 13.6). Let us assume that before some iteration i the current content of M is included in M_0. We need to show that the same happens after the ith iteration as well. If all the constraints are satisfied by M then M is returned, and our claim is true. So, let R_j be any constraint not satisfied by M. Then, by inductive assumption, since $M \subseteq M_0$, $M \cap X_j \subseteq M_0 \cap X_j$. But M_0 is a solution. Therefore $M_0 \cap X_j$ is a solution to R_j (i.e., the characteristic function of $M_0 \cap X_j$ is a row in R_j). Therefore there exists a solution to R_j that contains $M \cap X_j$ and the least such solution S is included in $M_0 \cap X_j$. So the set S considered within the **while** loop is included in M_0 and hence the new value of M is also included in M_0.

Let us finally look at the iteration where M does not grow. There must be such an iteration because the size of M is bound by the size of M_0. Then the only possibility is that all the constraints $R_1, \ldots, R_m$ are satisfied. Thus M is a solution. Since M is included in M_0, $M = M_0$ and it is what will be returned. This completes half of our argument; if a solution exists, it will be found and in fact the least solution will be returned.

If there is no solution, we will find ourselves in the situation where M will not grow, but there will be unsatisfied constraint. Then our algorithm will return the string 'No solution exists,' as it should. $\square$

We need to check the feasibility of the algorithm **HornCSP**. We established that there can be at most r iterations of the main loop, where $r = |X|$. Clearly $r \leq |\mathcal{P}|$. So let us see what happens within the main loop. There are several tasks that need to be performed. First, we need to establish the current content of the variable $\mathcal{T}$, that is, the set of currently unsatisfied constraints. This, clearly, can be done in polynomial time in the size of $\mathcal{P}$. Then, if we established that $\mathcal{T}$ is non-empty, we choose one unsatisfied constraint R_i and need to check if there is a way to satisfy R_i by means of some superset of $M \cap X_i$. This requires testing if $M \cap X_i$ is included in some set coded by a row of R_i. If this is the case, we need to find a smallest such set. All these tasks require polynomial number of operations. Finally, if a set U has been found, we need to add its elements to the current M. Certainly, this can be done in polynomial time. We conclude that **HornCSP** runs in polynomial time. As mentioned above a version of this argument can be used to establish that the dual Horn CSPs can also be solved in polynomial time.

We can now prove the promised "easier part" of Schaefer theorem.

PROPOSITION 13.9

If $\mathcal{F}$ is one of classes (1)–(6) above and a Boolean CSP problem $\mathcal{P}$ consists of relations that are all $\mathcal{F}$-relations, then the problem $\mathcal{P}$ can be solved in polynomial time.

Proof: The result is entirely obvious if all relations of $\mathcal{P}$ are **1**-satisfiable; for the valuation taking the value 1 on all variables is then a solution. The case of **0**-satisfiable tables is similar.

Here is how we proceed in the case of Krom relations. Given a table R_i of $\mathcal{F}$, we can find a set of 2-clauses, K_i such that R_i is the positive part of the truth table for R_i in polynomial time (Proposition 9.28). Indeed, the positive part of the truth table is, as noted above, nothing else but table representation of the set of satisfying valuations. So now we compute in polynomial time sets K_i of 2-clauses such that R_i is the positive part of the truth table for the formula $\bigwedge K_i$. Now, let us consider the set of 2-clauses K, where $K = \bigcup_i K_i$. Then by considering the fact that there was a single polynomial $f(\cdot)$ such that the size of K_i, $|K_i|$, is $f(|R_i|)$ we find that there is a single polynomial $g(\cdot)$ such that $|K|$ is bound by $g(|R_1| + \ldots + |R_m|)$, in particular $g(|\mathcal{F}|)$. Now, we can use a SAT solver algorithm to solve K; the satisfying valuations for K are precisely solutions to the problem $\mathcal{F}$ (Proposition 13.4). Thus we can test if $\mathcal{F}$ has a solution in polynomial time.

A very similar approach is used for the affine case. Here, instead of Proposition 9.28, we use Proposition 9.37. For each table R_i we identify an affine theory A_i so that rows of R_i are solutions to the system of linear equations A_i. Then we set $A = \bigcup A_i$. Clearly, solutions to the affine problem $\mathcal{P}$ are satisfying valuations for A which we can find (if any exists) using Gaussian elimination (see Section 9.7). Moreover, the size of A is polynomial in the size of $\mathcal{P}$.

Now, we have to handle the remaining two cases: Horn and dual Horn. In Proposition 13.8 we proved that the algorithm **HornCSP** solves the Horn case correctly. Then, in the discussion following Proposition 13.8 we showed that the algorithm **HornCSP** runs in polynomial time. This takes care of the Horn case. The other, dual Horn, is very similar, except that when in the Horn case we use the fact that the families of models of Horn theories are closed under intersections (here we represent valuations by sets of propositional variables), the dual Horn case is characterized by families closed under unions. □

13.5 Schaefer dichotomy theorem for Boolean constraint satisfaction

In Proposition 13.9 we proved that if $\mathcal{F}$ is included in one of six Schaefer classes, then the constraint satisfaction problem $CSP(\mathcal{F})$ is polynomial. That is, there is an algorithm running in polynomial time solving every instance of $CSP(\mathcal{F})$. Now, our goal is to prove a remarkable dichotomy theorem due to T. Schaefer, namely that if $\mathcal{F}$ is not included in any single one of Schaefer classes then the corresponding constraint satisfaction problem is NP-complete.

There are various proofs of the Schaefer Theorem, including a direct one. But the one we will give here, based on the argument of H. Chen [Ch09] (but grounded in work of others, see below), will be based on universal algebra techniques. Of course, this is not a book on universal algebra, and the argument we give below will be (almost) self-contained. But the point of view will have the algebraic flavor. The 'grand plan' of the argument uses a number of facts interesting on their own merit. We will look more deeply into the structure theory of Boolean functions. The first major step is to look at a kind of definability associated with the constraint satisfaction. This definability is, actually, a simplified version of SQL database query language (with much restricted WHERE clauses). The class of relations definable out of a class Γ of Boolean relations, called *primitively-positively* definable (pp-definable, for short), turns out to be closely related to the *polymorphisms* of Boolean relations.[3] We introduce two related classes of objects. One, on the side of Boolean functions: given a collection Γ of relations, we consider the class of all functions that are polymorphisms of Γ. This class is denoted $Pol(\Gamma)$. On the side of Boolean relations, given a collection $\mathcal{F}$ of Boolean functions, we look at all relations for which a collection $\mathcal{F}$ of functions are polymorphisms. This class of relations is denoted by $Inv(\mathcal{F})$.

Denoting by $\langle \Gamma \rangle$ the collection of pp-definable relations definable from Γ we cite without proof the following classical result due to D. Geiger, namely that $\langle \Gamma \rangle = Inv(Pol(\Gamma))$.[4] On the side of collections of Boolean functions we look at *clones*, families of functions containing projections and closed under compositions, identification of variables, permutations of variables and adding dummy variables. The reason why we do so is that the collection $Pol(\Gamma)$ is a clone. Then we come to the main point of the argument, a property of clones due to I. Rosenberg. The basic polymorphisms corresponding to classes of Horn, dual Horn, affine, and Krom classes of relations appear in that characterization. The argument we will show digs deeply into the very nature of Boolean functions. Once that characterization is obtained, the argument becomes easy, although still not trivial.

---

[3] Again, these are *not* polymorphisms as considered in object-oriented programming.
[4] This is the result in universal algebra, and the only point in the proof of the Schaefer theorem which is not proved here.

To realize the program outlined above, we need to be a bit formal. Let Γ be a collection of Boolean relations. A *constraint* over Γ is a table R from Γ together with an assignment of propositional variables to its columns. There is a delicate aspect to such constraint. Namely, we can assign to two (or more) columns the same variable. This means that we select out of the rows of R only those rows where the corresponding columns hold equal values. The reader familiar with SQL will note that this type of a constraint is just a query which is a selection (according to some equality condition). We will write $R(x_{i_1}, \ldots, x_{i_k})$ for such constraint I. Let Var be the set of propositional variables extending the set of those actually appearing as the names of columns. Then a valuation $v : Var \to Bool$ satisfies I if $\langle v(x_{i_1}), \ldots, v(x_{i_k}) \rangle \in R$. Now, given such constraint language (that is, a collection of relations) Γ, an *instance* of Γ is a finite set of constraints over Γ. Then $CSP(\Gamma)$ is the following decision problem: Given an instance S of Γ is there an assignment v that satisfies all the constraints in S?

We will now introduce the notion of primitive-positive definability out of a set of relations Γ. Here is what we do. By $x_i = x_j$ we mean a table with two columns labeled x_i and x_j, respectively, and two rows: $(0,0)$ and $(1,1)$. Such equality constraint is satisfied by a valuation v if $v(x_i) = v(x_k)$. Now, a relation S is *primitively-positively* definable from a set of tables Γ if S is a set of tuples satisfying the formula

$$\exists_{x_{i_1}, \ldots, x_{i_j}} F,$$

where F is a conjunction of constraints over Γ and of equality constraints.

For instance, if we have in Γ two relations R_1 and R_2 with, respectively, three and two arguments, then relation S with the definition

$$S(x_1, x_2) \equiv \exists_{x_3, x_4} (R(x_1, x_2, x_3) \wedge R_2(x_2, x_4))$$

is primitively-positively definable from Γ.

Let us observe that with this definition the problem of identical names for different columns disappears. For instance, relation (of two variables) $R_1(x_1, x_2, x_1)$ is primitively-positively definable by

$$\exists_{x_3} (R_1(x_1, x_2, x_3) \wedge x_1 = x_3).$$

We will call such elimination of repeated variable names *disambiguation*.

We define $\langle \Gamma \rangle$ as the collection of all relations (tables) primitively-positively (pp-) definable out of Γ.

A couple of simple observations which we will ask the reader to prove in the Section 13.6 Exercises, follow. For instance, whenever Γ contains at least one non-empty relation then $\langle \Gamma \rangle$ is infinite. Also the operator $\langle \cdot \rangle$ is a monotone idempotent operator. We now have the following fact due to P. Jeavons.

PROPOSITION 13.10
Let Γ, Γ' be two finite sets of Boolean relations such that $\Gamma' \subseteq \langle \Gamma \rangle$ then $CSP(\Gamma')$ polynomially reduces to $CSP(\Gamma)$.

Proof: Given an instance $\mathcal{S}'$ of $CSP(\Gamma')$ we need to find an instance $\mathcal{S}$ of $CSP(\Gamma)$ such that

1. $\mathcal{S}$ is (uniformly) computed out of $\mathcal{S}'$, and
2. $\mathcal{S}'$ has a solution if and only if $\mathcal{S}$ has a solution.

To this end, for each constraint R'_j in Γ' we fix a representation

$$R'_j(x_1, \ldots, x_{n_j}) \equiv \exists_{x_{n_j+1}, \ldots, x_{n_j+s_j}} C_j(x_1, \ldots, x_{n_j+s_j}).$$

Without the loss of generality we can choose existentially quantified variables in the definitions so that each such variable occurs in only one definition. Moreover, using disambiguation we can assume that each term in each conjunction C_j has different column names. We use equality constraints to enforce the equalities among column names. So now, the instance $\mathcal{S}'$ is nothing more than an existentially quantified conjunction of terms such that each of these terms itself is an existentially quantified conjunction of relations from Γ and equality constraints. Let us call the matrix of that formula C'. Thus we deal with a formula $\exists_{\vec{x}} C'$. With the disambiguation effort described above, the prenex normal form of this formula has the form

$$\exists_{\vec{x}, \vec{x}'} C''(\vec{x}, \vec{x}', \vec{x}'').$$

The variables $\vec{x}$ are existentially quantified variables of the instance $\mathcal{S}'$. The variables $\vec{x}'$ come from the conjuncts of the formula C'. Finally, variables $\vec{x}''$ are the variables of $\mathcal{S}'$ that are not quantified. So now, let us drop the quantifiers of the prenex of the formula $\exists_{\vec{x}, \vec{x}'} C''(\vec{x}, \vec{x}', \vec{x}'')$. Then, dealing with the quantifier-free formula resulting from dropping the quantifier, let us drop all conjunctions. Let us pause to see what is the result of this transformation. We get an instance $\mathcal{S}$ of $CSP(\Gamma)$. It should be clear that this new instance has a solution if and only if $\mathcal{S}'$ has a solution. The size of $\mathcal{S}$ is polynomial (in fact linear) in the size of $\mathcal{S}'$. There is still a minor problem: $\mathcal{S}$ contains equality constraints. But we can eliminate these constraints as follows: equality constraints generate an equivalence relation in the variables of Var, namely the least equivalence relation generated by these equalities. In each coset we select one variable and substitute it in all occurrences of variables in that coset. This does not change satisfiability, it does not increase size of the problem, and it eliminates equality constraints. In this way, we get an instance of $\mathcal{S}$ with the desired properties. □

We then get the following corollary.

COROLLARY 13.5
Let Γ be a finite collection of Boolean relations. Then $CSP(\Gamma)$ is solvable in polynomial time if and only if for all finite $\Gamma' \subseteq \langle \Gamma \rangle$, $CSP(\Gamma')$ is solvable in polynomial time.

Proof: The implication $\Rightarrow$ follows immediately from Proposition 13.10. The implication $\Leftarrow$ follows from the fact that $\Gamma \subseteq \langle \Gamma \rangle$. □

We will now discuss *polymorphisms*. In our studies of models of Horn theories in Chapter 9 we discussed "bitwise conjunction." This was an operation in $Bool^n$. Like ordinary Boolean conjunction it was a binary operation. But it computed the result coordinatewise. In our characterization of other classes of theories we encountered other polymorphisms: there was the "bitwise disjunction," "bitwise majority," "bitwise sum-of-three," and also bitwise constant functions. Now, we introduce polymorphisms in full generality.

Let $f : Bool^m \rightarrow Bool$ be an m-ary operation. We are lifting the operation f to $Bool^k$ as follows. Let us take m tuples in $Bool^k$,

$$(t_{11}, \ldots t_{1k}), \ldots, (t_{m1}, \ldots, t_{mk}).$$

We visualize this collection as a matrix

$$\begin{pmatrix} t_{11} & \ldots & t_{1k} \\ & \ldots & \\ t_{m1} & \ldots & t_{mk} \end{pmatrix}.$$

Each *column* of this matrix is a Boolean vector of length m. We can execute the operation f on this vector. Since we execute f exactly k times (once for each column), we get as a result a vector of length k. This is the result of applying f on rows $(t_{11}, \ldots t_{1k}), \ldots, (t_{m1}, \ldots, t_{mk})$. We call this resulting vector of length k the *bitwise execution* of f (on rows $r_1, \ldots, r_m$). Now, we say that an m-ary operation f is a *polymorphism* for a Boolean relation R if for any choice of m rows $r_1, \ldots, r_m$, of table R the result of bitwise execution of f on these rows is again a row in R. Let us look at a couple of examples (in Chapter 9 we had some examples already, but by now they must be forgotten).

Example 13.3

1. The operation $\wedge$ is a polymorphism for

$$\begin{pmatrix} 0 & 1 & 0 & 0 & 1 \\ 1 & 0 & 1 & 0 & 1 \\ 0 & 0 & 0 & 0 & 1 \end{pmatrix}.$$

2. The 7 row table

$$\begin{pmatrix} 0 & 0 & 1 \\ 0 & 1 & 0 \\ 0 & 1 & 1 \\ 1 & 0 & 0 \\ 1 & 0 & 1 \\ 1 & 1 & 0 \\ 1 & 1 & 1 \end{pmatrix}$$

(which is, of course, $Mod(p \vee q \vee r)$) has $\vee$ as a polymorphism but $\wedge$ is *not* its polymorphism, nor is the constant function 0 its polymorphism (but the constant function 1 is its polymorphism).

Let us recall that in Chapter 9 we proved six facts: that the constant function 1 is a polymorphism of a Boolean relation R if and only if R contains a row of 1s, that the constant function 0 is a polymorphism of R if and only if R contains a row of 0s, that the conjunction function is a polymorphism of R if and only if R is the set of models of a Horn theory, that the disjunction function is a polymorphism of R if and only if R is the set of models of a dual Horn theory, that the ternary sum operation (sum_3) is a polymorphism of R if and only if R is the set of models of an affine theory, and that the ternary operation maj is a polymorphism of R if and only if R is the set of models of a theory consisting of 2-clauses.

With these examples in mind it is natural (as the researchers in universal algebra do) to introduce two concepts. First, given a collection Γ of Boolean tables, we define

$$Pol(\Gamma) = \{f : \forall_{R \in \Gamma} f \text{ is a polymorphism of } R\}.$$

Next, we say that R is *invariant* under f if f is a polymorphism of R. Then we define, for a collection F of Boolean functions,

$$Inv(F) = \{R : \forall_{f \in F} R \text{ is invariant under } f\}.$$

Given a set of Boolean functions F, $[F]$ is the least clone containing F as a subset. Clearly, $[F]$ is the intersection of all clones containing F.

Here is a fundamental result due to D. Geiger which is, actually, more general. We state it version in Boolean case; it is valid without its limitation.

THEOREM 13.1 (Geiger theorem)

1. For every finite set Γ of Boolean relations

$$\langle \Gamma \rangle = Inv(Pol(\Gamma)).$$

2. For every family of Boolean functions F

$$[F] = Pol(Inv(F)).$$

The proof of the Geiger theorem, while can be done using SAT, belongs to universal algebra, not to our interests. We will ask the reader to prove an easier half of it in one of exercises. Let us also observe that Theorem 13.1 can be proved in generality (we do not have to limit it to the Boolean case) but in this book it is all we need. There is plenty to learn about relationships between Boolean tables and Boolean operations. There are thick tomes covering that theory and, more generally, theory of clones over algebras and relations over these algebras (for instance, [La06]) and we refer the reader to these books.

We now use Theorem 13.1 to get a useful corollary, which ties polymorphisms with the constraint satisfaction.

COROLLARY 13.6
Let Γ, Γ' be finite collections of Boolean relations. If $Pol(\Gamma) \subseteq Pol(\Gamma')$ then $\Gamma' \subseteq \langle \Gamma \rangle$ and therefore $CSP(\Gamma')$ polynomially reduces to $CSP(\Gamma)$.

Proof: If $Pol(\Gamma) \subseteq Pol(\Gamma')$ then every relation invariant under $Pol(\Gamma')$ is invariant under $Pol(\Gamma)$. Thus we have

$$\Gamma' \subseteq \langle \Gamma' \rangle = Inv(Pol(\Gamma')) \subseteq Inv(Pol(\Gamma)) = \langle \Gamma \rangle.$$

Therefore $\Gamma' \subseteq \langle \Gamma \rangle$ and the conclusion follows from Proposition 13.10. □

The connection of clones to constraint satisfaction is given by the following observation.

PROPOSITION 13.11
For every family Γ of Boolean relations, $Pol(\Gamma)$ is a clone. Conversely, every clone is the family of polymorphisms for a suitably chosen set of tables.

Proof: It is easy to verify the first part. The second part follows from Geiger theorem $[F] = F$. Namely, given a clone F, F is the family of all polymorphisms for $Inv(F)$. □

Let us recall that our goal in this section is to prove the Schaefer dichotomy theorem. Here is what we do now. First, we will formulate it (only half of it was formulated above.)

THEOREM 13.2
Let Γ be a collection of Boolean tables. If Γ is included in one of the Schaefer classes, then $CSP(\Gamma)$ is solvable in polynomial time. Otherwise, it is NP-complete.

We already proved the "easy half" of Theorem 13.2 in Proposition 13.9. Now, we will need to prove the "hard part."

In view of the connections between collections of Boolean relations and collections of Boolean functions, the Schaefer theorem can be expressed as a fact about polymorphisms. We do this now.

THEOREM 13.3 (Schaefer Theorem)
Let Γ be a collection of Boolean tables. If one of conditions (1)–(6) below holds, then $CSP(\Gamma)$ is solvable in polynomial time. Otherwise, it is NP-complete.

1. Constant function 1 is a polymorphism of all $R \in \Gamma$.
2. Constant function 0 is a polymorphism of all $R \in \Gamma$.
3. Binary function $\wedge$ is a polymorphism of all $R \in \Gamma$.
4. Binary function $\vee$ is a polymorphism of all $R \in \Gamma$.
5. Ternary function maj is a polymorphism of all $R \in \Gamma$.
6. Ternary function sum_3 is a polymorphism of all $R \in \Gamma$.

The proof of Theorem 13.3 will occupy us until the end of this chapter. Of course, we only prove the NP-completeness result. The rest was done in Proposition 13.9.
We need a couple of definitions, all pertaining to Boolean functions. A function $f : Bool^n \to Bool$ is *essentially unary* if there is a coordinate i, $1 \leq i \leq n$, and a unary function g such that $f(x_1, \ldots, x_n) = g(x_i)$, that is, f depends on at most one variable, x_i. Let us observe that the coordinate i is not necessarily unique. There are important essentially unary functions called *projections*. A projection is a function $\pi_j^m(x_1, \ldots, x_m)$ identically equal to x_i. Of course, projections are essentially unary functions. Recall that a *clone* is a collection of Boolean functions that contains all projections and is closed under composition (i.e., substitutions), permutations of variables, identifications of variables, and adding dummy variables. We stated above that for any collection Γ of Boolean relations the set $Pol(\Gamma)$ is a clone (we will ask the reader to prove it as an exercise).
Given a function $f : Bool^n \to Bool$, a unary function $\hat{f}$ is defined by $\hat{f}(x) = f(x, \ldots, x)$. Let us observe that $\hat{f}$ draws the information from two constant inputs to f, and that it is always one of four possible functions: constant 1, constant 0, x, and $\bar{x}$. Next, we say that a Boolean function f is *idempotent* if $\hat{f}$ is the identity function. Finally, we say that a Boolean function f *acts as a permutation* if $\hat{f}$ is non-constant (that is, it is an identity or it is the negation).
We will use the following main technical result, due to I. Rosenberg, in our proof of the Schaefer theorem.

THEOREM 13.4 (Rosenberg Theorem)
Let C be a clone. Then either C contains only essentially unary functions, or C contains at least one of functions $\wedge, \vee, maj,$ or sum_3.

Proof: Let us observe that if we had enough of information about Post lattice (the lattice of clones) we could just look at the lattice and see that Theorem 13.4 is true. But we did not study Post lattice (although it is present in many places of this book), and so we prove the assertion directly. Besides, the proof is esthetically appealing.
Let C be a clone, and let f be a non-essentially unary function, $f \in C$. Thus, the function f is an n-ary function, $n > 0$. There will be several cases to investigate.
Case 1: $\hat{f}$ is constant, $\hat{f}(x) \equiv 0$. Since f is not essentially unary there must be a Boolean vector $(a_1, \ldots, a_n)$ such that $f(a_1, \ldots, a_n) \neq 0$. The reason is that otherwise, if f is constantly 0 then every variable of f is witness to the fact that f is

essentially unary. So, we have a vector $(a_1, \ldots, a_n)$ such that $f(a_1, \ldots, a_n) = 1$. At least one of a_is must be 0 and at least one of them must be 1 (because otherwise our assumption that $\hat{f}$ is identically 0 is violated). So now, let us define a binary function g by setting

$$g(x_1, x_0) = f(x_{a_1}, \ldots, x_{a_n}).$$

The function g belongs to the clone $\mathcal{C}$, because all we do is substitute projections π_0^2 and π_1^2 into the function f. Now, let us look at g more closely. We find that

$$g(0,0) = f(0, \ldots, 0) = 0.$$

Indeed, what do we do? We substitute in f variables x_{a_1} for x_1, then x_{a_2} for x_2, etc., and then we put for each occurrence of x_0 and for each occurrence of x_1 the value 0. Thus we substitute the value 0 in every variable of f, and this gives us the value 0. In a very similar manner we find that $g(1,1)$ is also 0 (remember $f(1, \ldots, 1) = 0$). Next let us look at $g(1,0)$. Here we first substitute x_{a_i} for x_i and then we substitute 1 for x_1, and 0 for x_0. What do we get? $f(a_1, \ldots, a_n)$. The latter value is 1, thus $g(1,0) = 1$.

Now, there are two subcases:

Subcase 1.1: $g(0,1) = 1$. Then, clearly, g is the Boolean addition (+ function). But then $sum_3(x, y, z) \equiv g(x, g(y, z))$. The latter function is in $\mathcal{C}$, because $\mathcal{C}$ is closed under substitution.

Subcase 1.2: $g(0,1) = 0$. Then, clearly, g is one of the conjunction functions (because its table has three zeros, and a single one). Which one? It is easy to see that $g(x, y) \equiv x \wedge \neg y$. But then

$$g(x, g(x, y)) = x \wedge (\neg(x \wedge \neg y)) = x \wedge (\neg x \vee y) = x \wedge y.$$

Consequently, in this subcase $\wedge$ belongs to the clone $\mathcal{C}$.

Case 2: $\hat{f}$ is constant, $\hat{f}(x) \equiv 1$. The argument is similar to Case 1, except that in the first subcase we find that $g(x, y)$ is actually $x + y + 1$. It is obvious to see that sum_3 can be defined from this function. There is also a second subcase, where we find that $\vee$ belongs to the clone $\mathcal{C}$.

Case 3: $\hat{f}$ is not constant. We first show that our clone $\mathcal{C}$ contains an idempotent, not essentially unary function. Indeed, if $\hat{f}$ is not constant, then either $\hat{f}$ is identity, or it is negation. In the first case, when $\hat{f}$ is identity, f itself is an idempotent, non-essentially unary function in $\mathcal{C}$.

Thus we are left with the case when $\hat{f}$ is the negation (complement). Here is what we do. We define an auxiliary function g setting

$$g(x_1, \ldots, x_n) = \hat{f}(f(x_1, \ldots, x_n)).$$

Thus

$$g(x_1, \ldots, x_n) = \overline{f(x_1, \ldots, x_n)}.$$

Since $\mathcal{C}$ is a clone, and both f and $\hat{f}$ are in $\mathcal{C}$, $g \in \mathcal{C}$. But now:

$$\hat{g}(x) = g(x, \ldots, x) = \hat{f}(f(x, \ldots, x)) = \hat{\hat{x}} = \neg\neg x = x.$$

TABLE 13.7: Cases for $m = 3$

$g(x,x,y)$	$g(x,y,x)$	$g(y,x,x)$
x	x	x
x	x	y
x	y	x
x	y	y
y	x	x
y	x	y
y	y	x
y	y	y

Moreover, g is not essentially unary because, clearly,

$$g(x_1, \ldots, x_n) = \neg f(x_1, \ldots, x_n),$$

thus

$$f(x_1, \ldots, x_n) = \neg g(x_1, \ldots, x_n),$$

and if $g(x_1, \ldots, x_n) = h(x_i)$ then $f(x_1, \ldots, x_n) = \neg h(x_i)$ so f would be essentially unary contradicting the choice of f.

So what do we know at this point? We do know that $\mathcal{C}$ contains a function that is, non-essentially unary and idempotent. Let us choose such function g. What is the arity m of such function g? Certainly it is not 1.

Subcase 3.1 $m = 2$.
Could it be that $g(0,1) \neq g(1,0)$? If this was the case then as g is idempotent, $g(0,0) = 0$, and $g(1,1) = 1$. So, when $g(0,1) = 0$, then $g(1,0) = 1$. But then, as is easily seen, g is the projection on the first coordinate. In the case when $g(0,1) = 1$, and $g(1,0) = 0$, g is the projection on the second coordinate. Thus in either case g is essentially unary contradicting the choice of g.
So it must be the case that $g(0,1) = g(1,0)$. There are two subcases possible. When $g(0,1) = g(1,0) = 0$, then the function g is $\wedge$ (just check the table of g). And if $g(0,1) = g(1,0) = 1$ then g is $\vee$. Thus we proved that in the case when $m = 2$ the clone $\mathcal{C}$ must contain either $\wedge$ or $\vee$.

Subcase 3.2: $m = 3$.
Since the minimal arity function that is, idempotent and not essentially unary is 3 it follows that any identification of two variables in g must create a function that is essentially unary (let us observe that identification of variables preserves idempotence). There are three ways in which we can identify two variables in a three-argument function. With appropriate renaming we have $g(x,x,y)$, $g(x,y,x)$, and $g(y,x,x)$. As we are in the case when $m = 3$, it must be the case that each of these three functions is essentially unary. There are eight possible cases. We list them in Table 13.7.

We are sure that the reader would not survive if we computed here what g is for all eight cases (but we did, and the reader who does not trust us can get the calculations by mail). We will do the detailed calculation for the first, second, and fourth rows. The remaining ones will be explicated but not calculated explicitly.

First row. We will show that in this case the function g is maj. We use the fact that $Bool$ has just two elements (kind of obvious, right?).

$g(0,0,0) = g(x,x,y)\begin{pmatrix} x & y \\ 0 & 0 \end{pmatrix} = x\begin{pmatrix} x \\ 0 \end{pmatrix} = 0$

$g(0,0,1) = g(x,x,y)\begin{pmatrix} x & y \\ 0 & 1 \end{pmatrix} = x\begin{pmatrix} x \\ 0 \end{pmatrix} = 0$

$g(0,1,0) = g(x,y,x)\begin{pmatrix} x & y \\ 0 & 1 \end{pmatrix} = x\begin{pmatrix} x \\ 0 \end{pmatrix} = 0$

$g(0,1,1) = g(y,x,x)\begin{pmatrix} x & y \\ 1 & 0 \end{pmatrix} = x\begin{pmatrix} x \\ 1 \end{pmatrix} = 1$

$g(1,0,0) = g(y,x,x)\begin{pmatrix} x & y \\ 0 & 1 \end{pmatrix} = x\begin{pmatrix} x \\ 0 \end{pmatrix} = 0$

$g(1,0,1) = g(x,y,x)\begin{pmatrix} x & y \\ 1 & 0 \end{pmatrix} = x\begin{pmatrix} x \\ 1 \end{pmatrix} = 1$

$g(1,1,0) = g(x,x,y)\begin{pmatrix} x & y \\ 1 & 0 \end{pmatrix} = x\begin{pmatrix} x \\ 1 \end{pmatrix} = 1$

$g(1,1,1) = g(x,x,y)\begin{pmatrix} x & y \\ 1 & 1 \end{pmatrix} = x\begin{pmatrix} x \\ 1 \end{pmatrix} = 1$

Now, if only the reader recalls the table for the ternary majority function maj, she will see that g is, indeed, maj.

Second row. Here we will see that the result of the identification is the projection on the first coordinate, contradicting the choice of g.

$g(0,0,0) = g(x,x,y)\begin{pmatrix} x & y \\ 0 & 0 \end{pmatrix} = x\begin{pmatrix} x \\ 0 \end{pmatrix} = 0$

$g(0,0,1) = g(x,x,y)\begin{pmatrix} x & y \\ 0 & 1 \end{pmatrix} = x\begin{pmatrix} x \\ 0 \end{pmatrix} = 0$

$g(0,1,0) = g(x,y,x)\begin{pmatrix} x & y \\ 0 & 1 \end{pmatrix} = x\begin{pmatrix} x \\ 0 \end{pmatrix} = 0$

$g(0,1,1) = g(y,x,x)\begin{pmatrix} x & y \\ 1 & 0 \end{pmatrix} = y\begin{pmatrix} y \\ 0 \end{pmatrix} = 0$

$g(1,0,0) = g(y,x,x)\begin{pmatrix} x & y \\ 0 & 1 \end{pmatrix} = x\begin{pmatrix} y \\ 1 \end{pmatrix} = 1$

$g(1,0,1) = g(x,y,x)\begin{pmatrix} x & y \\ 1 & 0 \end{pmatrix} = x\begin{pmatrix} x \\ 1 \end{pmatrix} = 1$

$g(1,1,0) = g(x,x,y)\begin{pmatrix} x & y \\ 1 & 0 \end{pmatrix} = x\begin{pmatrix} x \\ 1 \end{pmatrix} = 1$

$g(1,1,1) = g(x,x,x)\begin{pmatrix} x & y \\ 1 & 1 \end{pmatrix} = x\begin{pmatrix} x \\ 1 \end{pmatrix} = 1$

Fourth row. Here we encounter a small complication. The function g itself is not any of the desired functions, but once we have that g we can compute (via suitably chosen substitution) the function maj.

$g(0,0,0) = g(x,x,y)\begin{pmatrix} x & y \\ 0 & 0 \end{pmatrix} = x\begin{pmatrix} x \\ 0 \end{pmatrix} = 0$

$g(0,0,1) = g(x,x,y)\begin{pmatrix} x & y \\ 0 & 1 \end{pmatrix} = x\begin{pmatrix} x \\ 0 \end{pmatrix} = 0$

$g(0,1,0) = g(x,y,x)\begin{pmatrix} x & y \\ 0 & 1 \end{pmatrix} = y\begin{pmatrix} y \\ 1 \end{pmatrix} = 1$

$g(0,1,1) = g(y,x,x)\begin{pmatrix} x & y \\ 1 & 0 \end{pmatrix} = y\begin{pmatrix} y \\ 0 \end{pmatrix} = 0$

$g(1,0,0) = g(y,x,x)\begin{pmatrix} x & y \\ 0 & 1 \end{pmatrix} = y\begin{pmatrix} y \\ 1 \end{pmatrix} = 1$

$g(1,0,1) = g(x,y,x)\begin{pmatrix} x & y \\ 1 & 0 \end{pmatrix} = y\begin{pmatrix} y \\ 0 \end{pmatrix} = 0$

$g(1,1,0) = g(x,x,y)\begin{pmatrix} x & y \\ 1 & 0 \end{pmatrix} = x\begin{pmatrix} x \\ 1 \end{pmatrix} = 1$

$g(1,1,1) = g(x,x,y)\begin{pmatrix} x & y \\ 1 & 1 \end{pmatrix} = x\begin{pmatrix} x \\ 1 \end{pmatrix} = 1$

The function g computed above is not maj (just look at the third row of the computation). But miraculously the function $h(x, y, z) \equiv g(x, y, g(x, y, z))$ is, actually, $maj(x, y, z)$. Since $\mathcal{C}$ is a clone and $g \in \mathcal{C}$, $maj \in \mathcal{C}$.

We will now tell the reader what happens in the cases of the third, fifth, sixth, seventh, and eighth rows. In the third row we found that g is the projection on the second coordinate (which contradicts the choice of g and so this case does not happen). The fifth row is the projection on the third coordinate and so it does not happen. The case of the sixth row is similar to the case of the fourth row, the function g defines function maj. In the case of the seventh row, the function g has the property that $g(g(x, y, z), y, z)$ is the function sum_3. In the case of the eighth row, we find that g is simply sum_3.

Subcase 3.3: $m > 3$.

This proof being quite long, it would be beneficial to see where we are. We have this parameter m, the least arity of idempotent non-essentially unary function in the clone $\mathcal{C}$. When $m = 2$ or $m = 3$ we found that $\mathcal{C}$ must contain one of $\wedge$, $\vee$, maj, or sum_3. Now, we will show that m cannot be bigger than three. So let us assume $m \geq 4$. Our assumption now is that a function g of arity bigger than or equal to 4 has a least possible arity among idempotent and non-essentially unary functions in $\mathcal{C}$. We have a couple of cases. The first case is when after the identification of x_i and x_j the result is projection to x_j. Then we have: $f(x_1, x_1, x_3, x_4, \ldots) = x_1$ and $f(x_1, x_2, x_3, x_3, \ldots) = x_3$. Then $f(0, 0, 1, 1, \ldots) = 0$ and simultaneously $f(0, 0, 1, 1, \ldots) = 1$ (we use here the fact that $m \geq 4$). This is an obvious contradiction. This leaves us with the case when identification of variables results in a projection, except that it is a projection to another variable. With appropriate renaming we can assume that

$$g(x_1, x_1, x_3, x_4, \ldots) \equiv x_4.$$

We claim that $g(x_1, x_2, x_1, x_4, \ldots)$ is also x_4. For there is x_j such that

$$g(x_1, x_2, x_1, x_4, \ldots) \equiv x_j.$$

Thus in particular $g(x_1, x_1, x_1, x_4, \ldots) \equiv x_j$. But $g(x_1, x_1, x_1, x_4, \ldots) \equiv x_4$ (as it is the result of "dummy" substitution into g of x_1 for x_3). But then $x_j \equiv x_4$ and so $j = 4$. By the same argument $g(x_1, x_2, x_2, x_4, \ldots)$ is also x_4. But $|Bool| = 2$. Therefore for any valuation v at least one of the equalities $v(x_1) = v(x_2)$, or $v(x_1) = v(x_3)$, or $v(x_2) = v(x_3)$ holds. This means that if we have a Boolean vector $(a_1, a_2, a_3, a_4, \ldots)$ then it belongs to the domain of one of the functions arising from identifications $g(x_1, x_1, x_3, x_4, \ldots)$, or $g(x_1, x_2, x_1, x_4, \ldots)$, or $g(x_1, x_2, x_2, x_4, \ldots)$. But each of these functions on its domain coincided with x_4! Thus for every Boolean vector $(a_1, a_2, a_3, a_4, \ldots)$, $g(a_1, a_2, a_3, a_4, \ldots) = a_4$, that is, g is the projection to the fourth coordinate. Thus $m \geq 4$ is, after all, impossible, and the proof is now complete. $\square$

We are closing on the proof of the Schaefer theorem, but we need the final push. Essentially, we need to see what happens when a clone is not included in any of the six clones of Schaefer. That is, we need to investigate what happens when none of six functions: constant 0, constant 1, $\wedge$, $\vee$, sum_3, maj is in the clone $Pol(\Gamma)$. By Rosenberg theorem, not having functions $\wedge$, $\vee$, maj, or sum_3 in the clone $Pol(\Gamma)$ implies that the clone $Pol(\Gamma)$ contains only unary functions. But since the clone $Pol(\Gamma)$ does not contain constant functions, all it can contain are essentially unary functions. Those are projections and negations of projections, thus functions that act like permutations.

PROPOSITION 13.12

If Γ is a finite set of Boolean relations such that $Pol(\Gamma)$ contains only unary functions that act as permutations then for every finite set of Boolean relations Γ', $CSP(\Gamma')$ polynomially reduces to $CSP(\Gamma)$. In particular, the problem $CSP(\Gamma)$ is NP-complete.

Proof: If $Pol(\Gamma)$ contains only projections then as projections are in every clone, $Pol(\Gamma') \subseteq Pol(\Gamma)$. Then we can use Corollary 13.5 directly. But there is another case, namely that Γ has another polymorphism, namely negation. We still need to reduce $CSP(\Gamma')$ to $CSP(\Gamma)$. The trouble is that negation does not have to be a polymorphism of Γ'. But the reader will recall how in Chapter 5 we made "mirror copies" of tables. We will do something similar here. Given a relation R' in Γ', we build a relation R'' that has one column more by setting

$$R'' = (\{0\} \times R') \cup (\{1\} \times \{(\bar{a}_1, \ldots, \bar{a}_n) : (a_1, \ldots, a_n) \in R'\}).$$

We then set $\Gamma'' = \{R'' : R' \in \Gamma'\}$. All relations in Γ'' are guaranteed to have negation (i.e., $\neg$) as a polymorphism. It is really easy to see. Yet another property is that polymorphisms of Γ' are polymorphisms of Γ'' as well. The reason is that the polymorphisms of Γ' are essentially unary acting as permutations (that is, projections or complemented projections) and these are, as is easy to see, polymorphisms of Γ''. But now it is guaranteed that all polymorphisms of Γ are polymorphisms of Γ''. This, by Proposition 13.10, implies that $CSP(\Gamma'')$ polynomially reduces to $CSP(\Gamma)$. So all we need to do is to polynomially reduce $CSP(\Gamma')$ to $CSP(\Gamma'')$. But this is easy. Let S' be an instance of $CSP(\Gamma')$. We select a new variable (say x_0), and for each constraint $R'(x_{i_1}, \ldots, x_{i_k})$ of S', we compute a new constraint $R''(x_0, x_{i_1}, \ldots, x_{i_k})$ by "mirroring," defined above. Now, the instance S'' of $CSP(\Gamma'')$ consists of these new constraints. It is obvious that the size of S'' is linear in the size of S' (a bit more than two times size of S'). Now, if v is a solution to S', then $v \cup \{(x_0, 0)\}$ is a solution to S''.

Conversely, let w be a solution of S''. Two cases are possible:

Case 1: $w(x_0) = 0$. Then the assignment $v = w \setminus \{(x_0, 0)\}$ is a solution to S'.

Case 2: $w(x_0) = 1$. Then the values of assignment w on variables different from x_0 must come from the "mirror part" of R'', $\{1\} \times \{(\bar{a}_1, \ldots, \bar{a}_n) : (a_1, \ldots, a_n) \in R'\}$. But then the assignment v, defined by

$$v(x_i) = \neg w(x_i),$$

is a solution to S'. Thus any algorithm for solving $CSP(\Gamma'')$ provides a way to solve $CSP(\Gamma')$ with only polynomial increase in the amount of work. Therefore $CSP(\Gamma')$ reduces polynomially to $CSP(\Gamma'')$ which, in turn, polynomially reduces to $CSP(\Gamma)$. But we established that there are constraint languages Γ' for which $CSP(\Gamma')$ is NP-complete, for instance the language consisting of eight tables for 3-clauses was such language. Thus $CSP(\Gamma)$ is NP-complete, as desired. This completes the argument. □

To sum up (we have repeated the argument already!), what does it mean when a clone $\mathcal{C}$ contains only unary functions that act as permutations? Again, let us stress that in

Knowledge representation and constraint satisfaction   317

view of Rosenberg theorem (Theorem 13.4) this means *precisely* that the clone C is not included in $Pol(\Gamma)$ for any of 6 Schaefer clones. Thus we get the following.

COROLLARY 13.7 (Schaefer theorem, "hard part")

If Γ is a set of Boolean relations that is not included in any of 6 Schaefer clones, then $CSP(\Gamma)$ is an NP-complete problem.

In the Exercises below we will have several interesting corollaries to Schaefer theorem. Here we only observe that it is amazing that there is a relationship between universal algebra constructions and the existence of algorithms solving constraint satisfaction problems. We note in passing that the fact that the domains of our CSPs were all two element was of critical importance in many places in the argument. In fact, Schaefer dichotomy has been lifted to three-element domains, but to date it is not known for larger finite domains.

13.6 Exercises

1. Instead of representing SEND-MORE-MONEY puzzle as a SAT problem, represent it as a Constraint Satisfaction problem over finite domains.

2. We could represent sudoku-solution problem (i.e., find all sudoku solutions) as a CSP with 81 variables each ranging over the domain (1..9). But such an attempt would be a disaster. Why?

3. Γ_{3SAT} is a collection of four Boolean relations: $Bool^3 \setminus \{(0,0,0)\}$, $Bool^3 \setminus \{(0,0,1)\}$, $Bool^3 \setminus \{(0,1,1)\}$, and $Bool^3 \setminus \{(1,1,1)\}$. The tables in this collection allow for expressing $3SAT$ as a constraint satisfaction problem. Find a pp-definition of the table

$$\begin{pmatrix} 0 & 0 & 1 \\ 0 & 1 & 0 \\ 1 & 0 & 0 \end{pmatrix}$$

using relations in Γ_{3SAT}.

4. Establish that the previous problem is trivial because *every* ternary Boolean relation is pp-definable from Γ_{3SAT}.

5. Find a pp-definition of the inequality relation, i.e.,

$$\begin{pmatrix} 0 & 1 \\ 1 & 0 \end{pmatrix}$$

using relations in Γ_{3SAT}.

6. Show that all functions are polymorphisms of the table:
$$\begin{pmatrix} 0 & 0 \\ 1 & 1 \end{pmatrix}.$$
What is the reason, and how does it relate to disambiguation?

7. When Boolean tables R_1 and R_2 have the same number of columns then the relation $R_1 \cap R_2$ is pp-definable from $\{R_1, R_2\}$.

8. Show that each of the Schaefer classes of relations is closed under pp-definability. For instance, when all relations in Γ are Horn, then any relation pp-definable in Γ is also Horn.

9. Using the above fact show that pp-definable relations are *not* closed under unions. Specifically, construct two relations R_1, R_2 such that R_1 and R_2 have the same number of columns, and $R_1 \cup R_2$ is not pp-definable from $\{R_1, R_2\}$.

10. Continuing the previous exercise, prove that, in general, the collection of relations pp-definable from Γ is not closed under complement.

11. On the other hand, there are classes Γ such that $\langle \Gamma \rangle$ is closed under complement. Construct one.

12. Prove that when Γ is a set of tables then $Pol(\Gamma)$ is non-empty.

13. Prove that when $\mathcal{F}$ is a set of Boolean functions then $Inv(\mathcal{F})$ is non-empty.

14. Let us strengthen the concept of definability to bring it a bit more in line with SQL database language, namely, we allow conjuncts of the form $x_i = \varepsilon$ with $\varepsilon \in Bool$. Show that *four* Schaefer classes, namely those of Horn, dual Horn, affine, and 2SAT, are closed under this concept of definability, but the remaining two classes (1-consistent and 0-consistent) are not closed under this concept of definability.

15. Prove the "easier half" of the first Geiger theorem, that is, $\langle \Gamma \rangle \subseteq Inv(Pol(\Gamma))$.

16. Prove that for every set $\mathcal{F}$ of operations there is a least clone that contains $\mathcal{F}$.

17. Prove the old but difficult theorem of E. Post: Every Boolean clone $\mathcal{C}$ has a finite basis, that is, there is a *finite* set of operations $\mathcal{F}$ such that $\mathcal{C}$ is the least clone containing $\mathcal{F}$.

18. Prove that Boolean clones form a lattice under inclusion.

19. Let R_{1in3} be the Boolean table
$$\begin{pmatrix} 0 & 0 & 1 \\ 0 & 1 & 0 \\ 1 & 0 & 0 \end{pmatrix}.$$
Prove that $CSP(\{R_{1in3}\})$ is NP-complete. Use the Schaefer theorem.

20. Introduce the table R_{2in3} and prove the same fact.

21. The table R_{NAE} (not-all-equal) is $Bool^2 \setminus \{(0,0,0),(1,1,1)\}$. Prove that like in the two above problems $CSP(\{R_{NAE}\})$ is NP-complete.

22. Compute explicitly the remaining five cases for $m = 3$ in the Rosenberg theorem. If you do not want to make the complete effort, do case eight.

Chapter 14

Answer set programming, an extension of Horn logic

14.1 Horn logic revisited 321
14.2 Models of programs 322
14.3 Supported models 323
14.4 Stable models 326
14.5 Answer set programming and SAT 329
14.6 Knowledge representation and ASP 333
14.7 Complexity issues for ASP 336
14.8 Exercises 337

In this chapter we introduce the reader to a formalism closely related to propositional satisfiability, namely answer set programming (ASP for short). ASP is an extension of Horn logic, i.e., logic based on an extension of the logic admitting only Horn clauses. The extension uses a nonstandard negation *not*. This negation is sometimes called *negation-by-failure*. As in other cases (viz. our discussion of relationship of SAT and integer programming) we will touch only "the tip of the iceberg." This area is well developed, and it has a number of monographs, for instance, [MT93, Ba03].

14.1 Extending Horn logic by nonstandard negation in the bodies of clauses

Let us recall that Horn clauses were clauses of the form $p \vee \neg q_1 \vee \ldots \vee \neg q_k$ and of the form $\neg q_1 \vee \ldots \vee \neg q_k$. The clauses of the first kind were called *program clauses*, the second kind – *constraints*. We could treat the program clauses as implications: $q_1 \wedge \ldots \wedge q_k \Rightarrow p$. Traditionally, such clauses, when written by ASP researchers, are written like this:

$$p \leftarrow q_1, \ldots, q_k.$$

Among many possible interpretations of such a clause is the following: "*Given that $q_1, \ldots, q_k$ all hold, p must hold, too.*"

There is a procedural aspect to such interpretation; once $q_1, \ldots q_k$ are all computed, we derive (conclude, compute) p.

Now, let us to allow on the right-hand side of the symbol ← terms of the form *not r*, where r is a propositional variable. In other words, we now allow program clauses of the form:

$$p \leftarrow q_1, \ldots, q_k, \text{not } r_1, \ldots, \text{not } r_m. \tag{14.1}$$

One intuition could be that if $q_1, \ldots, q_k$ were computed and $r_1, \ldots, r_m$ were not, then we accept p as computed. But it is pretty obvious that such interpretation raises issues. What if one of the r_js was not computed originally, but later on was computed? Surely there is something wrong with the accepted computation of p. It is our goal in this chapter to find a reasonable interpretation of "*were not computed*" which will make possible the extension of Horn clauses to clauses of the form (14.1). We will call clauses of the form (14.1) *normal program clauses*. A *normal program* is a set of normal clauses. Given a normal clause C of the form (14.1) we call p the head of C (and denote it by $head(C)$,) and the formula $q_1 \wedge \ldots \wedge q_k \wedge \neg r_1 \wedge \ldots \wedge \neg r_m$ the body of C and denoted by $body(C)$.

Our goal below is two-fold. First, we need to introduce a meaningful semantics for normal programs. Second, we need to see what could be done with programs consisting of clauses of the form (14.1). Thus the first problem is the issue in semantics. The other problem is the issue in knowledge representation. We need to get to the right semantics of programs in stages for we will see that there is more than one possibility of defining semantics of such programs.

14.2 Models of programs

The simplest semantics we give to programs is one where we interpret *not p* as $\neg p$. Here is what we do. We say (as we did before) that a set of atoms M satisfies p if $p \in M$, and that M satisfies *not p* if $p \notin M$. Then we say that $M \models body(C)$ if $q_1, \ldots, q_k \in M$, and $r_1, \ldots, r_m \notin M$ (here C is the clause like in (14.1).). Then we say that M is a model of the program P if for all clauses $C \in P$, whenever $M \models body(C)$ then $M \models head(C)$. Let us look at an example.

Example 14.1
Let P be the following program:

$p \leftarrow q, \text{not } r$
$p \leftarrow s$
$q \leftarrow \text{not } t.$

Clearly, $M_1 = \{q, r\}$ is a model of P. Indeed, since M does not satisfy the bodies of the first two clauses, we just do not care whether the heads are in M or not. For the third clause, $M \models \neg t$, and $M \models q$.
Let us observe that $M_2 = \{q\}$ is not a model of P (first clause fails) and $M_3 = \{r\}$ is also not a model of P (look at the third clause).

Now, let Q be the following program:

$p \leftarrow q, \text{not } r.$
$r \leftarrow \text{not } p.$
$q \leftarrow .$

Here both $\{p,q\}$ and $\{q,r\}$ are models of Q, but $\{p,q,r\}$ is also a model of Q. □

A minimal model of a program P is a model M of P such that no proper subset of M is a model of P. In our Example 14.1, $\{q,r\}$ was a minimal model of P, while $\{p,q,r\}$ was not a minimal model of Q.

Now, let P be a normal program. Given a normal program clause $C \in P$,

$$p \leftarrow q_1, \ldots, q_k, \text{not } r_1, \ldots, \text{not } r_m.$$

We assign to C the clause $p \vee r_1 \vee \ldots \vee r_m \vee \neg q_1 \vee \ldots \neg q_k$. We call that clause *propositional interpretation* of C and denote that clause by $pl(C)$. We define $pl(P) = \{pl(C) : C \in P\}$. We then have the following simple fact.

PROPOSITION 14.1
Let P be a normal logic program and M a set of propositional variables. Then:

1. *M is a model of P if and only if $M \models pl(P)$.*

2. *M is a minimal model of P if and only if $M \models pl(P)$ and no strictly smaller set N is a model of $pl(P)$.*

It should be clear that every program possesses a model (because $cl(P)$ contains no constraints, and so by Proposition 7.1 it is satisfiable). Likewise, Lemma 13.1 implies the second part of Proposition 14.1, that is, every normal program possesses a minimal model.

14.3 Supported models, completion of the program

When we look at the least model of a Horn program P (we denoted that model $lm(P)$) it is easy to see that the presence of an atom in that model is always justified by the clauses of the program. The reason is that the program P determines the operator T_P. That operator T_P, which is monotone, is then iterated (this is what we called the Kleene construction of the least fixpoint) and the sequence of iterations converges to the least fixpoint. What is important here is that the model is obtained as the fixpoint. So here we will extend the definition of T_P to the case when P is no longer Horn. The price to be paid is that the operator thus obtained is no longer monotone, and so existence of a fixpoint is not guaranteed. But if there are fixpoints, then such fixpoints are models with reasonable properties.

Here is how we define the operator T_P now. Let M be a set of propositional variables. We define

$$T_P(M) = \{head(C) : C \in P \text{ and } M \models body(C)\}.$$

When P is the Horn program, this definition coincides with the one we gave above. But in the general case we get a different behavior of T_P.

Example 14.2
Let P be the program

$p \leftarrow not\ q$
$q \leftarrow not\ p$

Then $T_P(\emptyset) = \{p, q\}$, $T_P(\{p\}) = \{p\}$, $T_P(\{q\}) = \{q\})$ and $T_P(\{p, q\}) = \emptyset$. Thus, in particular, T_P is not monotone, but both $\{p\}$ and $\{q\}$ are its fixpoints. Now, let P' be the program consisting of a single clause

$p \leftarrow not\ p$

Then $T_{P'}(\emptyset) = \{p\}$ and $T_{P'}(\{p\}) = \emptyset$. Thus the operator $T_{P'}$ possesses no fixpoints.

□

Loosely motivated by the fixpoint considerations discussed above, we define, for each propositional variable p, the *completion formula* with respect to program P, $comp_P(p)$, by

$$p \equiv \bigvee \{body(C) : C \in P \text{ and } head(C) - p\}$$

Next, we define $comp_P = \{comp_P(p) : p \in Var\}$. Let us observe that, in principle, some formulas in $comp_P$ may be infinitary. This will happen if for some variable p the program P contains infinitely many clauses with head p. When P is a *finite* normal program we will not have this problem as all formulas $comp_P(p)$ and the theory $comp_P$ will be finite. For that reason we will limit our attention to the case of finite normal programs, although with enough effort the theory of infinitary propositional formulas can be developed, and most of our theory put through.

Let us observe that under the finiteness assumption made above, $comp_P$ is a propositional theory. Moreover, transformation of $comp_P$ to clausal form can be done in linear time.

PROPOSITION 14.2
Let M be a set of propositional variables, and P a finite normal program. If $M \models comp_P$ then M is a model of P.

Proof: We need to prove that M is a model of $pl(P)$. So, let us assume that for some clause C of P, $M \models body(C)$. Let $p = head(C)$ and let ψ_p be the right-hand side of the formula $comp_P(p)$. Then, as $M \models body(C)$, $M \models \psi_p$. But as $M \models comp_P(p)$, $M \models p$, that is, $M \models head(C)$, as desired. □

But now, we can prove a result connecting the operator T_P with the formula $comp_P$.

PROPOSITION 14.3 (Apt and van Emden theorem)

Let M be a set of propositional variables, and P a finite normal program. Then M is a model of $comp_P$ if and only if M is a fixpoint of the operator T_P.

Proof: First, let us assume that $M \models comp_P$. We need to show two inclusions.
We first show $M \subseteq T_P(M)$. Let $p \in M$. Then $M \models \psi_p$ (see above for the definition of ψ_p). Therefore, for some clause $C \in P$ such that $head(C) = p$, $M \models body(C)$. Thus, by definition, $p \in T_P(M)$. As p is an arbitrary variable, inclusion $\subseteq$ follows.
Next, we show $T_P(M) \subseteq M$. Let $p \in T_P(M)$. Then for some clause C with head p, $M \models body(C)$. Thus $M \models \psi_p$ and since $M \models comp_P(p)$, $M \models p$, i.e., $p \in M$, as desired.
Second, let us assume that $M = T_P(M)$ and p be an arbitrary propositional variable. We need to show that $M \models comp_P(p)$. Again two cases require our attention.
When $p \in M$, i.e., $p \in T_P(M)$, then for some program clause C with head p, $M \models body(C)$. But then $M \models \psi_p$, and both sides of the equivalence $comp_P(p)$ are true in M, and so $M \models comp_P(p)$.
When $p \notin M$, i.e., $p \notin T_P(M)$, then for all program clauses C in P, $head(C) = p$, it must be the case that $M \not\models body(C)$. But then $M \not\models \psi_P$, and both sides of $comp_P(p)$ are false in M. Thus the equivalence $comp_P(p)$ is true in M, as desired. □

The results proved above justify the following definition: we call a set of variables a *supported model* of a normal program P if $M = T_P(M)$. We observed that not every program possesses supported models but that testing whether a program possesses a supported model may be easily reduced to SAT. Moreover, since supported models were models of a propositional theory, once a program possesses a supported model, it also possesses a model that is, supported and minimal. Let us observe that the least model of a Horn program, being a fixpoint of T_P, is supported. We state the properties of the least model of a Horn program formally.

PROPOSITION 14.4

The least model $lm(P)$ of a Horn program P is minimal (in fact, the only minimal model of P) and supported.

14.4 Stable models of normal programs

We saw in Section 14.3 that the least model of a Horn program is supported and minimal. Let us observe that the nice aspect of the least model M of a Horn program is that, as a least fixpoint of a monotonic operator, it is computed in layers. In other words, we can assign to every propositional variable p in M its *level*, the least integer i such that p belongs to $(T_P)^i(\emptyset)$.

But what does it mean? This means that we accept p as belonging to M on the basis of the membership of other variables with *strictly lower* levels. Therefore, there is never a circular dependence of p on itself in such computation. This is, of course, an intuitive observation only.

It is natural to express the property of dependence of the presence of a variable in the model in terms of a certain graph, called the *call-graph* G_P, associated with program P. In this graph, the vertex labeled with p points to all vertices labeled with q_i whenever a clause $p \leftarrow q_1, \ldots, q_k$ belongs to P. It is now clear that we accept sinks in this graph unconditionally, and when for some clause C as above, all q_i, $1 \leq i \leq k$ are accepted, then p is accepted. Once we admit negation *not* in the bodies of clauses, the perspective changes slightly. We can consider a variety of graphs associated with the program. One of these graphs, where a vertex points only to the vertices labeled with variables occurring positively in the body, is called the *positive call-graph*. We will denote it by G_P, too.

Once we admit negation in the bodies of clauses there is no simple way to exclude multiple intended models (in the Horn case we had just one intended model, the least one). Let us look at an example.

Example 14.3
Let P be a program:

$p \leftarrow not\ q$
$q \leftarrow not\ p$

Then, clearly, there are just two supported models of P, $\{p\}$ and $\{q\}$. There is no way to say which one is "better" (unless we have a preference relation on the set of variables). □

Example 14.3 tells us that there is no way to avoid multiple intended models. Therefore, we need to choose the properties of the least model of the Horn program that we want to preserve. There are several: the least model, supportedness, existence of levels. We cannot have all of them; we need to drop something. The definition below, due to Gelfond and Lifschitz, provides a way to define models which are minimal, supported, and have levels. Not every program will have such models (as there are programs that do not even possess supported models) but when they do, these models have a nice and natural behavior.

Answer set programming

Before we define this class of models, *stable models*, we need a bit of terminology. Specifically, we need to split the body of a normal program clause into positive and negative parts. So, let

$$C : p \leftarrow q_1, \ldots, q_k, not\ r_1, \ldots, not\ r_m$$

be a program clause. The formula $q_1 \wedge \ldots \wedge q_k$ will be called the positive part of the body, $posBody(C)$. We will abuse notation and write $posBody$ in clauses as a sequence of variables, and not as a conjunction. The negative part of the body will be the set of variables $\{r_1, \ldots, r_m\}$. We will denote it by $negBody$. While those objects are, in principle, of different type, it will be convenient to have them in this form.

Now, let C be a normal program clause, and let M be a set of propositional variables. We define

$$C^M = \begin{cases} \textbf{nil} \text{ if } M \cap negBody(C) \neq \emptyset \\ head(C) \leftarrow posBody(C),\ otherwise. \end{cases}$$

So what do we do in the computation of C^M? When M directly contradicts one of the negative literals in the body of C, we eliminate C altogether. Otherwise, we eliminate in C all negative literals, We now define $P^M = \{C^M : M \in P\}$. This program P^M is called *the Gelfond-Lifschitz reduct* of P by M, and is denoted by $GL_P(M)$. Let us observe that the program P^M is always a Horn program, and so it possesses its least model $lm(P^M)$. This model depends on M. Let us denote it by N_M. When P is fixed, the assignment $M \mapsto N_M$ is an operator in $\mathcal{P}(Var)$. This operator is called *Gelfond-Lifschitz operator* associated with P. Before we introduce stable models of programs, we will prove two useful properties of the Gelfond-Lifschitz operator.

PROPOSITION 14.5
The operator $GL_P(\cdot)$ is antimonotone. That is, for all $M_1 \subseteq M_2$, $GL_P(M_2) \subseteq GL_P(M_1)$.

Proof: If $M_1 \subseteq M_2$ then $P^{M_2} \subseteq P^{M_1}$ because we prune more with M_2, and thus leave less. For word-processing reasons let us denote P^{M_1} by Q, and P^{M_2} by R. Then $R \subseteq Q$, and clearly, for all N, $T_R(N) \subseteq T_Q(N)$. Therefore, by a simple induction, for every non-negative integer i:

$$(T_R)^i(N) \subseteq (T_Q)^i(N).$$

But then this holds for $N = \emptyset$ and so the least fixpoint of the operator T_R is included in the least fixpoint of the operator T_Q, as desired. □

To some extent, Proposition 14.5 is bad news. The reason is that antimonotone operators do not have to possess fixpoints. In fact, we have this situation for GL_P for some P.

Example 14.4
Let P be the program:

$$p \leftarrow q$$
$$q \leftarrow not\ p$$

The operator GL_P has no fixpoints. Four cases have to be considered: $M_1 = \emptyset$, $M_2 = \{p\}$, $M_3 = \{q\}$, and $M_4 = \{p, q\}$. We hope the reader will compute all four of these cases and see that there are no fixpoints. □

We also need the following lemma.

LEMMA 14.1
Let P be a normal program, and let N be a model of P. Then $GL_P(N) \subseteq N$.

Proof: We first show that for every $N' \subseteq N$, $T_{P^N}(N') \subseteq N$. Indeed, let $p \leftarrow q_1, \ldots, q_k$ be a clause in P^N, and $q_1, \ldots, q_k \in N'$. Then for some $r_1, \ldots, r_m$, all out of N, the clause $p \leftarrow q_1, \ldots, q_k, not\ r_1, \ldots, not\ r_m$ belongs to P. Since all q_i, $1 \leq i \leq k$ belong to N and none of r_j, $1 \leq j \leq m$ belongs to N, $N \models body(C)$ and since $N \models C$, $p \in N$, as desired.
But now, by an easy induction on i we show that $(T_{P^N})^i(\emptyset) \subseteq N$ for all non-negative integers i, and so $GL_P(N) \subseteq N$. □

Now, let us define a set M of variables *stable* for a program P if $GL_P(M) = M$. Sets stable for P are commonly called *stable models* for P or *answer sets* for P. To justify the first of these namings we prove a fundamental theorem on sets stable for programs.

THEOREM 14.1 (Gelfond and Lifschitz theorem)
Let P be a normal program, and let M be stable for P. Then:

1. M is a model of P.
2. M is a supported model of P.
3. M is a minimal model of P.
4. M has levels.

Proof. We show (1). Let $C \in P$, $C : p \leftarrow q_1, \ldots, q_k, not\ r_1, \ldots, not\ r_m$. If $M \cap \{r_1, \ldots, r_m\} \neq \emptyset$, then $M \not\models body(C)$ and thus $M \models C$. If $M \cap \{r_1, \ldots, r_m\} = \emptyset$, but for some i, $1 \leq i \leq k$, $q_i \notin M$, then again $M \not\models body(C)$ and thus $M \models C$. Finally, if $M \cap \{r_1, \ldots, r_m\} = \emptyset$, and for all i, $1 \leq i \leq k$, $q_i \in M$, then $p \leftarrow q_1, \ldots, q_m$ belongs to P^M and since M is a model of P^M, $p \in M$, as desired.
We show (2). Let $p \in M$. We need to find support for p, that is, we need to find a clause C with head p such that $M \models body(C)$.
But $M = lm(P^M)$. As the model M is a supported model of P^M, there is a clause $D = p \leftarrow q_1, \ldots, q_k$ in P^M with $q_1, \ldots, q_k$ all in M. But this clause D is E^M for

some $E \in P$. Now we found the desired C; it is E. It is easy to see that its body is satisfied by M. As $head(C) = p$ we are done.

We show (3). Let us assume M is stable for P, $N \models P$, $N \subseteq M$. Then we have:

$$GL_P(N) \subseteq N \subseteq M = GL_P(M) \subseteq GL_P(N).$$

Indeed, the first inclusion follows from Lemma 14.1 as N is a model of P. The second inclusion is our assumption. The equality following it is another assumption: M is stable for P. Finally, the last inclusion follows from antimonotonicity of the operator GL_P. Thus we "squeezed" N and M into equality, as desired.

To see (4) we observe that the levels of $lm(P^M)$ are levels for M. □

14.5 Answer set programming and SAT

We will now discuss the possible use of the SAT solver as a back-end engine for ASP computation. We know at this point that if all we care about are supported models, then a SAT solver can be easily used for testing if a finite propositional normal program has such models. But could something be done for stable models computation? In this section we will actually do this; we will show how a SAT solver could be used for such purpose.

The problem we face is that on the one hand stable model computation is similar to least model computation, but on the other hand there are all these negative literals in the body which somehow have to be taken care of. To handle the problem of negative literals, we introduce the concept of *guarded unit resolution* derivation. We will essentially simulate as much as possible the unit resolution derivations (see Figure 9.1 for a graphic presentation of unit resolution). We first introduce a slightly different (but it is just a "syntactic sugar") representation of normal program clauses. Instead of $p \leftarrow q_1, \ldots, q_k, not\ r_1, \ldots, not\ r_m$ we will write our clause as

$$p \leftarrow q_1, \ldots, q_k : \{r_1, \ldots, r_m\}.$$

When $k = 0$, we drop $\leftarrow$ and write $p : \{r_1, \ldots, r_m\}$. This notation highlights the fact that positive and negative literals in normal program clauses play a different role (under stable semantics). Namely, we want to *compute* $q_1, \ldots, q_k$ while keeping $r_1, \ldots, r_m$ out of the putative stable model. Now we introduce guarded unit resolution rule as follows. Both of the input are guarded clauses, but one of the inputs is a guarded variable (unit). Here is the guarded unit resolution rule:

$$\frac{p \leftarrow q_1, \ldots q_k : \{r_1, \ldots, r_m\} \quad q_j : \{s_1, \ldots, s_n\}}{p \leftarrow q_1, \ldots q_{j-1}, q_{j+1}, \ldots, q_k : \{r_1, \ldots, r_m, s_1, \ldots, s_n\}}.$$

What happens here is that we eliminate one literal in the body of the first argument. But there is a cost involved: we grow the guard. The goal in the guarded resolution derivations is the same as in the derivation of variables using unit resolution. The difference is that we do not derive variables alone; we derive a variable *and* its guard. That guard is a collection of variables. The intuition is that the guard is telling us what variables must be *outside* of the model if we want to use the derivation to compute an element in the model. When we discharge all the positive variables in the body we take the symbol ← as well. We get a *guarded variable*.

Example 14.5
Let our program P be:

$p \leftarrow not\ q$
$q \leftarrow r, not\ p$
$r \leftarrow$

The first clause has guard $\{q\}$. The second clause has guard $\{p\}$. The third clause also has a guard – it is empty. We save on word processing and do not write empty guards. Now, here is a derivation of the guarded unit (variable) $q : \{p\}$ from our program.

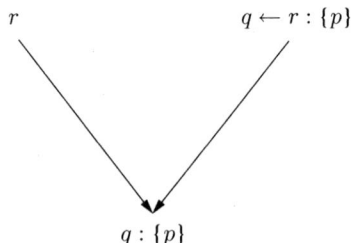

FIGURE 14.1: The derivation of a guarded unit

□

Now we have to relate guards, derivations, and the Gelfond-Lifschitz operator. First, we need a bit of terminology. A derivation of a guarded variable $p : \{r_1, \ldots, r_m\}$ is *admitted* by a set of variables M if $\{r_1, \ldots, r_m\} \cap M = \emptyset$. Then, we say that a variable p is guarded-derivable from P if for some set $\{r_1, \ldots, r_m\}$ the guarded variable $p : \{r_1, \ldots, r_m\}$ is derived from P. The guarded resolution tree is the witness of the derivability, but all that matters is the guard of p. The reason is that the guards only grow, and so the guard in the root is the inclusion-largest guard in the tree. Here is a characterization of $GL_P(M)$ in proof-theoretic terms.

PROPOSITION 14.6
The set $GL_P(M)$ consists precisely of those variables p such that there is a

guarded derivation of the guarded unit $p : \{r_1, \ldots, r_m\}$ *that is admitted by* M.

Proof: Two inclusions need to be proved: first that $GL_P(M)$ is included in the set of variables that have guarded derivation admitted by M, and the converse inclusion. Both are proved by induction: the first on the level of variables in N_M, the other one by induction on the height of the derivation tree. □

Proposition 14.6 shows that, in reality, the Gelfond-Lifschitz operator is a proof-theoretic construct. Here is one nice consequence.

COROLLARY 14.1
Let P be a propositional normal logic program, and let M be a set of propositional variables. Then M is a stable model for P if and only if

1. *For every $p \in M$ there is a guarded derivation of p admitted by M and*
2. *For every $p \notin M$ there is no guarded derivation of p admitted by M.*

The proof-theoretic characterization of stable semantics given in Corollary 14.1 leads to a useful result with practical consequences.

PROPOSITION 14.7 (Fages theorem)
Let P be a finite normal propositional program such that the positive call-graph G_P of P is acyclic. Then every supported model of P is a stable model of P.

Proof: Let P be a program as in the assumptions and let M be a supported model of P. In particular, M is a model of P (Proposition 14.2) and hence by Lemma 14.1, $GL_P(M) \subseteq M$. Therefore, all we need to prove is that $M \subseteq GL_P(M)$. This means, in view of Proposition 14.6, that we need to exhibit, for every $p \in M$, a guarded derivation of $p : Z$ such that $M \cap Z = \emptyset$.

Our assumption is that the graph G_P is acyclic. Therefore, we can topologically sort it. This results in the ordering $\preceq$ of variables occurring positively in the bodies of clauses of P. Let us look at that ordering, and let us inspect the elements of M as they appear in the order $\preceq$ restricted to variables occurring positively in P. We actually inspect the elements of M from the last to the first.

Let p be the last element of M in the order $\preceq$. Since M is a supported model of P, there is a clause $C : p \leftarrow q_1, \ldots, q_k, \text{not } r_1, \ldots, \text{not } r_m$ such that $body(C)$ is satisfied by M. In particular, all q_is must belong to M. But then q_is must occur later in the ordering $\prec$ since in the call-graph for P, G_P, p points to them. As p is the last, k must be 0. Thus that clause with head p that supports the presence of p in M must have only negative literals in the body. It is, then, $C : p \leftarrow \text{not } r_1, \ldots, \text{not } r_m$. But then, clearly $p : \{r_1, \ldots, r_m\}$ is a guarded unit admitted by M.

Now we execute the inductive step. Say we have a variable p and let us assume that for every variable q which belongs to M and is later in the ordering $\preceq$, q has a guard Z such that $q : Z$ is a guarded unit that is, admitted by M. As $p \in M$

and M is supported, there is a clause $C : p \leftarrow q_1, \ldots, q_k, \text{not } r_1, \ldots, \text{not } r_m$ such that $body(C)$ is satisfied by M. In particular, all q_js are in M and so there are Z_js, $1 \leq j \leq k$ so that $q_j : Z_j$ is a guarded variable admitted by M. Now we proceed as we did in combining unit resolution in our discussion of the least model of the Horn programs (Chapter 9). The only issue is that we have to combine the guards too. But, fortunately, all Z_js are disjoint from M and also $\{r_1, \ldots, r_m\}$ is disjoint from M. Therefore, setting Z as

$$Z_1 \cup \ldots \cup Z_k \cup \{r_1, \ldots, r_m\},$$

we see that $M \cap Z = \emptyset$. Therefore, the guarded unit $p : Z$ is admitted by M, and the (guarded) resolution tree described above and with $p : Z$ in the root is the desired derivation admitted by M and so p belongs to $GL_P(M)$. As p is an arbitrary element of M, we are done. □

It is easy to test if the graph G_P of a program P is acyclic. All we need to do is to try to topologically sort G_P. We trust the reader has taken an undergraduate data structures course.

The net effect of the Fages theorem (Proposition 14.7) is that if we find that the graph G_P is acyclic, then in order to find stable models of P all we have to do is to compute completion, transform it to a CNF, and submit it to the SAT solver. If a satisfying assignment is returned, we have a stable model. If the resulting CNF is unsatisfiable, we conclude that there are no stable models.

But, unfortunately, the graph G_P may possess cycles (in fact, in interesting cases this will happen). What options we have in such situation? At least two possibilities occur (both are used in ASP solvers).

1. We could use SAT solver as a "candidate generator."
2. We could extend the completion of program P so to eliminate those supported models which are not stable.

Idea (1) is quite simple. We modify the SAT solver (not that difficult) so it returns one by one all satisfying assignments for the completion of P. As those satisfying assignments are returned, we test each of them for stability. If there is a stable model of P it will, eventually, be returned among the supported models of P and then we will easily establish the fact that it is stable. This idea is in the spirit of so-called satisfiability modulo theories, SMT, a currently intensely studied paradigm for use of SAT.

Idea (2) explores a different possibility: we can generate additional formulas and add them to the $pl(P)$. If suitably chosen, such additional formulas would prune the candidates for stable models, leaving only those that really are stable.

We will explore now a technique, due to Lin and Zhao[LZ04], that realizes the possibility (2). To this end, let L be a subset of variables that occur in heads of clauses of P. We call such set a *loop* if L is strongly connected in G_P, that is, for any two vertices p and q in L there is a cycle H consisting of elements of L such that both p and q are in H.

Now, recall that for a normal program clause P the $posBody(C)$ was the conjunction of all variables occurring positively in the body of C. With a slight abuse of notation we can also think of $posBody(C)$ as a set of propositional variables. Given a loop L we define $P^-(L)$ as the set of those clauses C in P that have the following property: $posBody(C) \cap L = \emptyset$. Now, let Φ_L^- be the following disjunction:

$$\bigvee_{C \in P^-(L)} body(C).$$

Let us observe that we take disjunction of entire bodies, not just their positive parts. We now define the loop-formula $LZ_P(L)$ associated with the loop L:

$$(\bigvee_{p \in L} p) \Rightarrow \Phi_L^-.$$

Given a normal propositional program P, let $LZ(P)$ be the theory consisting of $LZ_P(L)$ for all loops L of P. We then have the following characterization of stable models which we will not prove; instead we will discuss it in the Section 14.8 Exercises.

PROPOSITION 14.8 (Lin and Zhao theorem)
Let P be a finite propositional program. Then a set of variables M is a stable model of P if and only if M is a supported model of P and $M \models LZ(P)$.

Proposition 14.8 provides a way to realize (2). First we compute all loop formulas for P, add them to the completion of P, transform to the CNF, and drop the result into the SAT solver. If the solver returns a satisfying assignment, that assignment is a stable model of P.
But there can be plenty of loops, in fact, exponentially many in the size of the program. What can we do then? We can use Proposition 14.8 as guidance for the technique (1) discussed above. The point is that we could add *some* loop formulas to the completion of P, not all of them at once. When we submit the resulting theory to the SAT solver we prune *some* non-stable models of the completion.

14.6 Knowledge representation and ASP

In this section we discuss some applications of ASP to knowledge representation. All we can do is to show the rudiments; a serious book on this topic is [Ba03].
We will show two examples of the use of ASP in knowledge representation, both pretty theoretical. One is a description of three-coloring of graphs, the other is a description of Hamiltonian cycles using ASP. The common feature of these solutions is the *separation of problem from data*. We will have the description of instances

(graphs in our case), but the description of the solution is uniform. We use variables in our program clauses. This presupposes availability of a grounder that compiles a program with variables into a propositional program. We will assume that the grounder recognizes at least some system predicates. In our case only the inequality predicate $\neq$ is used, but the existing systems allow for other predicates such as the ternary predicate $X + Y = Z$ (true when the variables are assigned numerical values x, y, z, resp., and $x + y = z$) or comparison predicates.

14.6.1 Three-coloring of graphs in ASP

Here we represent finite graphs by means of sets of propositional variables. Specifically, for a graph $G = (V, E)$, we assign to G an *extensional database*, Ext_G. It consists of variables $vtx(a)$ for every $a \in V$, and variables $edge(a, b)$ for all $(a, b) \in E$. The predicates $vtx(\cdot)$ and $edge(\cdot, \cdot)$ will never occur in the heads of clauses that we will write in our intentional database. This *automatically* implies that we will make a closed world assumption on these predicates, as in stable semantics we can only compute variables when they occur in the heads of clauses. This is what makes ASP attractive in knowledge representation; taking care of closed world assumption (when needed) is very easy (unlike in SAT, where we either have to add the full diagram of a relation, or implement a CWA-aware grounder).

In addition to the two predicates discussed above, we have an additional one, to describe our three colors: *red, blue,* and *green*, abbreviated r, b, and g. The predicate $clr(\cdot)$ will be true on these three constants only.

To write a program that colors the graph we will need three clauses only. One of these is a constraint, that the had of the clause is empty. This is a small problem only, Using a simple trick we can make sure that heads are nonempty. Specifically, given a clause $C :\leftarrow B$ we select a new variable $x \notin Var$ and replace C by $x \leftarrow B$, not x. We assume that the solver will do this for us. So here is the program (with variables) *3Col*:

$color(X, Y) \leftarrow vtx(X), clr(Y), not\ othercolor(X, Y)$
$othercolor(X, Y) \leftarrow color(X, Z), Y \neq Z$
$\leftarrow color(X, Y), color(Z, Y), edge(X, Z), vtx(X), vtx(Z), clr(Y)$

The union of these two programs (graph description and three-coloring description), when grounded, has the following property.

PROPOSITION 14.9

There is a one-to-one and "onto" correspondence between the stable models of the grounding of the program $Ext_G \cup 3Col$ and three-colorings of the graph G.

Thus if we have a grounder (several are available, one is is a part of the package *lparse*, c.f. [SNS02]) and an ASP solver then we can use them to solve a 3-coloring problem for graphs. Also, the reader should realize that the 3-coloring is incidental here; we can write four-coloring program by adding one additional color, say *yellow*.

14.6.2 Hamiltonian cycles in ASP

We will now show how the problem of finding the Hamiltonian cycle can be represented in ASP.

This is, again, a graph problem. We will represent finite graphs exactly as we did in Section 14.6.1. To model a Hamiltonian cycle in the graph G we will need four additional predicates: one of those, $in(\cdot,\cdot)$, will be the interpretation of the Hamiltonian cycle. The other, $out(\cdot,\cdot)$, will be interpreted by edges which are not on the cycle. The third one, which is unary and denoted as $reached(\cdot)$, will be interpreted by vertices that are on the cycle. Finally, the last one, $initialvtx(\cdot)$, simulates a constant, one vertex which is the beginning and the end of the cycle. So, our extensional database (graph description) is as in our description of 3-coloring. As before, we denote it as Ext_G. The problem description (independent of any specific graph) consists of eight clauses (with variables). We denote this program by *Ham*. Here it is:

$initialvtx(a_1)$
$in(X,Y) \leftarrow edge(X,Y), not\ out(X,Y)$
$out(X,Y) \leftarrow edge(X,Y), not\ in(X,Y)$
$\leftarrow in(X,Y), in(Z,Y), X \neq Z$
$\leftarrow in(X,Y), in(X,Z), Y \neq Z$
$reached(X) \leftarrow in(X,Y), reached(Y)$
$reached(X) \leftarrow in(X,Y), initialvtx(Y)$
$\leftarrow vtx(X), not\ reached(X)$

The first clause defines the beginning (and the end) of the putative cycle. The second and third clauses choose the edges that are going to be on the cycle. The fourth and fifth clauses enforce the condition that the interpretation of the predicate *in* will be a cycle. This task is shared, actually, by the sixth and seventh clauses. They compute the vertices that are on the putative cycle. The last clause makes sure that all vertices are on the cycle (observe the fact that the initial vertex must also be reached!).

As above we have the following fact.

PROPOSITION 14.10
There is a one-to-one and "onto" correspondence between the stable models of the grounding of the program $Ext_G \cup Ham$ and the Hamiltonian cycles of the graph G.

Thus, given a graph G, an ASP solver, and a grounder, we can submit the graph description of G, $Ext(G)$, and the eight clauses of the program *Ham* and, given enough resources, the solver will return a Hamiltonian cycle in G if there is one.

14.7 Complexity issues for ASP

The satisfiability problem for ASP is the language SM consisting of finite propositional programs that possess a stable model. To see the complexity of the problem SM let us first observe that the three tasks involved in checking if a given set of variables M is a stable model of a program P are all polynomial; computation of the reduct P_M can be done in linear time (each program clause is scanned once and then tested if it should generate a clause in P^M). Then the computation of the least model of the reduct is done in linear time in the size of P using the Dowling-Gallier algorithm. Finally, we need to check if M coincides with N_M, again an easy task. Therefore, the language SM is in the class NP.

Next, we will show a polynomial time reduction of SAT to SM. Here is how we do this. Let T be a finite clausal theory. We assume that T does not contain tautologies. For each variable p we choose two new atoms which we denote as $in(p)$ and $out(p)$. We also choose one new additional propositional variable x. We will now construct a program P_T. The program P_T consists of two parts. The first of these depends only on the set of variables of T, not T itself. We will call it the *generator* program, G. The second part, which we call the *checker* program, Q_T, depends on T. In fact, the idea of the generator program is very similar to one behind the Turing machine for testing satisfiability.

The generator program consists of $2n$ program clauses (here n is the number of variables in Var_T). They are:

$in(p_j) \leftarrow not\ out(p_j)$
$out(p_j) \leftarrow not\ in(p_j)$

We add such a pair of program clauses for each p_j, $1 \leq j \leq n$.
Then for each clause $D : q_1 \vee \ldots \vee q_k \vee \neg r_1 \vee \ldots \vee \neg r_m$ of T we add a program clause C_D:

$$x \leftarrow out(q_1), \ldots, out(q_k), in(r_1), \ldots, in(r_m), not\ x$$

(the same x is used in all these clauses). Then we form the checker part of our program Q_T as $\{C_D : D \in T\}$. Finally, we set $P_T = G \cup Q_T$. Clearly the computation of P_T can be done in linear time in the size of T. We now have the following fact.

PROPOSITION 14.11
The program P_T possesses stable models if and only if T is satisfiable. In

fact, the assignment $M \mapsto N_M$ where

$$N_M = \{p : in(p) \in M\} \cup \{\neg q : out(q) \in M\},$$

establishes one-to-one and "onto" correspondence between stable models of P_T and satisfying assignments for T.

Corollary 14.2 follows immediately.

COROLLARY 14.2
The problem SM is an NP-complete problem.

14.8 Exercises

1. Prove that supported models of a program P can contain only variables that occur in heads of clauses of P. Conclude that the same property holds for stable models of P.

2. Find all supported models and all stable models for the program P.

    $p \leftarrow q, r$
    $q \leftarrow p$
    $r \leftarrow not\ s$

3. The operator GL_P is antimonotone. Therefore, if we iterate it twice, that is, if we execute $M \mapsto GL_P(GL_P(M))$ we get a monotone operator. Prove it.

4. So now we know that GL_P^2 is a monotone operator. By the Knaster-Tarski fixpoint theorem it possesses the least and largest fixpoints M_1 and M_2. Prove that $GL_P(M_1) = M_2$ and $GL_P(M_2) = M_1$.

5. Prove that every fixpoint N of GL_P satisfies the inclusions

    $$M_1 \subseteq N \subseteq M_2.$$

6. Use Gelfond and Lifschitz theorem to show that the stable models of a logic program form an antichain. That is, whenever M_1, M_2 are stable models of a logic program P and $M_1 \subseteq M_2$ then $M_1 = M_2$.

7. Let us modify the Gelfond-Lifschitz reduct by pruning more. This construction is due to Truszczyński. Namely, we eliminate all clauses with bodies not satisfied by M (clearly, the Gelfond-Lifschitz reduct prunes less). Then as in the second part of the reduct procedure we drop the negative part. We get a smaller Horn program. Let us call it P^{M-}. Now, we get a modified operator; we assign

to M the least model of P^{M^-}. Prove that the fixpoints of this operator and the fixpoints of the Gelfond-Lifschitz operator coincide.

8. Prove the Dung theorem: For every program P there is a program P' such that bodies of clauses of P' have no positive literals, and stable models of P' and of P coincide.
9. The "easy" part of the Lin and Zhao theorem is the part that says that stable models of P are supported models of P that also satisfy $LZ(P)$. Prove that easy part.
10. The "hard" part of Lin and Zhao Theorem is the other implication. Prove it.
11. Prove the validity of the reduction of SAT to SM given in Proposition 14.11.

Chapter 15

Conclusions

Logic has been a domain of human intellectual activity for thousands of years. Every major civilization needed to create tools for humans to reason. The need for precision of argumentation, for legal reasoning, and for creating the order in the world that surrounds humans forced codification of the principles of reasoning.

Surprisingly, the natural "consumer" of precise thinking, namely mathematicians, did not pay much attention to the formalization of rules of reasoning. Even today mathematicians are often biased against those who want to overformalize the way mathematics is produced. There is something in such an attitude. After all, mathematicians practiced their craft for as long a time as logic has been around, or even longer. The arguments of Euclid are as valid today as they were over two thousand years ago. Many constructions, say in number theory, in geometry, and in calculus, are grounded in the experience of mathematicians who lived long ago. Moreover, modulo the use of concepts and techniques developed recently, many results of recent past and present could be easily explained to the mathematicians of past centuries. So, the issue of correctness of the reasoning is not the first priority for a working mathematician.

The raise of ineffective mathematics of late nineteenth century, especially the abstract concepts of sets, and specifically of abstract functions (earlier mathematicians did not accept ineffective constructions of functions), forced mathematicians and philosophers to look at the principles of defining and manipulation of such objects. The emergence of paradoxes such as the Russell paradox (the issue of existence of the set consisting of sets that do not belong to themselves), of Cantor paradox (the set of all sets), and others forced the development of formal methods dealing with the issues such as the logic used by mathematicians, the question of semantics of formal languages of logic, and other related issues such as definability, axiomatizability, and other concepts.

While the mathematicians and philosophers struggled with the nature of mathematical reasoning, the new issues of the effectiveness of reasoning and the decidability of fragments of mathematics became more urgent. One can only say that a premonition of the advent of computers as we know them resulted in works by Hilbert, Gödel, Turing, Church, Kleene, Mostowski, and Post that proposed some variants of the computability theory that persist even today.

But it was the invention of digital computers during World War II that changed everything. Perhaps not immediately, but soon after the introduction of computers, it became clear that they could deal with some reasoning problems. Suddenly, computers could solve problems that humanity could deal with "in principle" but not "in practice." Of course, one needs to program computers, but once computers with stored programs were built, a new perspective opened.

Most of the material covered in this book relates to the techniques developed *after* it became clear that humanity has, in computers, a collection of tools that change the rules completely. The question of effectiveness of constructions, of algorithms that *really* could be implemented, became urgent.

One can think about computers as tools for dealing with problems that can be suitably formalized, but on another level, a computer is just an electronic device. The underlying combinational and sequential circuits need to be verified and tested. The software that runs on the computers also needs to be verified and tested. It did not take long for computer scientists to realize that computers themselves can be used to test the designs of both hardware and software. This *self-reference* of computers is, on reflection, amazing. Computer science deals with this problem by considering a universal Turing machine. But we cannot realize a physical implementation of such universal device, we can only approximate it with the real hardware and software with their limitations resulting from the fact that everything we deal with must be finite.

Of course, as we build stronger computers, with bigger CPUs, much larger designs have to be tested and verified. Clearly, sometimes the laws of physics will start to intervene. In the meanwhile we certainly can enjoy the availability of computers in their many roles: of computing devices, of controlling devices, and of communication devices.

The many roles of computers resulted in an explosion of formalisms which often are called *logic*. Many of these systems provide the community with a language that can be used to describe a task at hand (say communication of processes, or description of communication protocols, or execution of programs). Very abstract languages such as modal logics devised by philosophers to describe intentional concepts such as *certainty* or *possibility* suddenly became practical, allowing us to describe properties of abstract machines. Even more, the need to verify properties of such machines resulted in *model checking algorithms* (a topic entirely absent in this book), that allows for testing models of various kinds of automata (a gross simplification) for various properties.

Very often the reasoning tasks of these new logics are reduced to reasoning tasks of classical propositional logics. What it means is that various reasoning tasks of those logics are reduced to other reasoning tasks for classical logic provers or solvers. In effect, programs performing reasoning tasks in classical propositional logic serve as *back-end systems* for these new logical systems. Model checking is one example of such a situation.

The quest for *artificial intelligence* and in particular for formalization of *commonsense reasoning* resulted in great progress in logic. We touched on this (but only the "tip of the iceberg") in Chapter 14.

The majority of topics covered in this book (but not all) has been invented since the advent of computers (i.e., in the 1950s or later). The *roots* of the theory presented in this book were created by the great logicians of the past, especially Boole, De Morgan, Frege of the nineteenth century and Russell, Hilbert, Tarski, Bernays and others in the early twentieth century. But as we said, the premonition of computers, visible already in the work of Gödel, Turing, and others, was the true impetus for the parts of logic treated here. The fundamental tools used here, such as the Knaster-Tarski fixpoint theorem, were also developed at the same time. But their real power became apparent only after they were applied in computer science. Similarly, the work of Post on Boolean functions (we presented several aspects, but not the deeper aspect of his work on the so-called *Post lattice*) had to wait for its most important applications until computers became ubiquitous.

We believe that anyone who wants to study seriously the foundations of algorithms that are used in reasoning systems (solvers, provers) needs most of the material covered in this book. We do not claim that this book prepares the reader to do the research in the area of computational logic (or more precisely of applications of computational propositional logic). But we are sure that anyone researching computational logics will have to deal with issues at least partly similar to those presented and studied in this book.

References

[ARSM07] F. Aloul, A. Ramani, K. Sakallah, and I. Markov. Solution and optimization of systems of pseudo-Boolean constraints, *IEEE Transactions on Computers*, 56, pp. 1415–1424, 2007.

[Apt03] K.R. Apt. *Principles of Constraint Programming*, Cambridge University Press, 2003.

[Ba03] C. Baral. *Knowledge Representation, Reasoning and Declarative Problem Solving*, Cambridge University Press, 2003.

[Ch09] H. Chen. A Rendevous of Logic, Complexity and Algebra. To appear in *ACM Computing Surveys*.

[Co71] S.A. Cook. The complexity of theorem proving procedures, *Proceedings of the 3rd ACM Symposium on Theory of Computing*, pp. 151–158, 1971.

[CCT87] W. Cook, C. Coullard, and G. Turan. On the complexity of cutting-plane proofs, *Discrete Applied Mathematics*, 18 pp. 25–38, 1987.

[Cr57] W. Craig. Linear Reasoning, A new form of Herbrand-Gentzen theorem, *Journal of Symbolic Logic*, 22, pp. 250–268, 1957.

[CKS01] N. Cregignou, S. Khanna, and M. Sudan. *Complexity Classifications of Boolean Constraint Satisfaction Problems*, SIAM Press, 2001.

[DP60] M. Davis and H. Putnam. A computing procedure for quantification theory, *Journal of the ACM*, 7, pp. 201–215, 1960.

[DLL62] M. Davis, G. Logemann, and D. Loveland. A machine program for theorem proving, *Communications of the ACM*, 5, pp.394–397, 1962.

[DM94] G. De Micheli. *Synthesis and Optimization of Digital Circuits*, McGraw-Hill, 1994.

[De89] R. Dechter. Enhancement schemes for constraint processing: backjumping, learning and cutset disposition, *Artificial Intelligence*, 41, pp. 273–312, 1989/1990.

[DGP04] H.E. Dixon, M.L. Ginsberg, and A.J. Barkes. Generalizing Boolean satisfiability I: background and survey of existing work, *Journal of Artificial Intelligence Research*, 21, 193–243, 2004.

References

[DG84] W.F. Dowling and J.H. Gallier. Linear-time algorithms for testing satisfiability of propositional Horn formulae, *Journal of Logic Programming*, 1, pp. 267–284, 1984.

[DLMT04] D.M. Dransfield, L. Liu, V.W. Marek, and M. Truszczynski. Satisfiability and computing van der Waerden numbers, *Electronic Journal of Combinatorics*, 11:#R41, 2004.

[ET01] D. East and M. Truszczyński. Propositional satisfiability in answer-set programming, *KI 2001*, Springer Lecture Notes in Computer Science 2174, pp. 138–153, Springer-Verlag, 2001.

[ET04] D. East and M. Truszczyński. Predicate-calculus based logics for modeling and solving search problems. *ACM Transactions on Computational Logic*, 7, pp. 38–83, 2006.

[EIMT06] D. East, M. Iakhiaev, A. Mikitiuk, and M. Truszczyński. Tools for modeling and solving search problems, *AI Communications*, 19, pp. 301–312, 2006.

[Fit90] M. Fitting. *First-Order Logic and Automated Theorem Proving*, Springer-Verlag, 1990.

[Fr95] J.W. Freeman. Improvements to Propositional Satisfiability Search Algorithms, Ph.D. dissertation, University of Pennsylvania, 1995.

[GJ79] M.R. Garey and D.S. Johnson. *Computers and Intractability; a Guide to the Theory of NP-Completeness*, W.H. Freeman, 1979.

[Ha85] A. Haken. Intractability of resolution, *Theoretical Computer Science*, 39, pp. 297–308, 1985.

[HHLM07] P.R. Hervig, M.J.H. Heule, P.M. van Lambalgen, and H. van Maaren. A new method to construct lower bounds for van der Waerden numbers, *Electronic Journal of Combinatorics* 14:#R6, 2007.

[Ho00] J. Hooker. *Logic-Based Methods for Optimization*, Wiley, 2000.

[Kr67] M. Krom, The decision problem for segregated formulas in first-order logic, *Mathematica Scandinavica*, 21, pp. 233-240, 1967.

[KM84] K. Kuratowski and A. Mostowski. *Set Theory*, North-Holland, 1984.

[La06] D. Lau. *Function Algebras on Finite Sets*, Springer-Verlag, 2006.

[LT06] L. Liu and M. Truszczyński. Properties and applications of monotone and convex constraints. *Journal of Artificial Intelligence Research*, 27, pp. 299–334, 2006.

[LZ04] L. Liu and M. Truszczyński. Properties and

References

[LM05]   F. Lin and Y. Zhao. ASSAT: Computing answer sets of a logic program by SAT solvers. *Artificial Intelligence* 157, pp. 115–137, 2004. structures for backtrack search SAT solvers, *Annals of Mathematics and Artificial Intelligence*, 43, pp. 137–152, 2005.

[Mak87]   J.A. Makowsky. Why Horn formulas matter in computer science: initial structures and generic examples, *Journal for Computer and System Sciences*, 34, pp. 266–292, 1987.

[MT93]   V.W. Marek and M. Truszczyński. *Nonmonotonic Logic – Context-Dependent Reasoning*, Springer-Verlag, 1993.

[MS95]   J.P. Marques-Silva. Search algorithms for satisfiability problems in combinatorial switching circuits, Ph.D. dissertation, University of Michigan, 1995.

[MS99]   J. Marques-Silva, K. Sakallah. GRASP: A search algorithm for propositional satisfiability, *IEEE Transactions on Computers*, 48, pp. 506–521, 1999.

[NS93]   A. Nerode and R.A. Shore. *Logic for Applications*, Springer-Verlag, 1993.

[Qu52]   W.V. Quine. The problem of simplifying truth functions, *American Mathematical Monthly*, 59, pp. 521–531, 1952.

[Ro65]   J.A. Robinson. A machine-oriented logic based on the resolution principle. *Journal of the ACM*, 12, pp. 23–41, 1965.

[Sch78]   T.J. Schaefer. The complexity of satisfiability problems, *Proceedings of the Tenth Annual ACM Symposium on Theory of Computing*, pp. 216–226, 1978.

[Sh67]   J.R. Shoenfield. *Mathematical Logic*, A.K. Peters/Association of Symbolic Logic (reprint), 2001.

[SNS02]   P. Simons, I. Niemelä, and T. Soininen. Extending and implementing the stable model semantics, *Artificial Intelligence*, 138, pp. 181–234, 2002.

[Sip06]   M. Sipser. *Introduction to the Theory of Computation*, Second Edition, Thomson, 2006.

[Sl67]   J.R. Slagle. Automated theorem proving with renameable and semantic resolution, *Journal of the ACM*, 14, pp. 687–697, 1967.

[Sm95]   R. Smullyan. *First-Order Logic*, Dover, 1995.

[SP05]   S. Subbarayan and D.K. Pradhan. NiVER - non increasing variable elimination resolution for preprocessing SAT instances, in: *SAT-2004*, Springer Lecture Notes in Computer Science 3542, pp. 276–291, 2005.

[Ta55]   A. Tarski. A lattice-theoretical fixpoint theorem and its applications, *Pacific Journal of Mathematics*, 5, pp. 285–309, 1955.

[WB96]   H.P. Williams and S.C. Brailsford. Computational logic and integer programming, in: J.E. Beasley (ed.), *Advances in Linear and Integer Programming*, pp. 249–281, Clarendon Press, 1996.

[ZS00]   H. Zhang and M.E. Stickel. Implementing the Davis-Putnam Method, in: *SAT-2000*, I. Gent, H. van Maaren, and T. Walsh (eds.), pp. 309–226, IOS Press, 2002.

[ZM+01] L. Zhang, C.F. Madigan, M.H. Moskewicz, and S. Malik. Efficient conflict driven learning in a Boolean satisfiability solver, *Proceedings of IEEE ICCAD*, pp. 279–285, 2001.

Index

2CNF, 185

Algorithm A_+, 180
Algorithm A_-, 181
Algorithm *DPLLdec*, 159
Algorithm *DPLLsearch*, 160
Algorithm *DPsearch*, 151
Algorithm *DPtest*, 150
Algorithm *Dptest*, 150
Antimonotone constraints, 276
Autark set, 124
Autarky, 23
    existence, 243
    for Horn theories, 176
    for Krom theories, 186

Basis for a set of clauses, 114
BCP, 155
Bitwise conjunction, 169
Bitwise disjunction, 182
Bitwise execution of a function, 308
Body of a program clause, 171
Body of the program clause, 322
Boolean Algebra, 4
    complete, 5
Boolean Constraint Propagation, 155
    for L^{cc}, 270
Boolean function
    essentially unary, 311
    idempotent, 311
    projection, 311
Boolean polynomials, 79
Branch
    open, 139

Call-graph
    positive, 326
Canonical tableau, 140

Canonical valuation, 109
Cardinality clauses, 269
Cardinality constraint, 267
    lower-bound, 268
    upper-bound, 268
Change set, 193
Clausal logic, 101
Clause, 12, 23, 102
    constraint, 102
    Horn, 167
    Krom, 185
Clone, 311
Closed-world Assumption, 286
Closure under resolution, 108
Compactness of Propositional Logic, 95
Complete set of clauses, 98
Complete set of functors, 75
Complete set of literals, 17
Completion formula, 324
Consistent set of formulas, 15
Consistent set of literals, 20
Constraint, 306
Constraint satisfaction problems, 292
    solutions, 292
Continuity of Cn operator, 99
Convex constraints, 281
Cook-Levin Theorem, 236
Craig lemma, 66
    strong form, 69
Crossing out variables, 128
CSP, 285
    "easy cases", 300
    complexity, 298
    Horn case, 300
    locally finite, 296
Cutting-plane rule, 208

CWA-consistent theory, 287

Davis-Putnam Algorithm, 144
Davis-Putnam lemma, 118
Davis-Putnam reduct, 117
De Morgan functors, 73
Deduction Theorem, 40
Diagram of a relation, 287
DIMACS format, 254
Disambiguation of variables, 306
Dowling-Gallier Algorithm, 174
Dual-constraint, 179
Duality, 38

Entailment, 16, 39

Fages theorem, 331
Family of sets closed under intersections, 167
Finitely satisfiable set of formulas, 98
Finitely satisfiable theory, 95
Fixpoint, 6
Formula, 11
    affine, 199
    nontautological, 24
    positive, 165
    symmetric, 38
Formulas
    equivalent, 32
    signed, 137
Function, 2
    idempotent, 311
Functor, 11, 73
    0-consistent, 83
    1-consistent, 83
    linear, 83
    monotone, 83
    self-dual, 83
    Sheffer-like, 88

Geiger theorem, 309
Gelfond-Lifschitz operator, 327
Gelfond-Lifschitz reduct, 327
Graph coloring as SAT, 259

Head of a program clause, 171, 322

Herbrand base, 262
Herbrand structure, 262
Herbrand-consistent theory, 264
Hintikka set, 135
Hintikka Theorem, 136
Hit set, 132, 279
Horn Theorem, 167
HornCSP algorithm, 302

Idempotent function, 311
Instance of a constraint, 306
Interpolant, 63
    strong, 69
Interpolation lemma, 66

König lemma, 94
König Theorem, 105
Kleene logic, 18
Kleene ordering, 19
Kleene theorem, 22
Kleene Theorem for L^{cc}, 271
Kleene Theorem for L^{wtc}, 276
Kleene three-valued logic, 18
Knaster-Tarski Theorem, 6
Krom clause, 185
Kullmann theorem, 243

Language L^{cc}, 269
Lattice, 4
    Axioms, 4
    complete, 4
    distributive, 4
    monotone operator, 6
    ordering, 4
Lean set, 124
Least fixpoint theorem, 6
Least model of Horn formula, 170
Level of the variable, 326
Lewis Theorem, 197
Lindenbaum Algebra, 32
Lindenbaum theorem, 98
Linear-combination rule, 207
Literal, 12
    dual, 35
    pure, 118

Localization Theorem, 15
Loop, 332

Maximal satisfiable subset of a clause set, 130
Minimal resolution consequence, 110
Minimally unsatisfiable set of clauses, 103
Minterm, 56
Model of a formula F, 40
Model of a program, 322
    minimal, 323
    stable, 327
    supported, 325
Monomial, 79
Monotone constraints, 276
Monotone operator, 6

Negation, nonstandard, 321
Normal Form, 45
Normal form, 45
    canonical, 53
    complete, 56
    conjunctive, 50
    disjunctive, 50
    negation, 46
    reduced, 54

Occurrence of a variable, 47
Operation bcp_F, 155
Operation Mod, 39
Operation Th, 39
Operation BCP, 155
Operator
    fixpoint, 6
Operator T_H, 172
Ordinal number, 5
    limit, 6
    successor, 6

Partial assignments, 18
Partial valuation, 18
Partial valuations, 18
Permutation
    consistent, 35
    of literals, 35
Permutation of literals
    shift, 195
Permutation of variables, 195
Plain clause in a set of clauses, 126
Polarity of a literal, 17
Polymorphism, 308
Polynomials, 80
Poset, 2
    Bounds, 3
    Chain in a poset, 3
    chain-complete, 9
    classification of elements, 3
    well-founded, 5
Post classes, 83
Post criterion for completeness, 85
Post Theorem, 85
Postfixpoint of an operator, 6
Power set of X, 1
pp-definability, 306
Predicate Logic, 253
Prefixpoint of an operator, 6
Primitive-positive definability, 306
Program clause, 167
Program clauses
    normal, 322
Propositional interpretation of a program, 323
Propositional Schemata, 254
Propositional variable, 11
Pseudo-Boolean constraint, 267
Pseudo-Boolean inequality, 206, 267
Pure literal, 23, 118

Quine Theorem, 110

Rank of a formula, 12
reduct, 146
Reduct by a valuation, 146
Relation
    equivalence, 2
    ordering, 2
Relation $\sim$, 32
Relation, extensional, 289
Relation, intentional, 289

Relations
    classification, 2
Renameable-K formula, 196
Resolution
    completeness, 110
    semantic, 119
    soundness, 109
    Unit, 174
    unit, 156
Resolution consequence, 110
Resolution derivation of a clause, 108
Resolution refutation, 113
Resolution rule of proof, 107
Resolvent, 108
Robinson Theorem, 16
Rosenberg Theorem, 311
Rule
    structural, 210

Satisfaction, 15
Satisfiable set of formulas, 15
Schaefer Theorem, 310
    "easy part", 304
Semantical Consequence, 39
Shift permutation, 195
Sign pattern of a matrix, 218
Solutions of constraint satisfaction problems, 292
Solver, 253
Stable models, 327
Standard polynomials, 80
Substitution Lemma, 30
Subsumption rule, 111
Supported model, 325

Table, 13, 75
    of a formula, 75
Table method for testing satisfiability and validity, 134
Tableau, 137, 139
    branch, 139
    canonical, 140
    finished, 139
Tableau rules, 138
Tail of a polynomial, 81

Tarski propositional fixpoint theorem, 52
Tautology, 23, 28
Theory
    finitely satisfiable, 95
Three-valued truth function, 20
Touching, 23
Transversal, 279
Tree, 94
Truth tables, 13
Turing machine, 228
    coding runs, 232
    run, 229

Unit resolution, 156
Unsatisfiable set of clauses, 103

Valuation, 13, 14
    partial, 13, 18
Variable
    assignment, 14
    guarded, 330
VER, 144

Weight constraints, 273
    lower-bound, 273
    upper-bound, 273

Zhegalkin polynomials, 80
Zorn lemma, 3